Technology Development for a Cosmic Microwave Background Probe of Inflation

Journal of Physics: Conference Series Volume 155

Boulder, Colorado, USA
25-28 August 2008

ISBN: 978-1-61738-378-6

Printed from e-media with permission by:

Curran Associates, Inc.
57 Morehouse Lane
Red Hook, NY 12571

Printed by Curran Associates, Inc. (2010)

For permission requests, please contact the Institute of Physics
at the address below.

Institute of Physics
Dirac House, Temple Back
Bristol BS1 6BE UK

Phone: 44 1 17 929 7481
Fax: 44 1 17 920 0979

techtracking@iop.org

Additional copies of this publication are available from:

Curran Associates, Inc.
57 Morehouse Lane
Red Hook, NY 12571 USA
Phone: 845-758-0400
Fax: 845-758-2634
Email: curran@proceedings.com
Web: www.proceedings.com

TABLE OF CONTENTS

PREFACE

In late 2007 NASA called for proposals to fund Astrophysics Strategic Mission Concept Studies. The goal was to generate concept studies for key future missions, which would be forwarded to the Astro2010 astrophysics decadal review committee for prioritization. Under the guidance and orchestration of the Primordial Polarization Program Definition Team, a NASA committee chartered to coordinate the activities of the cosmic microwave background (CMB) community, a CMB proposal aiming to represent the consensus of the entire community was submitted. A CMBPol Mission Concept Study grant was awarded in early 2008. Under the grant we reviewed the entire activities of the CMB community and proposed a path for the next decade. We also assessed the case and recommended a path for a future CMB polarization satellite.

The grant funded three community-wide workshops that were held over the summer of 2008. The goal of the first workshop, held at Fermilab, was to discuss the theoretical foundation of inflation and its signature on the CMB, as well as the theoretical aspects of other polarimetric signatures observable at millimeter wavelengths. Volume 1141 of the American Institute of Physics conference proceedings summarizes the results of this workshop. The second workshop, held at Annapolis, Maryland, centered on expected systematic effects in polarimetric experiments and their potential mitigation. The third workshop, held at the NIST facility at Boulder, Colorado, focused on the technology requirements necessary to make incisive CMB polarization measurements and what was needed to advance the technology to the readiness level required for a start of a space-borne mission. The electronic proceedings presented here are the result of this third workshop.

In preparing for the workshop the organizers assigned topical-editors for each technology topic. Each of them solicited white paper contributions from experts in their respective areas. The white papers were distributed to all participants prior to the workshop and were the basis for presentations and further discussions during the workshop. After the workshop the white papers were assembled to the summaries presented in these proceedings.

The papers give the current technology state-of-the-art with some projections for further development in the future. They all represent the immense technological progress in our field.

The prevailing opinion during the workshop was that from a technology aspect alone, the community is confident in its ability to begin a space mission in the next few years.

Journal of Physics: Conference Series **155** (2009) 012001 doi:10.1088/1742-6596/155/1/012001

Direct Detectors for the Einstein Inflation Probe

Dominic Benford[1], **Harvey Moseley**[1] **and Jonas Zmuidzinas**[2]

[1] Observational Cosmology Laboratory, NASA/GSFC, Greenbelt, MD 20771

[2] Downs Laboratory, Department of Physics, California Institute of Technology, Pasadena, CA

E-mail: Dominic.Benford@nasa.gov
Harvey.Moseley@nasa.gov
jonas@submm.caltech.edu

Abstract.
Here we review the principles of operation, history, present status, and future prospects for the primary candidate detectors for Cosmic Microwave Background (CMB) polarization studies. The three detector types we will discuss are semiconductor-based bolometers, superconducting transition edge sensor (TES) bolometer, and Microwave Kinetic Inductance Detectors (MKIDs). All of these detector types can provide the sensitivity to permit background-limited measurements of the CMB, but the ultimate selection of detectors will be largely determined by the ease of production and reliability of large arrays of such detectors. This paper describes the present state of development of these detectors, efforts to integrate them into large arrays, and the detector system developments necessary to enable a space CMB polarization mission.

1. Introduction

The B-mode polarization of the CMB is espected to be a very weak signal; to put limits on the tensor to scalar ration of primordial fluctuations $R \sim 0.01$, we must measure a B-mode signal as small as $\sim 10^{-8}$ of the CMB power. The measurement of such a weak signal over the full sky requires arrays of 10^3 to 10^4 single mode background limited detectors with detective quantum efficiency (DQE) near unity operating for a year, with backgrounds dominated by the CMB itself. Beyond sensitivity, the performance requirements for the detectors, such as response time and the $\frac{1}{f}$ noise corner depend on the design of the experiment, so the ultimate selection of the"best" detector must wait for a system design.

At millimeter wavelengths, many optical functions of the instrument, such as beam formation and filtering, are being implemented in planar superconducting circuitry integrated with the detectors. Different groups are including these beam forming and spectral filtering elements into the focal plane elements [1–8]. The filtered radiation must be coupled into the sensor element for detection. In bolometers, the component which absorbs the power can be different from the sensor element and can be optimized separately, but must be thermally connected to it. The ultimate design of a focal plane for a space mission to measure CMB polarization is not known, but the design tools and production processes required to design it are largely in place.

In this paper, we will focus on the development of the sensor elements and the multiplexing systems required to read out these integrated focal planes. The planar superconducting microwave circuits required for beam forming and filtering are treated separately in this report. Here, we review the state of the art for three classes of detectors which are candidates for use in a CMB polarization mission. All three types can in principle be coupled to planar circuits. The ultimate choice of a detector type will depend on a number of things:

(i) Performance: The detector must provide background limited sensitivity and detective quantum efficiency near unity observing the CMB at relevant wavelengths.

(ii) Readout: The detector must be compatible with a low noise readout system which can be easily integrated with the array.

Journal of Physics: Conference Series **155** (2009) 012001 doi:10.1088/1742-6596/155/1/012001

(iii) Practical issues:
   (a) Fabrication process for detector must be compatible with that of other focal plane elements.
   (b) Array should be simple to integrate into the optical system.
   (c) Power consumption should be low to be compatible with coolers.

In this report, we review the state of the art for superconducting transition edge sensors (TES) thermal sensors, semiconductor thermal sensors and microwave kinetic inductance detectors (MKIDs), which are sensitive to photon number. Since the field is changing rapidly, the choice of detectors will depend on the rate of development of the different detectors and multiplexing systems and the requirements of the specific instrument implementation, so this review should be seen as a progress report.

2. Transition Edge Sensor Bolometers for CMB Polarimetry

by Dominic Benford

2.1. Introduction
A superconducting transition-edge sensor (TES) bolometer is a highly sensitive detector of light that can resolve individual photons at energies above the near-infrared and can operate at all wavelengths between the radio and gamma rays. The TES is a thermal sensor that measures a total power rate or energy deposition by measuring with great precision the increase in the resistance of a superconducting material that is biased within the superconducting-to-normal transition region. Here I describe the technology, including a technical description, a description of its state of maturity, and discuss its benefits and disadvantages from a statistical and systematic viewpoint. I analyze the technological readiness level of this option (see attached description of NASA's TRL definitions), and of the cost and timescale that would be required for the community to bring the technology to TRL 5, which would be a level appropriate to respond to an Announcement of Opportunity for a satellite mission.

2.2. A Broad History of Bolometers
Going back into history, the bolometer can be said to have originated a very long time ago. Arguably, the Greeks[1] and Incas[2] understood how to extract heat from light, so perhaps bolometry has existed since antiquity. (The more pedantic might argue that, first of all, the above examples are not designed to measure the heat in light in any precise way, and hence the "-metry" portion of the term is invalid, and additionally, it appears likely that the ancients believed that mirrors and lenses could be used to ignite fires by capturing and concentrating fire received from the sun, hence implying that there was no concept at the time of measuring the heat content of light to begin with.) I will begin this didactic history of the bolometer by referencing Frederick William Herschel [9], who in the waning days of the 18th century figured out

[1] Archimedes is said to have used mirrors to concentrate sunlight onto Roman warships during the siege of Syracuse (214-212 BC) to have set them on fire; while the veracity of this story has long been in dispute, the Greeks were certainly aware of the heat content of sunlight, and this is mentioned by Aristophanes in his play The Clouds of 424 BC: "Good chance but you have noted / A pretty toy, a trinket in the shops, / Which being rightly held produceth fire / From things combustible – / A burning glass, / Vulgarly call'd" (translation of Thomas Mitchell, 1822).

[2] For instance, at the Feast of Raymi, a fire was ignited by using a concave metal mirror focused onto cotton wool; the failure of the sun, a key deity of the Incan religion, to ignite said fire was taken as an ill omen (see, e.g, History of the Conquest of Peru by William Prescott, 1855). In a fictional account (Prisoners of the Sun, 1949), Hergé suggests that the pyres of the executed might be ignited by mirror, thereby placing the responsibility for carrying out executions on the sun god.

that the infrared exists by means of no more simple equipment than a prism and a thermometer. Of course, the quantitative measurement of the solar spectrum was quite inaccurate since he was unaware of the change in dispersion with wavelength in the infrared, but the idea of using a thermometer to measure the total amount of energy in a beam of light was still pretty clever. This led, the better part of a century later, the scientific polyglot Samuel Pierpont Langley to invent the bolometer [10] for his investigations of the solar spectrum (a problem not fully solved during the intervening eighty years). His bolometer design is shown in Figure 1.

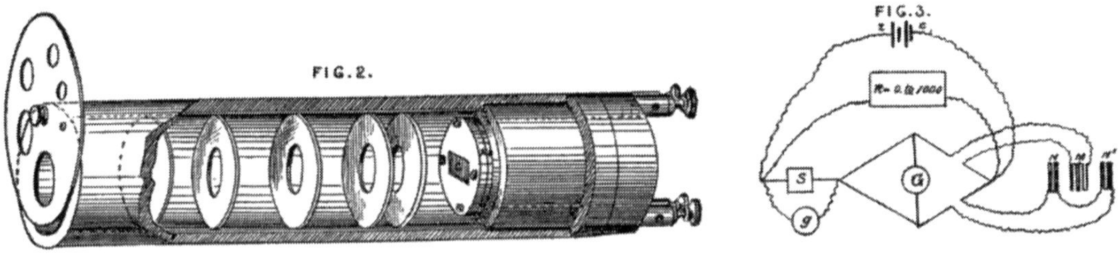

Figure 1. Bolometric instrument of S. P. Langley.

The next major advances in bolometry keep the core of Langley's original principles: an absorber converting light into heat, and a thermistor converting heat into an electrical signal. Superconducting bolometers made an early appearance, separately suggested by Goetz [11] and developed by Andrews et al. [12], with a composite structure consisting of a blackened aluminum foil absorber attached to a tantalum thermistor. It is of historical note that the Andrews group [13] found excess noise in their CbN (now known as NbN) superconducting bolometer for which they could ascertain no cause; this problem seems to plague researchers still today. However, no excess noise was reported in the more systematic study of Fuson [14] at the same institution. An important but logical improvement (see, e.g., Boyle and Rodgers' 4 K carbon resistor bolometer [15]) was made by Low [16] in 1961 by cooling the bolometer with liquid helium, increasing the sensitivity tremendously (by at least a factor of ten, from a noise equivalent power (NEP) of $\sim 5 \times 10^{-12}$ W/$\sqrt{\text{Hz}}$ [15] to $\sim 5 \times 10^{-13}$ W/$\sqrt{\text{Hz}}$ [16]). It is in 1977 that the term "transition edge" makes its way into the literature [17], with the seminal works of Clarke et al. [18] (Figure 2).

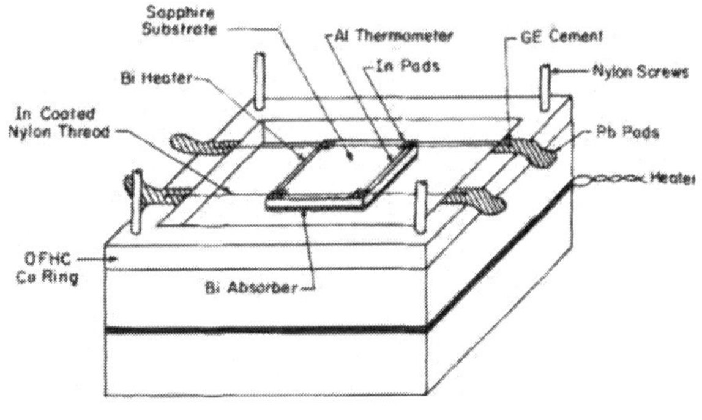

Figure 2. Superconducting bolometer demonstrated by Clarke et al. [17].

The idea of multiplexing large arrays of TES bolometers [19] goes back at least as far as 1990, although that line of research apparently ended shortly thereafter. The modern – an imprecise word, to be sure – TES bolometer period begins, in my opinion, with the mid-1990s work of the Berkeley group on voltage-biased superconducting bolometers [20], although the salient ideas

Journal of Physics: Conference Series **155** (2009) 012001 doi:10.1088/1742-6596/155/1/012001

were presented first by Kent Irwin [21] for particle detection. I therefore end the history of superconducting bolometers in 1996.

No introduction to bolometers would be complete without at least some mention of the role of fabrication techniques and operational approaches. For instance, the idea of making bolometers by sputtering metals onto thin films has been around for at least eighty years [22]. A bolometer operating below the thermodynamic limit for 300 K radiation existed in 1946 [14]. Co-deposition of patterned thermistor and absorber films on substrates goes back at least forty years [23] . An array of bolometers – five pixels – was described by De Waard at the same time [24]. Pushing to subKelvin temperatures [25] brought another order of magnitude improvement in the noise equivalent power (NEP) of a factor of ten over the previous [16] work to $\sim 3 \times 10^{-14}$ W/$\sqrt{\text{Hz}}$. The concept of the monolithic bolometer – entirely produced by microlithographic techniques – was a breakthrough in 1984 by Downey et al. [26], although much of that had been done earlier [27]. Cooling bolometers with an adiabatic demagnetization refrigerator (ADR) was put forth by Britt and Richards in 1981 [28] with an eye to space flight use. The performance of a 200 mK bolometer was another two orders of magnitude better [29], at an NEP of $\sim 2 \times 10^{-16}$ W/$\sqrt{\text{Hz}}$. The Berkeley group proposed the TES-based hot electron bolometer [30] – an antenna-coupled device – which would provide a 100 mK NEP of $\sim 2 \times 10^{-18}$ W/$\sqrt{\text{Hz}}$, yet another two orders of magnitude and below the photon limit for most imaging applications except for the darkest sky in space.

There is also a wealth of historical discussion of the theory of bolometer performance. The WWII-era work of Milatz and Van Der Velden [31] on the noise components of bolometers is familiar even today. Billings et al. [32] worked out the sensitivity and temporal properties of metallic bolometers in 1947. The prolific R. Clark Jones developed explicit treatments for semiconducting bolometers [33] the ultimate sensitivity of Lambertian detectors [34], figures of merit for detectors [35], and culminating in his description of nonequilibrium bolometer performance in 1953 [36], three decades before the similar well-regarded work by John Mather [37]. This dynamic impedance view was extended by Y. Urano [38], who established a thermoelectric dynamic version resulting in stability predictions. At a similar time, investigations into superconducting bolometers pursued still-relevant topics such as the effect of bias current on bolometer behavior [39] and the excess noise on the transition [40]. An equivalent-circuit technique for superconducting bolometers was given by Maul and Strandberg [41], who were also involved in a noise-mitigation technique using external magnetic fields [42], and who claimed to have reached the thermodynamic limit by that means. Bertin and Rose [43] compared superconducting bolometers to the then-state-of-the-art semiconducting bolometers and concluded that superconducting bolometers were superior. Too bad this information was largely disregarded. An effort was made to model $1/f$ noise [44] and phase-slip shot noise [45] in bolometers. The award for not-quite-insightful-enough goes to the ironically-named Sherlock and Wyatt for their general analysis of self-heating in bolometers which mentioned in passing that for superconducting bolometers, the positive electrothermal feedback induced by the current bias could be reduced to near zero if the load resistor were set to a value of roughly the TES resistance [46].

2.3. A Brief Theoretical Description of the TES Bolometer

A very thorough discussion of the theory of TES bolometers, in particular derivations of or at least lists of relevant equations, has been written by Irwin and Hilton [47]. I will not try to reproduce that weighty tome here, but will rather summarize the equations using their notation to facilitate comparison. The linear differential equations for the electrothermal behavior of a TES bolometer are:

$$L\frac{\mathrm{d}I}{\mathrm{d}t} = V - IR(T,I) \text{ and } C\frac{\mathrm{d}T}{\mathrm{d}t} = P_{\text{bath}} + P_{\text{J}} + P, \qquad (1)$$

Journal of Physics: Conference Series **155** (2009) 012001 doi:10.1088/1742-6596/155/1/012001

where the electrical parameters refer to the inductance L, the Thevenin equivalent bias voltage V, the measured detector current I, the Thevenin equivalent bias load resistor R_L, and the TES resistance $R(T, I)$, and the thermal parameters are the heat capacity C and TES temperature T, P_{bath} is the power cooling the TES to the heat bath, P_J is the Joule power dissipation from the bias, and P is the optical signal power to be found. In detail, the resistance function depends on other physical variables such as the external magnetic field, but for simplicity it is easier to consider the constant-field case. The shape of $R(T, I)$ can be modeled theoretically and measured directly (Figure 3).

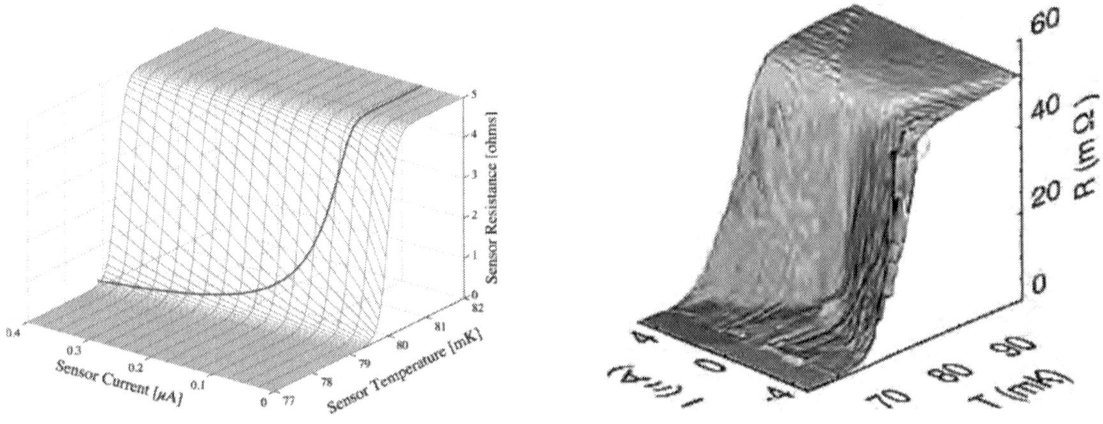

Figure 3. (Left) A modeled resistance function $R(T, I)$, showing the path followed during a bias sweep; adapted from a figure by Blas Cabrera [48]; (Right) Measurements of the $R(T, I)$ surface by Lindeman et al. [49], showing the shape of the transition in detail; note the similarity to the gross features of the model, given that the data have bipolar current.

The resistance can be expanded about nominal values of R, T, and I in the small-signal limit:

$$R(T, I) \approx R_0 + \left.\frac{\partial R}{\partial T}\right|_{I_0} \partial T + \left.\frac{\partial R}{\partial I}\right|_{T_0} \partial I \tag{2}$$

which has terms reminiscent of the dimensionless sensitivity of the resistance with temperature and current:

$$\alpha = \frac{T_0}{R_0} \left.\frac{\partial R}{\partial T}\right|_{I_0} \text{ and } \beta = \frac{I_0}{R_0} \left.\frac{\partial R}{\partial I}\right|_{T_0} \tag{3}$$

Substituting in a large number of similar small-signal approximations providing linear perturbations about equilibrium, the overall coupled differential equations relating the current and temperature to the voltage and power become:

$$\frac{\mathrm{d}}{\mathrm{d}t} \begin{bmatrix} \partial I \\ \partial T \end{bmatrix} = - \begin{bmatrix} \frac{1}{\tau_{el}} & \frac{\mathscr{L}_I G}{I_0 L} \\ -\frac{I_0 R_0 (2+\beta_I)}{C} & \frac{1}{\tau_I} \end{bmatrix} \cdot \begin{bmatrix} \partial I \\ \partial T \end{bmatrix} + \begin{bmatrix} \frac{\partial V}{L} \\ \frac{\partial P}{C} \end{bmatrix} \tag{4}$$

where the variable τ_{el} and τ_I refer to the electrical and thermal time constants, respectively, and the DC loop gain $\mathscr{L}_I$ under constant current I is given by:

$$\mathscr{L}_I = \frac{P_{J_0} \alpha_I}{G T_0} \tag{5}$$

Solving Equation 4 and integrating yields two eigenvalues for the reciprocal of the time constants involved in the TES bolometer response:

$$\frac{1}{\tau_+} = \frac{1}{2\tau_{\rm el}} + \frac{1}{2\tau_I} + \frac{1}{2}\sqrt{(\frac{1}{\tau_{\rm el}} - \frac{1}{\tau_I})^2 - 4\frac{R_0}{L}\frac{\mathscr{L}_I(2+\beta_I)}{\tau}} \tag{6}$$

$$\frac{1}{\tau_-} = \frac{1}{2\tau_{\rm el}} + \frac{1}{2\tau_I} - \frac{1}{2}\sqrt{(\frac{1}{\tau_{\rm el}} - \frac{1}{\tau_I})^2 - 4\frac{R_0}{L}\frac{\mathscr{L}_I(2+\beta_I)}{\tau}} \tag{7}$$

When a varying power is applied to a bolometer, the measured response will follow the power, but with a responsivity that varies with frequency according to the two poles defined by the time constants in Equations 6 and 7. If a small, sinusoidal power is applied to the bolometer, the solution to Equation 4 can be explicitly solved, and contains the time constants τ_+ and τ_-. Resubstituting, one can derive the small-signal response in current to a given power:

$$s_I(\omega) = -\frac{1}{I_0 R_0}\left[\left(\frac{L}{\tau_{\rm el} R_0 \mathscr{L}_I}\right) + \left(1 - \frac{R_l}{R_0}\right) + i\omega\frac{L\tau}{R_0\mathscr{L}_I}\left(\frac{1}{\tau_I} - \frac{1}{\tau_{\rm el}}\right) - \frac{\omega^2\tau}{\mathscr{L}_I}\frac{L}{R_0}\right]^{-1} \tag{8}$$

Typically, a TES bolometer is run with a strong voltage bias ($R_{\rm L} \ll R_0$) and strong electrothermal feedback (high $\mathscr{L}_I$), and so the DC responsivity simplifies to:

$$s_I(0) = -\frac{1}{I_0 R_0} \tag{9}$$

In a similar fashion to the above, the complex impedance of a TES bolometer can be calculated from Equation 4 by applying a voltage $V(\omega)$ and a response current $I(\omega)$. The result is a total circuit impedance of

$$Z(\omega) \equiv \frac{V(\omega)}{I(\omega)} = R_L + i\omega L + R_0(1+\beta_I) + \frac{R_0\mathscr{L}_I}{1-\mathscr{L}_I}\frac{2+\beta_I}{1+i\omega\tau_I} \tag{10}$$

which features the impedance of the bias circuit and the TES, and hence we can write

$$Z_{\rm TES}(\omega) = R_0(1+\beta_I) + \frac{R_0\mathscr{L}_I}{1-\mathscr{L}_I}\frac{2+\beta_I}{1+i\omega\tau_I} \tag{11}$$

In addition to the performance parameters above, a prediction of the noise of a TES bolometer is an absolute necessity. This has been done by many in the past, and Kent Irwin has added a nonlinear, nonequilibrium, first and second order calculation [50] to the existing theory [47]. The total noise is a combination of the thermal fluctutation noise in the bolometer thermal isolation, Johnson noise in the thermistor, noise from the bias circuit, noise from the readout amplifier, and excess noise arising from one or more of a variety of potential nonideal performance aspects. In this order, the total noise is:

$$\begin{aligned} S_{\rm total}(\omega) =& 4k_{\rm B}T_0^2 G F_{\rm thermal} + 4k_{\rm B}T_0 R_0 I_0^2 (1+2\beta_I)\frac{1+\omega^2\tau^2}{\mathscr{L}_I^2} + \\ & 4k_{\rm B}T_0 R_L I_0^2 \frac{(\mathscr{L}_I - 1)^2}{\mathscr{L}_I^2}\left(1+\omega^2\tau_I^2\right) + \frac{S_{I_{\rm amp}}}{|s_I(\omega)|^2} + S_{\rm Excess}(\omega) \end{aligned} \tag{12}$$

The thermal link function $F_{\rm thermal}$ can be calculated in the specular phonon or scattered phonon limit for a thermal link with temperature dependence T^n [51]:

$$F_{\rm thermal} = \frac{\left(\frac{T_{\rm bath}}{T_0}\right)^{n+2} + 1}{2}\ \text{(specular) or } F_{\rm thermal} = \frac{n+1}{2n+3}\frac{\left(\frac{T_{\rm bath}}{T_0}\right)^{2n+3} - 1}{\left(\frac{T_{\rm bath}}{T_0}\right)^{n+1} - 1}\ \text{(scattered)} \tag{13}$$

The thermal link function represents the behavior of fluctuations in a thermally isolating structure where a significant conductivity gradient exists across the structure. The conductance G is calculated at the warm end (T_0), and hence F_{thermal} reaches a maximum of unity at that point.

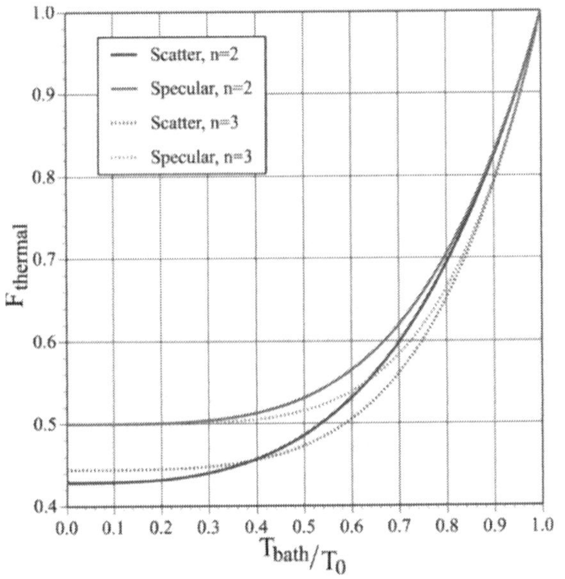

Figure 4. Thermal link functions as given by equations for specular- or scattering-dominated links in Dan McCammon's review [51].

Figure 4 shows the functional dependence for a reasonable range of the thermal conductivity index n. In Equation 12, there are two analytic unknowns. One is the amplifier noise, which, while it can be subjected to some amount of analysis, is more likely to be a quantity of known (measured) value in current noise units, hence the notation $S_{I_{\mathrm{amp}}}(\omega)$. The other is a quantity of excess noise in power units $S_{\mathrm{Excess}}(\omega)$. This is a subject of much investigation, and will be summarized later.

2.4. State of Maturity of TES Bolometer Development (or, how did we get here?)
My thesis advisor, Tom Phillips, once told me that the best papers to write are those that are the first in their field or those that are the last in their field. In keeping with this notion, and in the earnest expectation that the last papers on TES bolometer aspects are far from written, this section will summarize developments in TES bolometers in the first decade following the advent of the voltage-biased approach, what I will term the Decade of Development.

Since the Berkeley group introduced the voltage-biased superconducting bolometer [20] (although with a nod to Stanford [21]), this is a good point of departure to discuss the decade of 1996-2006. Their early efforts led to useful theory and experiment in the strong-feedback limit [52]. They also introduced a TES bolometer manufactured entirely with photolithography [53] – see Figure 5 –a step forward in fabrication technique that enables very large format arrays – which turned out to be the next thing they demonstrated, although only mechanically [54]. This same trio of authors also developed a theoretical treatment of the excess thermal fluctuation noise arising from undercoupled or distributed heat capacities on the bolometer [55].

The Stanford group achieved one early and notable milestone: the first astronomical application of TES bolometers. Using a small number of tungsten TES hot electron bolometers, Romani et al. [56] used photon-counting pixels that produced arrival time ($\delta t = 100\ ns$) and energy ($\delta E_\gamma \leq 0.15\ eV$) resolved measurements of the Crab pulsar from the near-IR through the near UV. This result is such a good indicator of the novel aspects of TES bolometers that it

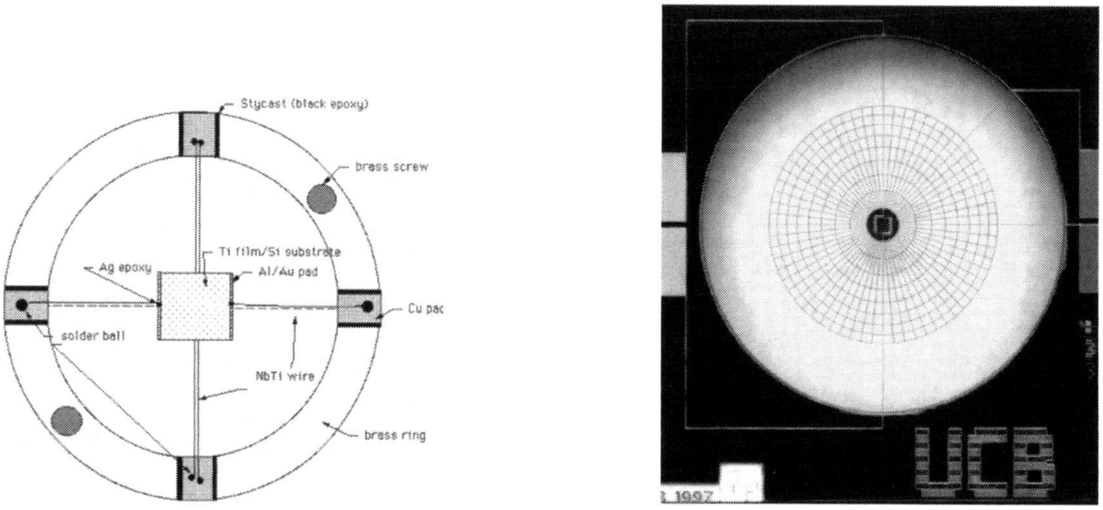

Figure 5. (Left) The layout of an early voltage-biased superconducting bolometer (e.g., [20]) is reminiscent of that in Figure 2 . (Right) The all-lithographic version (e.g., [53]) made large-format arrays possible.

represents an entirely new (although not CMB-related) approach to astronomical observations. Figure 6 shows the resultant photon count rate vs. phase and wavelength. As an aside, it

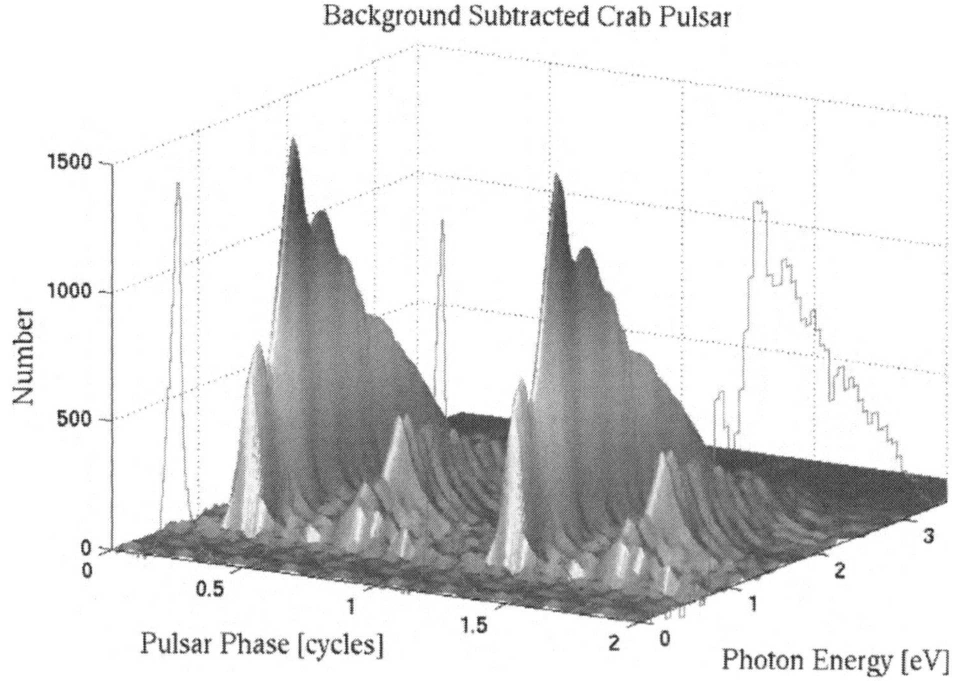

Figure 6. The first astronomical application of TES bolometers was to measure the brightness of the Crab Pulsar as a function of wavelength and phase [56].

is worthy of mention that this may be the only astronomical instrument yet fielded using

TES bolometers without a multiplexed SQUID readout. The NASA/GSFC group has been developing TES bolometers for long wavelength purposes since 1998 [57], in strong collaboration with NIST/Boulder. Developments intended for Herschel resulted in the first demonstration of a TES bolometer system operating using SQUID multiplexers [58] including the optical detection of light while multiplexing. Staguhn et al. [59] demonstrated that TES bolometers could be used with SQUID multiplexers and still retrieve Johnson-noise-limited readout. Later, ongoing efforts resulted in a robust design for near-phonon-noise-limited bolometers [60, 61]. Benford et al. fielded the first astronomical instrument using multiplexed TES bolometers [62]and produced the first astronomical images using an array of multiplexed TES bolometers [63]. This is shown in Figure 7, since it may be the first astronomical demonstration relevant to CMB polarimetry.

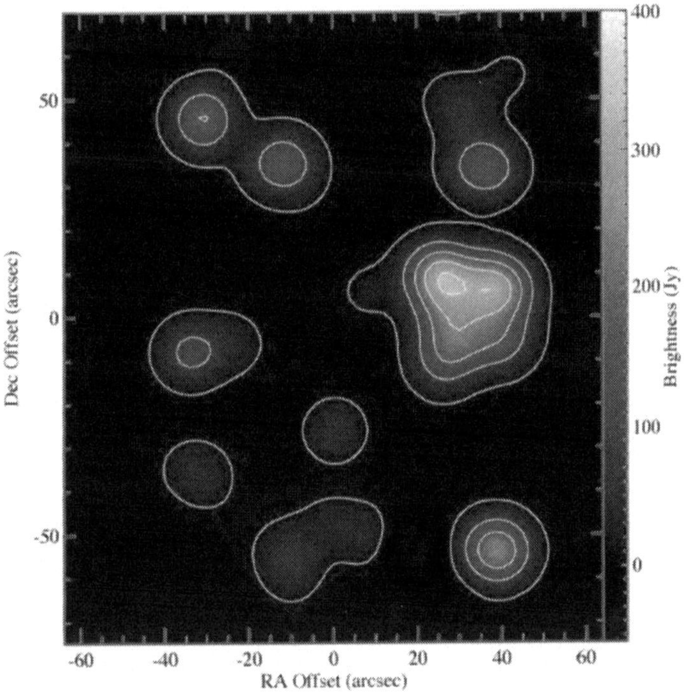

Figure 7. TES bolometer array 350 m image of the high-mass star formation region G34.3+0.2 (Benford et al. [63]).

The closing of the Decade of Development at Goddard was summarized by the realization that the TES capability available at the close of 2005 was robust enough to pursue astronomical instruments, rather than technology developments. In Section 4, I will cover the issue of technology readiness level – but in advance, I note that this decision marked the completion of TRL 4 ("Component and/or breadboard validation in laboratory environment"). A separate group at Goddard engaged in the development of TES-based X-ray microcalorimeters pursued the precise measurement and theoretical understanding of detector parameters. These can be typified by, for example, the interesting plot of $\beta(V)$ by Saab et al. [64], the hinting-at-some-underlying-physics of the excess noise vs. resistance of Lindeman et al. [65] and the optimization modeling of Bandler et al. [66].

NIST/Boulder has had a strong history in the SQUID multiplexers used by many groups, in particular the time-domain devices that have been incorporated into past and ongoing instrument developments. As a result, most of their focus was on the maturing of the readouts, resulting in the 8-element [67] 32-element [68], and 1,280-element [69] time domain SQUID

Journal of Physics: Conference Series 155 (2009) 012001 doi:10.1088/1742-6596/155/1/012001

multiplexers. They also investigated the excess noise problem in TES bolometers using both geometry (in conjunction with NASA/GSFC [70]) and magnetic fields to mitigate the effect [71].

In contrast to the statement a few paragraphs ago, the Decade of Development did see the beginning of several TES-based millimeter-wavelength instruments. One of them was developed at the University of Pennsylvania, a 3 mm camera for the Green Bank Telescope [72] started under the name "Penn Array Receiver." (At the time of first observations, it was renamed to MUSTANG, but that was a later event). This camera is based on an 8×8 TES array [73] in a fully planar, close-packed format – the first such array designed for long wavelength astronomy. At the same time, the Princeton group began work on the Atacama Cosmology Telescope [74], which featured a three-band camera in the millimeter wavelength range. In a far-reaching vision, each band would feature a 32×32 array for true multikilopixel operation [75], based on the NASA/GSFC pop-up-detector architecture [76]. During the middle part of the 2000s, detector array development and, perhaps more importantly, detector system development proceeded on these two instruments. However, it was a time of much promise and little actual use of TES arrays in actual astronomical applications.

At Caltech/JPL, attention to TES bolometers started late (in part because of that august institution's well-established semiconducting bolometer array work receiving the majority of the focus), and so their most notable contribution – and quite relevant to CMB polarimetry – concerns antenna-coupled TES bolometers. One notable example is a membrane-isolated TES bolometer coupled via a microstrip to a twin-slot antenna [77]. While not strictly TES-related, the wideband or dual-polarization antenna networks of Goldin et al. [78] have definite application to them, as a TES version appeared the following year [79].

I will close this section with a return to the Berkeley group, who in addition to pursuing technology development (e.g., their own twin-slot TES bolometer [80], took the first step into the next generation: large-scale engineering of TES bolometer arrays. Their APEX-SZ camera achieved first light during an engineering run in December 2005 [81], with a small fraction of their bolometer complement (13%) operational. This instrument features monolithically-fabricated superconducting TES bolometer arrays in a hex-packed spiderweb geometry, read out by frequency-domain SQUID multiplexers. The first light results (Figure 8) point the way to mature, kilopixel-scale TES instruments, and are therefore a suitable closing to the Decade of Development of the TES bolometer.

2.5. Recent Papers and Developments (or, here is where we are.)

2.5.1. TES Excess Noise and Geometry Over the last few years, the quantification of "excess noise"[3] has become a burgeoning field. One source of such, "β noise" [82], is a consequence of rigorous nonequilibrium theory [50], and manifests as an increased Johnson noise. A similar but unrelated "β noise" is postulated by Ullom [71], but not duplicated by Hoevers [83]. Instead, they consistently find an effective white voltage noise [84, 85] Other noise sources include phase slip shot noise or flux-flow noise [86, 87], percolation noise [88], internal thermal fluctuation noise (ITFN) [52, 89], and the rather empirical-sounding fluctuation superconductvity noise [90] A technique for reducing ITFN (high resistivity TES with high conductivity substrate) appeared also to suppress the white voltage noise [91]. This makes it seem like a valid source of noise. On the other hand, an analysis on several points led Brandt et al. to conclude that both phase slip and percolation noise are present [92]. The detailed thermal models of Kinnunen et al. show that white noise and ITFN are both present [93]. As an assignment to reflect on, note that the electron-phonon-decoupling device of Barrentine et al. [94] might show no excess noise and, more importantly, a virtually featureless spectrum from 10 Hz – 1 kHz. Several groups have

[3] I shall continue to use this term despite what Kent Irwin said in 2006 [50], tongue presumably lodged firmly in cheek, that there is no such thing as excess noise, just "noise" and the ratio between it and what you expected.

Journal of Physics: Conference Series **155** (2009) 012001 doi:10.1088/1742-6596/155/1/012001

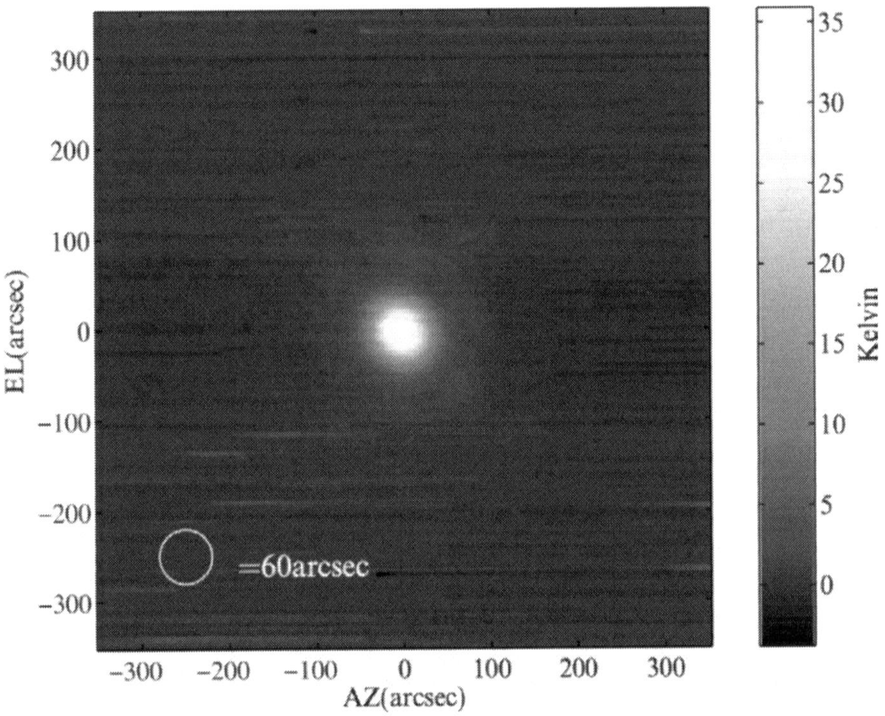

Figure 8. TES bolometer array 2 mm wavelength image of Jupiter using APEX-SZ (from Dobbs et al. [81]).

approached the excess noise as a geometrical problem. In this view, the reason(s) for the noise can all be reduced to a TES geometry problem, such as edge effects, percolation state space, etc. Adding or changing boundary conditions such as normal metal bars, stripes, or dots can change these effects and influence the excess noise. Among the first efforts of this sort were bars and stripes [60, 61, 71]. Dots are preferred by others [85]. A very clever circular geometry[4] has been in use for several years [95]. All of these appear to have excess noise, and where it is suppressed, other detector parameters have also changed. I would therefore conclude that to first order, geometry is not the determining factor in excess noise. In summary, while there are many competing and apparently valid (both theoretically and experimentally) models for the "excess noise", the only robust conclusion that can be drawn is that what gives rise to it is a poorly understood function of many (perhaps hidden) variables in the design of a TES bolometer.

2.5.2. Antenna-coupled devices Antenna-coupling provides a convenient method for controlling the solid angle of response of a TES bolometer, and hence limits the stray light it receives. There are additional benefits that can be realized in packaging (such as space to run wiring), although a price is paid relative to the efficiency of a filled array for wide-field mapping [96]. The simplest antenna-coupled TES bolometer array would be the effective equivalent of a bow-tie antenna array [1], but more sophisticated approaches are possible. Waveguide coupling includes feedhorns that can significantly improve the sidelobe response [2, 3]. The microstrips that are

[4] While I have called it a circular geometry, it is in fact annular. Due to its relation with the geometry of the Corbino effect (the Hall effect in a circular plate, named for Orso Mario Corbino), this is referred to as the Corbino TES geometry.

often used in an antenna-coupled TES bolometer can be used to define the wavelength response in an array, even allowing multicolor arrays [4] Alternatively, the microstrip can be used to combine the power of multiple antennas, such as in making a beam-forming antenna in two polarizations [5, 6]. Or, in a more integrated approach, functions such as polarization switching, band defining, and beam defining can be achieved simultaneously[7, 8].

The figures of merit for arrays of antenna-coupled TES bolometers are fairly straightforward. Depending on which of these functions are being defined by the antenna-coupling approach, they include: the repeatability of band definition and out-of-band suppression; the forward efficiency and sidelobe rejection of the spatial response function; the degree of cross-polarization and its repeatability; and the efficiency of using on-chip polarization modulation as opposed to full aperture modulation. For each approach, there is an increasing maturity of the numerical techniques for predicting antenna-coupled performance, but there is as yet a dearth of well-analyzed experimental effort on large data sets to extract the precise proof of their function.

2.5.3. Sensitivity Given the brightness of the CMB, it could be claimed that the sensitivity of the TES bolometer is not a technologically challenging aspect. This is, of course, a relative measure as compared to the sensitivity required for the detectors on a cryogenic space telescope operating at wavelengths of around 200-400 μm. A calculation of the components contributing to sky brightness in near-Earth space is shown in Figure 9, along with an estimate for the optical power and photon NEP that would be likely for a CMB polarization experiment in space. The NEP requirement for a CMB polarization detector is around 2×10^{-18} W/$\sqrt{\mathrm{Hz}}$. Prototype far-infrared TES bolometers such as those for SAFIRE with an NEP requirement of 10^{-19} W/$\sqrt{\mathrm{Hz}}$ [97], BLISS [98] on SPICA with a best NEP of $\sim 6\times10^{-20}$ W/$\sqrt{\mathrm{Hz}}$ [99], SAFARI on SPICA with an NEP of $\sim 10^{-18}$ W/$\sqrt{\mathrm{Hz}}$ [100], and even technology developments for the SAFIR observatory with an NEP approaching 10^{-20} W/$\sqrt{\mathrm{Hz}}$ [101, 102].

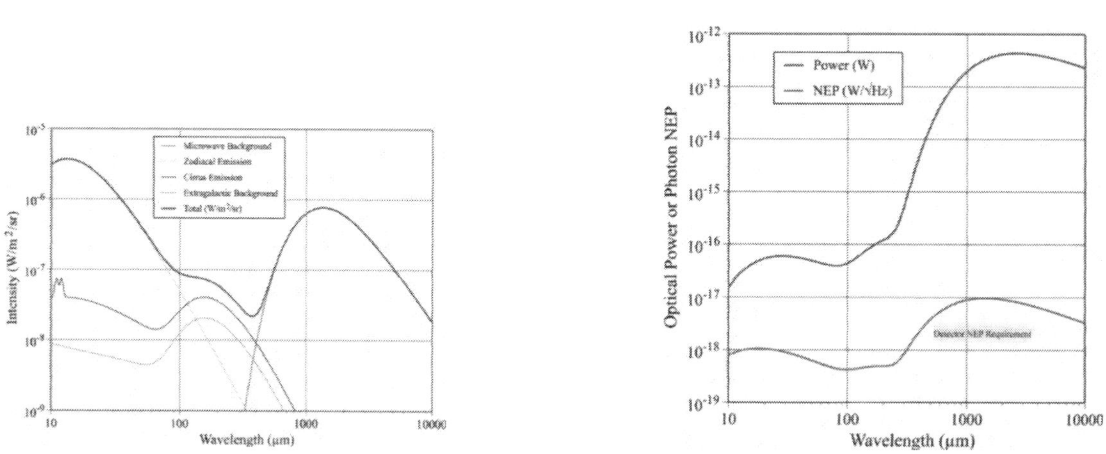

Figure 9. (Left) The calculated components of the celestial background shows that the CMB bands are significantly brighter than the far-infrared bands, and so sensitivity will be driven by far-IR instruments. (Right) A calculation of the optical power and photon NEP for a representative CMB polarization mission shows that the detector NEP requirement is not extreme.

2.5.4. Instruments As mentioned above, the Atacama Cosmology Telescope features in its Millimeter Bolometer Array Camera [103] one of the largest TES array sets currently in production or in use. Featuring three arrays of 1,024 pixels each, it has only recently been operated in its complete configuration. At the time of this writing, the first light images consisted only of the data from one array (145 GHz) [104] from late (October-December) 2007. A photo of an ACT array and a map of the Sunyaev-Zel'dovich effect in the Bullet Cluster are shown in Figure 10.

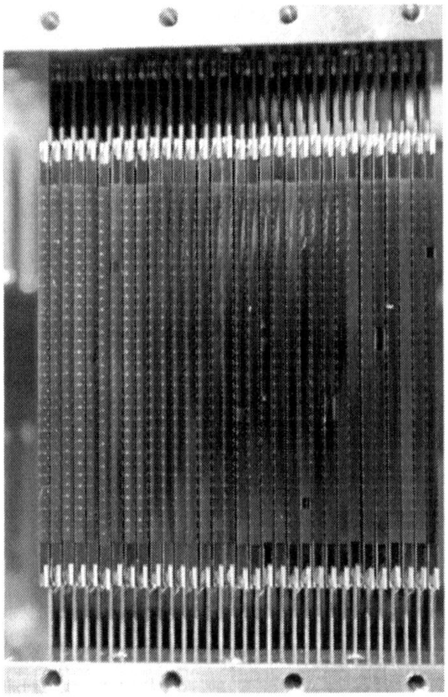

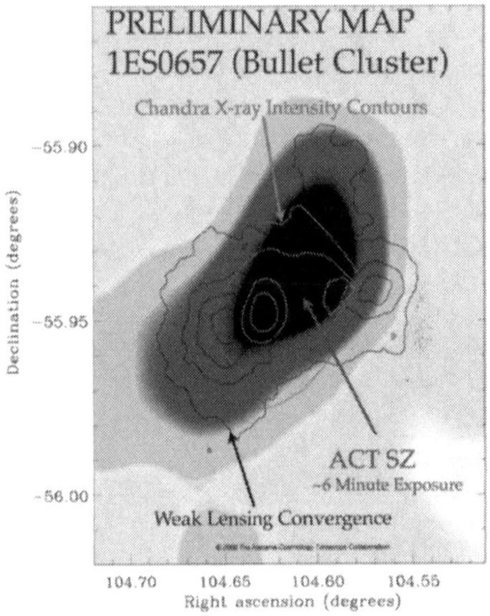

Figure 10. (Left) a photo of the 32×32 array of TES pop-up detectors produced by Princeton and NASA/GSFC. (Right) A 6 minute exposure at 2mm resulting in a Sunyaev-Zel'dovich effect map of the Bullet Cluster [105] proves the power of large arrays.

The MUltiplexed Superconducting TES Array at Ninety Gigahertz (MUSTANG), a collaborative effort between the University of Pennsylvania, NASA/GSFC, NIST/Boulder, NRAO, and Cardiff, is an instrument designed to take advantage of the high angular resolution possible at 3 mm wavelength at the Green Bank Telescope (GBT). The pixels in its 8×8 array subtend around $4''$ each, which, combined with the large scan patterns possible (indeed, necessary) on the GBT, produces large maps with fine detail. This can be seen in Figure 11, where the map width of $\sim 400''$ contains around 3,000 individual beam areas. MUSTANG achieved its first light image on Saturn on September 26, 2006, making it the first operational fully-sampled array at a telescope, and is one of the first to produce astronomical results [106]. Also, while not relevant for satellite or balloon-borne applications, it is worth pointing out that MUSTANG was also the first instrument to collect scientific data with a TES array cooled by a pulse tube cooler.

In August 2005, the NASA/GSFC group kicked off work on a new camera, the Goddard IRAM Superconducting 2 Millimeter Observer (GISMO), a large-field-of-view imager for the 2 mm wavelength band. Optimized for operation at the IRAM 30m telescope [107], it could achieve greater sensitivity to very high redshift sources than cameras operational or proposed at shorter wavelengths [108]. Its 8×16 array of close-packed TES bolometers [109] covers an extent of roughly $2' \times 4'$ by using fast optics providing near-optimal survey speed and source position/flux recovery [110]. The bolometer array is the first fielded array of the backshort-under-grid (BUG)

Journal of Physics: Conference Series **155** (2009) 012001 doi:10.1088/1742-6596/155/1/012001

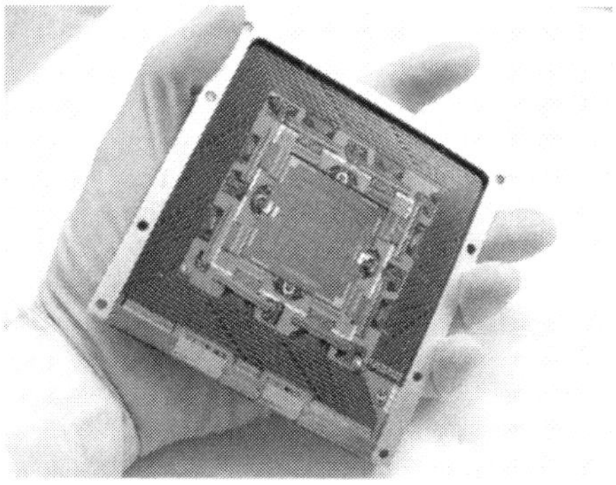

Figure 11. (Left) a photo of the 8×8 array of planar, close-packed TES bolometers fabricated at NASA/GSFC for the U. Penn MUSTANG instrument. (Right) An image at 3mm wavelength of the Crab nebula shows a wealth of structure revealed by the high angular resolution of the GBT.

architecture [111–114], a technology developed to enable and mature versatile, space-qualifiable kilopixel TES arrays with NIST large-format SQUID multiplexers. GISMO achieved first light in November 2007([see Figure 12).

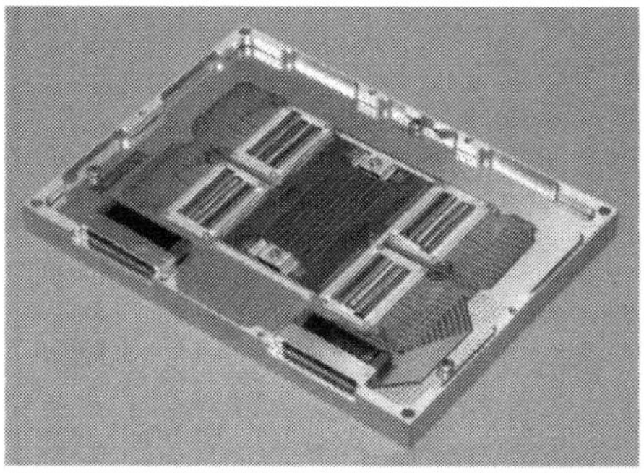
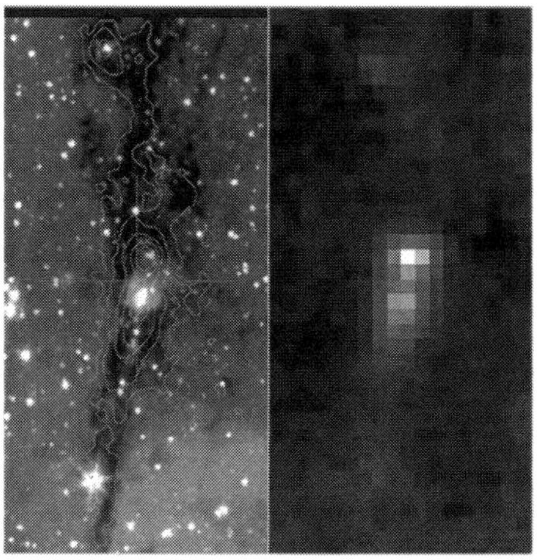

Figure 12. (Left) a photo of the 8×16 array of planar, close-packed TES bolometers fabricated at NASA/GSFC for the GISMO instrument. (Right) An IRAC image with 1.3 mm MAMBO contours shown to the left of the GISMO 2 mm image of the same field: a dense core known as IRDC43.

The South Pole Telescope is another Sunyaev-Zel'dovich-optimized telescope like ACT, but rather larger (10 m as opposed to 6 m) but with fewer detectors (960 as opposed to 3,072). Its first light on a cluster, in April 2007, narrowly beat ACT; however, significant data analysis remains before achieving the theoretical performance [115]. A similar future instrument is CLOVER, again a tri-band system, but aimed at polarimetry of the CMB. CLOVER's compact design [116] is better-suited to detecting B-mode polarization. Its TES bolometer array [117] is made up of individual detectors that are microstrip-coupled to antennas, providing the possibility of packing the array rather differently than would be possible with a monolithic array.

Figure 13. (Left) photos of the 39-element horn-coupled array of TES bolometers fabricated at IPHT/Jena for the MPIfR SABOCA instrument. (Right) A 350 μm image of the NGC6334 star forming region.

A group led by MPIfR at Bonn has developed the SABOCA 350 μm camera, using TES arrays produced at the Institute of Photonic Technology (IPHT) of Jena. Using 39 bolometers coupled to a conical horn array, the instrument saw first light at APEX in May 2008. Photos of its design and an early image are shown in Figure 13. One notable feature of SABOCA relevant to CMB polarimetry is its measured stability (after subtracting correlated sky noise): a noise power spectrum shows no $1/f$ drift down to 0.0025 Hz [118]. Stability on minute timescales makes it possible for even very slow mechanical polarizers to complete a cycle within a bolometer's stability time.

In addition to the above, there are several TES array-based instruments currently in development. I present briefly[5] some sense of what is to come in the next few years, since these may push the technology readiness level (see next sections) better than anything else could before 2010. The most advanced of these is SCUBA-2 [119], which is already at the JCMT and awaiting first light [120]. While it has only single commissioning-grade 32 × 40 TES arrays in it currently, a set of science-grade arrays is expected in early 2009. The EBEX [121]balloon experiment will fly its $\sim$ 1500 TES bolometers for engineering in 2008 and for science by 2010 [122]. Similarly, SPIDER [123], with its $\sim$ 2600 TES bolometers [124], is scheduled for its first flight in April 2010 [125, 126] Finally, in around 2012, the SAFIRE instrument for SOFIA [127], with its low background kilopixel bolometer arrays [128] for suborbital far-infrared spectroscopy [129], will demonstrate detectors that have passed some of the rigors of design for space flight.

2.6. Technology Readiness Level Assessment (or, where is here anyway?)

This section will cover the Technology Readiness Level (TRL[6]), a minefield of perceptions. Since this assessment is mine and mine alone, it is incumbent upon me to explain in detail why I rate things as I choose to do. A representative definition of the TRL scale is given in Figure 14. The definitions of mid-level TRLs as put forth by John Mankins [130] and as interpreted by this author are provided in Table 1. My scale can be illustrated by the following example: if you

[5] When I say "briefly", I mean that I give almost no details – this is what the cited references are for, if you're curious – and that I cover only a fraction of what is likely to be extant in the coming five years or so. If I tried to be more complete, it would only increase the offense to the ever-shrinking group of experiments that I had omitted, forgotten, was unaware of, or had decided belligerently to ignore.

[6] Note: in speaking, people often refer to the 'TRL level' of a technology. At the risk of arrogant pedantry, I point out that the acronym 'TRL' already contains the word 'level'. Hence, the appropriate phrase is to refer to 'the TRL of a technology.'

intend to produce, starting from scratch, a working large-format TES bolometer array camera for a major project, then your TRLs can be defined as follows: TRL 3 occurs when you have your first working TES devices in your lab cryostat; TRL 4 is reached when a prototype of the flight array is working in the lab; TRL 5 is marked by successfully operating the prototype array on a ground-based telescope and understanding the details of its performance; TRL 6 can be achieved when your completed suborbital instrument flies and/or when your engineering models of the flight array have undergone qualification testing.

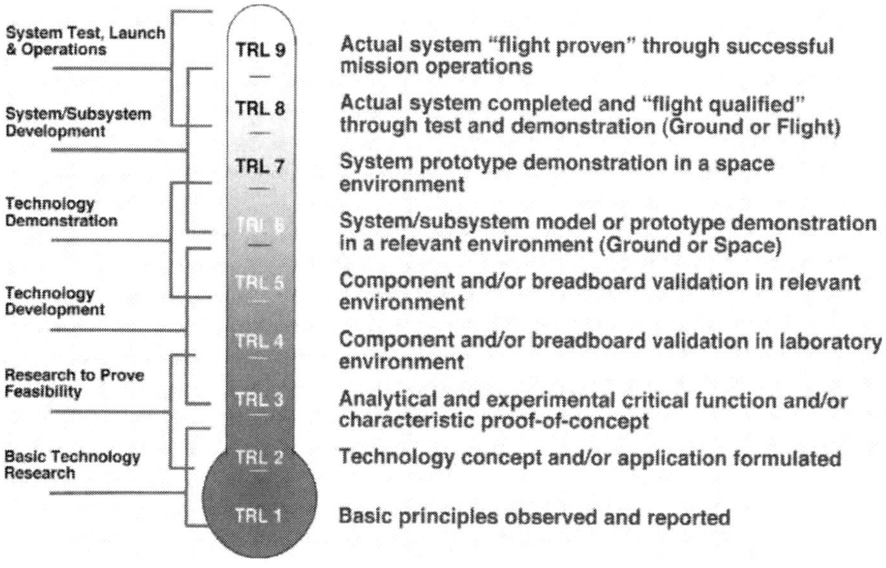

Figure 14. This TRL chart gives the flavor of technology readiness, but interpretation of readiness allows for a latitude of around ± 1 when technologies are evaluated by different people. Chart is from NASA's HRST Technology Assessments Technology Readiness Levels [131]

To summarize the positive aspects of what has been achieved to date, I will offer the following off-the-cuff statements that pull together the most important achievements. First of all, several instruments (e.g., APEX-SZ, ACT, MUSTANG, GISMO, SABOCA) exist. That, in the space of less than a decade, the instrumental community has been able to go from barely-proof-of-principle detectors to kilopixel arrays operating near fundamental noise limits is astonishing. People should feel rightly proud of this phenomenal progress. Part of this success concerns the pixel count: SCUBA-2 engendered the 1,280-pixel multiplexer, so an array of around this size is now both state-of-the-art and surprisingly commonplace. Furthermore, this pixel count is probably close to what is needed for a CMB Einstein Inflation Probe. The sensitivity of the detectors listed above is not fabulous (close to an NEP of 10^{-17} W/$\sqrt{\text{Hz}}$ for all of them), because it doesn't need to be. However, this is approaching what is needed for CMB polarimetry from a balloon, and there are efforts ongoing to reach the greater sensitivities needed for CMB polarimetry from space. Finally, one other positive aspect of recent work is that the readouts – varieties of SQUID multiplexers – all seem to function well. Their performance is excellent, and they are generally easy to interface to detector arrays (at least, at the 10^2 pixel level.) In the near-infrared, multiplexers seem to have their own unique quirks that change from generation to generation. My experience with the NIST time domain multiplexers is that they generally continue to become more sophisticated while retaining the same footprint, interfaces, etc. Perhaps future generations of multiplexers will be different, but for the time being, upgrading has been nearly transparent for around five years.

TRL Number & Name	Description	TES Examples
3 Analytical and experimental critical function and/or characteristic proof-of-concept	At this step in the maturation process, active research and development (R&D) is initiated. This must include both analytical studies to set the technology into an appropriate context and laboratory-based studies to physically validate that the analytical predictions are correct. These studies and experiments should constitute "proof-of-concept" validation of the applications/concepts formulated at TRL 2.	Conducting TES noise modeling and experiments. Determining scaling functions of performance parameters. Produce small TES arrays and verify dark performance in the lab.
4 Component and/or breadboard validation in laboratory environment	Following successful "proof-of-concept" work, basic technological elements must be integrated to establish that the "pieces" will work together to achieve concept-enabling levels of performance for a component and/or breadboard. This validation must be devised to support the concept that was formulated earlier, and should also be consistent with the requirements of potential system applications. The validation is relatively "low-fidelity" compared to the eventual system: it could be composed of ad hoc discrete components in a laboratory.	Develop working system for a ground-based instrument and operate it in a laboratory environment. Demonstrate scalability for large-format arrays. Conduct detailed performance tests and evaluate performance under a variety of conditions.
5 Component and/or breadboard validation in relevant environment	At this, the fidelity of the component and/or breadboard being tested has to increase significantly. The basic technological elements must be integrated with reasonably realistic supporting elements so that the total applications (component-level, sub-system level, or system-level) can be tested in a simulated' or somewhat realistic environment. From one- to-several new technologies might be involved in the demonstration.	Deploy a TES instrument to a ground-based platform and produce scientific results. Produce a working integrated large-format detector array system in a laboratory environment.
6 System/subsystem model or prototype demonstration in a relevant environment (ground or space)	A major step in the level of fidelity of the technology demonstration follows the completion of TRL 5. At TRL 6, a representative model or prototype system or system which would go well beyond ad hoc, patch-cord' or discrete component level breadboarding would be tested in a relevant environment. At this level, if the only relevant environment' is the environment of space, then the model/prototype must be demonstrated in space. Of course, the demonstration should be successful to represent a true TRL 6. Not all technologies will undergo a TRL 6 demonstration: at this point the maturation step is driven more by assuring management confidence than by R&D requirements. The demonstration might represent an actual system application, or it might only be similar to the planned application, but using the same technologies. At this level, several-to-many new technologies might be integrated into the demonstration.	Produce an engineering / flight qualification model of a large format TES bolometer array. Successfully qualify mounts, electronics, etc. in appropriate facilities for shock, vibe, EMI, radiation, etc. Operate TES instrument in a fully-remote or autonomous mode. Deploy a TES instrument to a suborbital platform and produce scientific results.

Table 1. Mid-Level Technology Readiness Levels as applied to TES Bolometers

Journal of Physics: Conference Series **155** (2009) 012001 doi:10.1088/1742-6596/155/1/012001

Rosy as the picture above may be, there are significant tasks yet to be accomplished. None of the list I give here is a Herculean task, but often recently it has felt as if the funding available was equivalent to being given a whiskbroom to clean the Augean stables. The first item in priority is the most basic: the maturity of observing with TES bolometers needs to be significantly advanced. When there are facility instruments with TES bolometers, a new community of astronomers will learn to rely on them. When experiments produce reams of data to be analyzed by a science team that has little direct contact with the instruments producing the data, they will learn to extract the best result despite peculiar effects. The wealth of experience with cryogenic semiconducting bolometers (dating back almost 40 years [16]) was a major contributor to the decision to use them on Herschel/SPIRE.

From a theoretical and experimental standpoint, it is essential that the excess noise problem be worked out. That ITFN exists and is calculable is fairly well-established. Other sources of noise such as phase slip or white voltage noise are less well-understood and well-characterized. This should continue to be investigated until some quantifiable prediction of which combination of effects makes up "excess noise".

Another nascent field is that of detector polarimetry. On-chip polarimeters that read out both polarizations, and which might even be able to switch them, are not a mature technology. A ground-based or balloon-borne camera with an appropriate polarimeter using such TES bolometers would be a big step forward.

Finally, there is the issue of space flight qualification. Suborbital experiments such as SAFIRE on SOFIA or the balloon-borne experiments EBEX or SPIDER go a long way to ensuring that a technology is robust enough to be used in a harsh and remote environment, but they do not compare to the rigors of qualification for space flight. This includes designing a detector subsystem for a demanding set of requirements and then conducting a variety of EMI, thermal, and vibration testing. This is expensive work, and probably will not happen without directed CMB polarimetry funding. However, it will eventually be necessary and the sooner it can be done, the better.

Summarizing, my perception is that the TES bolometer array for CMB polarimetry is a technology (really a combination of many technologies) that is close to achieving a TRL of 5. If allowed to use TRL as a continuous scale, I would rank TES bolometer arrays as TRL=4.8. I note that large arrays (kilopixel scale) are just now being fielded, and very little scientifically useful data from TES arrays has been acquired. We have not yet achieved a thorough understanding of the noise properties of real TES bolometers. However, advances are rapid and making headway on all fronts, and perhaps TRL=5.0 will be achieved as the first astronomical data publications emerge from the ACT and SPT collaborations [132, 133].

2.7. TES Detectors in Context (or, what else is here?)

Despite the focus devoted to the transition edge sensor bolometer, it should not be inferred from the above that this is the only detection technology – nor necessarily the best – for a CMB polarization application. It is worth considering the context of other technological approaches, and indeed the overall systems aspects of TES detectors. In considering other technologies, it is important to remember some guiding points about detectors for the CMB. Most importantly, at frequencies near $\sim$ 100 GHz, several competing technologies are close to or below the detector noise equivalent power requirement set by the CMB photon noise in space. Mapping speed is thus improved principally by increasing the number of detectors that can be feasibly deployed in a focal plane array. The detector response time is also important, as is its stability, inasmuch as these constrain the viable scan patterns and polarization modulation schemes that will be used to limit systematics.

The highest maturity direct detector for CMB astrophysics is the semiconducting bolometer, which has been used for decades in the measurement of the spectrum and anisotropy of the CMB, including in the Nobel-prize-winning space-flight COBE/FIRAS instrument [134]. Semiconducting bolometers are discussed below (Section 3). Their thermal and electrical properties are well-established both experimentally and theoretically (e.g., [37, 135]). The most recent semiconducting bolometers for long wavelength applications have employed advanced absorbers such as the well-known spiderweb bolometers [136, 137]. Large semiconducting arrays exist in ground-based instruments such as SHARC II [138] and BOLOCAM [139], and the balloon-borne BLAST [140]. However, with few exceptions, large scale multiplexing is not possible for such arrays, and hence individual JFET amplifiers must be connected to each pixel, with the large power dissipation and high temperatures inherent from such amplifiers. The low magnitudes of α for semiconducting thermistors result in slower bolometers for a given design than would be the case with a TES thermistor. Also, methods for optical coupling with semiconducting detectors are essentially limited to absorber-coupled (as opposed to antenna-coupled) methods. While none of these are insuperable obstacles, the combination enhances the attractiveness of TES bolometers.

Coherent amplifiers provide another competitive alternative for CMB detection, at least at frequencies up to $\sim$ 100 GHz. The quantum noise of such devices dominates at shorter wavelengths, and in fact extant devices have best performance several times the quantum limit. Thus, there is certainly some cause to consider a mission that has direct detectors in the focal plane as well. As one consideration, the foreground removal requires spanning the wavelength range where synchrotron, free-free, CMB, and dust emission change significantly relative to each other. From the recent summary of Dunkley [141], this will require detectors operating at $\lambda = 1.5$ mm or less, where coherent detectors have a noticeable raw sensitivity disadvantage compared to direct detectors. However, in favor of coherent detectors, it must be admitted that common amplifiers need to be cooled only to $\sim$ 20 K rather than the $\sim$ 100 mK required for bolometers. Mitigating this advantage, however, is the need to provide a much (orders of magnitude) larger cooling power for amplifiers at the higher temperature than would be required for the refrigerators cooling bolometers. Other aspects in favor of coherent detectors are operational simplifications such as wide dynamic range, low nonlinearity, and insensitivity to cosmic rays and microphonics. One disadvantage is the difficulty of producing large format arrays of coherent amplifiers, although this may now have been solved to some degree by the QUIET experiment [142]. It has been noted that coherent amplifiers offer great flexibility in signal processing after amplification, when both amplitude and phase are available, but before power detection, when only amplitude is available. However, TES bolometers can readily be fed by antennas through microstrip (e.g., [4–8]), permitting many of the same capabilities. Overall, given that covering all the frequency bands necessary for CMB polarization measurement with a sufficient number of pixels at sufficient sensitivity is of paramount importance, it seems that coherent detectors require further development to enable a CMB polarization space mission.

On the topic of further development, it is worthwhile to note that bolometers as a whole have made great strides in sensitivity over a very long history (Figure 15). Even within the limited realm of NASA-relevant work, there are non-CMB applications for bolometers (the far-infrared has been noted above in several instances, and the closely-related x-ray microcalorimeter detectors are fundamentally little different). As a result, per-pixel sensitivity has now achieved the sensitivity requirement of CMB (Figure 15) applications in space. Further strides in overall mapping speed are now being made by the investment in large format bolometer array development. This is where CMB-targeted technology maturation is needed.

To summarize some of the advantages of TES bolometers as compared to other relevant technologies, there are relatively few points. Their sensitivity permits background-limited operation at all wavelengths needed for CMB polarization measurements. Large format arrays have been demonstrated using SQUID multiplexers, which yield excellent mapping speed. The accelerated response time permits a range of viable scan patterns and polarization modulation

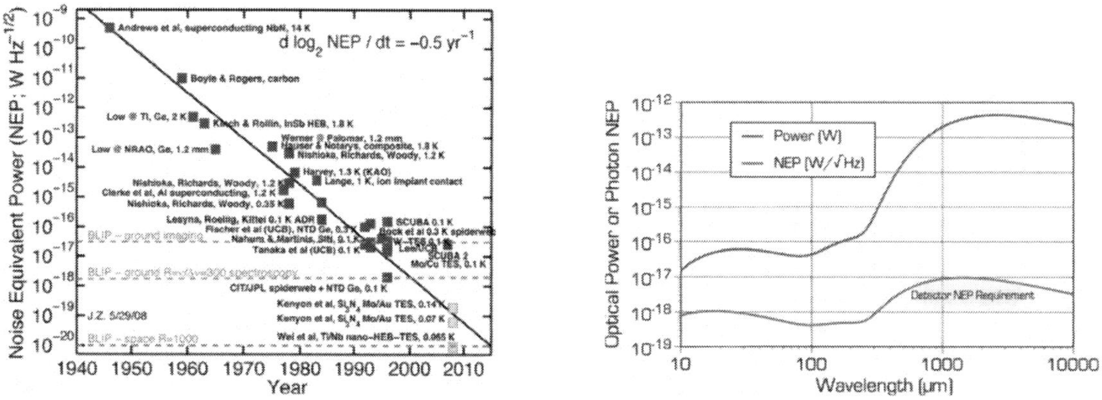

Figure 15. (Left) The NEP of bolometers has improved by a factor of two every two years for more than half a century (figure courtesy of Jonas Zmuidzinas). (Right) The natural sky backgrounds in space require detectors of a sensitivity that is now easily reached by bolometers. Larger format arrays continue to improve the effective mapping speed of fielded instruments where background-limited sensitivity has been achieved.

timescales. Some improvement may be realized in dynamic range and stability. Multiple groups now have experience in fabricating, testing, and fielding TES bolometer arrays, and so their heritage is substantial enough that development for space applications is appropriate.

2.8. TES Development Plan (or, From Here to Launch!)

This highly-speculative section approaches the task of how to prepare and mature the TES component technology for the cosmic microwave background polarimetry. Thus far, very little has had any focus on this one mission, instead being an overview of the field as it related to superconducting bolometers at long wavelengths in general. Measurements of CMB polarization promise to allow us to distinguish among models of the first instants of our universe. Advancing detector array technology is a stepping-stone to the goal of arrays of polarization sensitive detectors required for missions such as the Einstein Inflation Probe (when used with specific coupling and modulation structures under development elsewhere). Further in the future, even larger arrays are a requirement for the efficient use of large cryogenic telescopes in space such as SAFIR [143] and SPIRIT [144].

In my consideration, I take it as a governing assumption that TES bolometer arrays must be perceived to be mature in order for them to form the basis of a CMB polarization Einstein Probe mission. While Probe-class missions (∼ \$600 M cost class) could, in principle, provide their own technology development initiatives, it seems equally likely that an insufficiently mature technology could, conversely, hinder the initiation of funding for a Probe. Hence, the perception of TRL=6 may be required in order to bring about a successful Einstein Inflation Probe. How can this be achieved?

The most effective way to rapidly increase the perceived maturity of TES bolometers will be to deploy larger, higher performance detector arrays in existing instruments or on existing telescopes. These will then need to be used regularly for major experiments or facility instruments. Through scientific use, we discover the interactions of the detectors with other elements of the experiments, and techniques for their mitigation. The technologies to build and read out multi-kilopixel arrays are just now well-established enough that there are multiple possible approaches and facilities for their production. Having a wealth of astronomical data and – more importantly – understanding of TES bolometers by astronomers[7] is critical. Facility instruments on ground-based telescopes both (1) allow a bigger 'credibility pool' for the technology and (2) bring more minds to bear on teasing out effects in the data.

A second, and I would hope simultaneous, path is to pursue the theoretical explanations for excess noise. This issue does not appear to be firmly settled, and a combination of theorists (to quantify predictions, estimate observables, and explain how to distinguish between the original causes) and experimentalists (whose job it is to provide high quality – and by that I mean both accurate and precise – data in order to disagree robustly with as many theories as possible) should continue this work. Before proposing a major space mission, it is certainly a good idea and might effectively by required, that we know how to predict the performance of a given TES bolometer correctly *ab initio*.

As a third item – in sequence rather than priority – is that it is imperative a collaboration successfully conducts a suborbital CMB investigation using a TES-based instrument. This is not news to anyone, but it's worth pointing out that while this is the venue that might provide TRL=6 for TES bolometers, modification of the array design approach that is flown on a balloon reduces this to the extent that there is an insufficiently rich body of knowledge at the TRL=1 to 5 level. Thoroughly prosecuting the above two elements will result in a more concrete estimate of the technology readiness – and will, truly, mature the technology.

2.9. Acknowledgements
In the writing of this article, I found help and guidance from many publications and people, the latter class containing: William Duncan, Lyman Page, Giorgio Siringo, Harvey Moseley, Adrian Lee, Jamie Bock, and Nikhil Jethava. I trust Kent Irwin and Gene Hilton to such a degree that I pretty much lifted equations from their paper to write my TES theory section. This research has made use of NASA's Astrophysics Data System Bibliographic Services.

3. Semiconductor Bolometers for Cosmic Microwave Background Polarization Studies

by Harvey Moseley

3.1. Abstract
I will review the state of development of semiconducting bolometers, their prospects for future Cosmic Microwave Background (CMB) polarization experiments, and the effort required for their successful deployment.

3.2. Introduction
Bolometers are thermal detectors which are comprised of a radiation absorber, a thermometer which senses the temperature rise in the detector due to the absorption of radiation, and a conducting link through which the detector is held in thermal contact with a heat sink (Figure 16). Benford 2.2 has reviewed the history of their development in theory and practice.

Semiconducting bolometers have been the mainstay of CMB experiments which use direct detectors (COBE/FIRAS, COBE DIRBE, MSAM,IRTS, Boomerang, MAXIMA, e.g)[145–152]. The Ge or Si-based impurity conduction thermometers have predictable temperature coefficients of resistance, noise, limited by thermodynamic fluctuations and Johnson noise, which agrees well with Mather's [37] theory, and can be designed and optimized for background limited

[7] By "astronomers", I mean here those people derisively referred to as "God damn user astronomers" by a notable Caltech professor, by which he meant that class of astronomers that can acquire great skill at observing using an instrument about which they understand next to nothing.

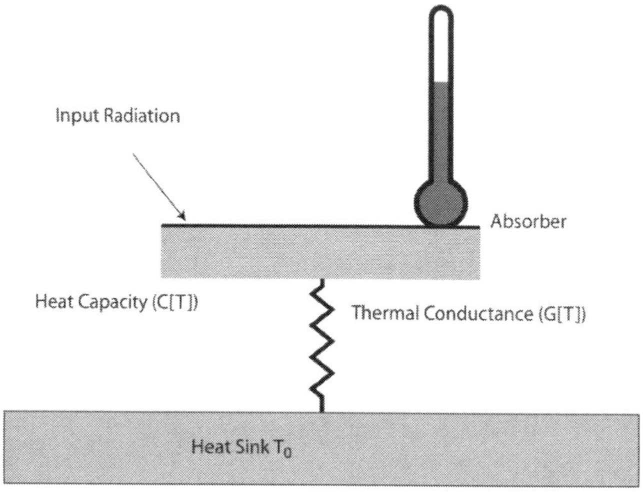

Figure 16. The basic components of a bolometer are shown. The response and minimum noise of such a system can be determined from thermodynamic quantities [37]

performance over a wide range of conditions covering proposed CMB experiments. Typically these thermometers are thermally connected to an absorbing element, which can be separately optimized, commonly using a resistive metallic film to absorb the incident radiation.

With a change of optical instrumentation and absorber structures, these detectors have been used for for CMB polarization measurements (Boomerang,Maxipol, e.g). In this paper, I will discuss the present state of semiconducting bolometers, their suitability for CMB studies, and consider their future prospects

3.3. Development

While Ge:Ga bolometers were introduced into infrared astronomy by Low [16], it was in the early 1980s that full promise of these devices was realized and developed. Mather [37] developed a theory which accurately describes the noise and response of such a detector, and provided clear insight into the scaling of the performance of the detector with temperature, conductance, heat capacity, and thermometer sensitivity. In his 1984 paper [135], he describes the optimization of such a detector. Following his procedure, we assume that for a given application, the radiant power is given, and there is a minimum detector heat capacity required to provide absorption and thermometry in the detector. Taking the absorbed power, signal modulation frequency range, detector heat capacity, and the thermal sensitivity

$$\alpha = \frac{dLog(R)}{dLog(T)} \tag{14}$$

of the detector as givens, the optimization of the detector consists of simultaneously adjusting the thermal conductance between the detector and its heat sink and the amount of bias power

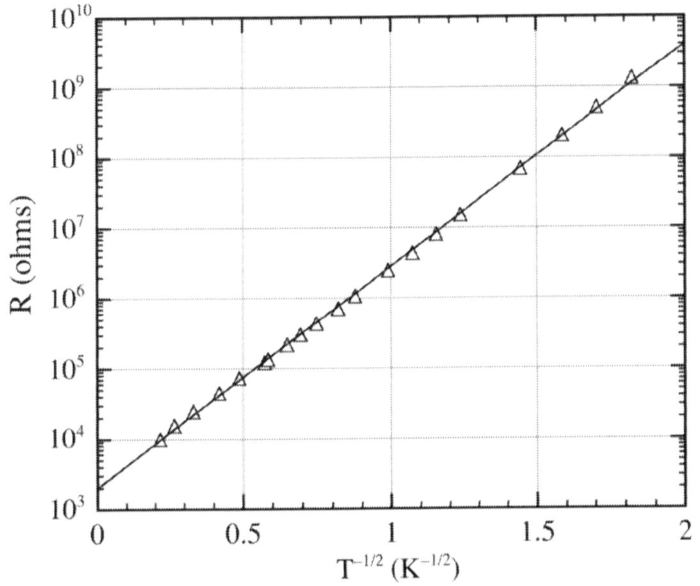

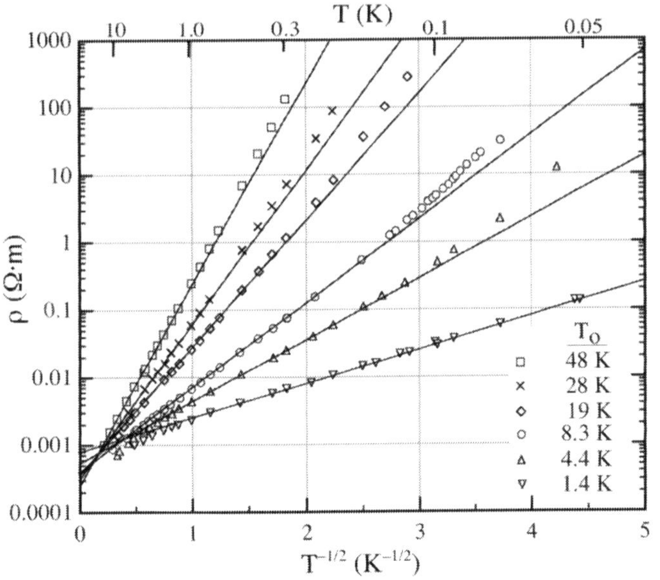

Figure 17. (Left) The resistance of impurity conduction thermometers in Si is well described over five orders of magnitude in resistance by the variable range hopping model. (Right) This theory accurately models impurity conduction of Si in the low temperature regime over a wide range of resistances and impurity concentrations [51]

dissipated in the detector to minimize system noise. The result of this optimization, shown as a function of α is seen in figure 18. Several general results arise from this simple analysis:

(i) For a thermal conductance $G \propto T^3$, the detector Noise Equivalent power (NEP) scales at $T^{\frac{5}{2}}$

(ii) For a given background power and wavelength, the detector noise can be designed to be below the background noise by lowering its operating temperature sufficiently

(iii) Higher values of the thermometer sensitivity, α, results in more response bandwidth.

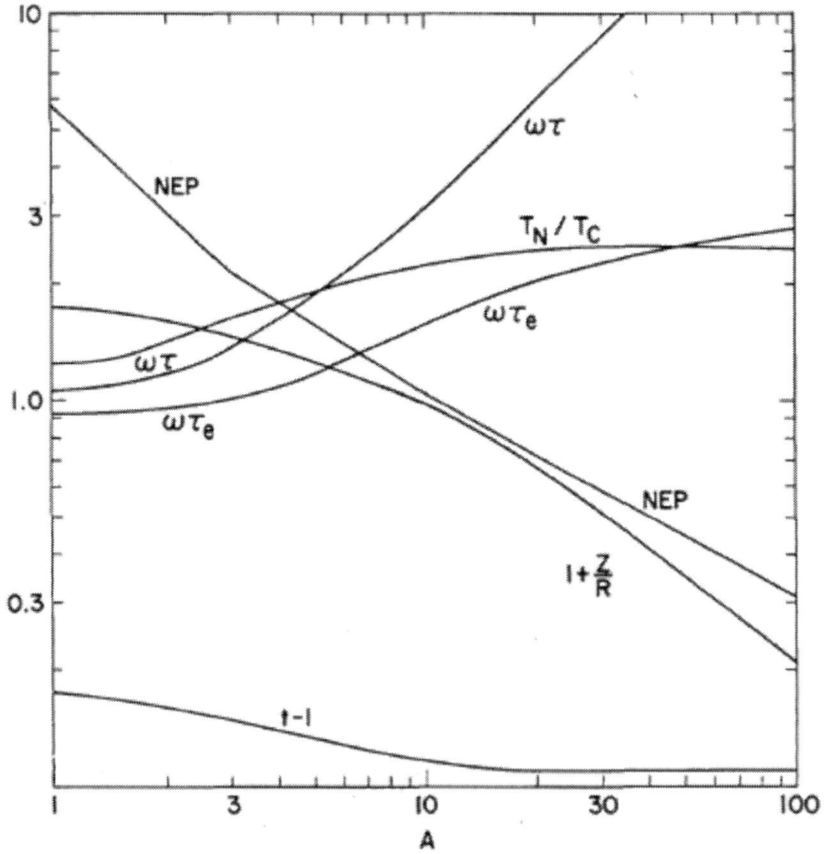

Figure 18. Using his thermodynamic model for bolometer performance in the linear regime, Mather [135] shows the optimal performance and detector operating conditions as a function of thermometer sensitivity α.

At the time of this work, semiconductor processing techniques were, for the first time, being applied to the development of bolometers, both in Si and Ge [26, 153]. The application of ion implanted contacts eliminated these elements as a source of noise. The advent of micromachining techniques permitted the development of very low heat capacity detectors, and the demonstration of the thermometers at $T \sim 0.1\ K$ brought the sensitivity required for a wide range of experiments, including CMB studies, within the available design space. The reproducibility of the devices arising from the improved fabrication techniques made large ($\sim$ 100 elements) arrays practical.

Journal of Physics: Conference Series **155** (2009) 012001 doi:10.1088/1742-6596/155/1/012001

3.4. Semiconductor Thermometers

The conduction mechanism in these Si and Ge impurity conduction bolometers is variable range hopping, seen in semiconductors doped near the metal-insulator transition. While the detailed physics of this transport process is complicated, experimenters have explored the conduction and noise properties of these thermometers as a function of impurity concentration, temperature, and current density [154, 155]. These measurements provide the basis for the design and optimization of bolometers for a wide range of experimental conditions The performance of Si and Ge thermometers are similar, with Ge having an advantage over Si at low temperature based on the ratio of the thermometer sensitivity to its heat capacity. But for the purpose of producing detectors for CMB polarization, which is a high background power application, the thermometer decision can be made on the basis of ease of fabrication. McCammon [51] has written a comprehensive review of the properties of semiconducting thermometers, and discusses the detector design process.

3.5. Amplifiers and Readout

The primary shortcoming of these semiconductor bolometers has been the lack of suitable amplifiers and multiplexers which can be integrated with them at the focal plane temperature. The noise of the detectors is typically double the Johnson noise of the thermometer, so amplifiers with noise temperatures lower than the physical temperature of the detector are required. The most convenient high performance amplifiers are JFETs, which require thermal ionization of carriers to function, and thus typically operate in the 70 to 150 K range. At this temperature, they can provide equivalent noise temperatures $\sim 0.1\,\mathrm{K}$, adequate for most present bolometer applications. Typically, detectors have been operated with a single JFET per detector, which has required significant cryogenic/electrical engineering to produce low noise, low power dissipation system. Modules in which the JFETs are suspended on thin Si_3N_4 membranes have been developed for the SPIRE instrument on ESA's Herschel. Each module, with 16 JFETs, dissipates 5 mW of power, or about $300\,\mu$W per channel, typical of other designs [156]. The XRS instrument on Astro-E2 used a two stage isolation system for the FETs, so that much of the power could be dumped into the intermediate heat sink at 16 K. This resulted in a 32 channel system in which a total of less than $100\,\mu$W was sunk at the 2 K heat sink [157]. While both these systems were successfully engineered, systems with an order of magnitude more pixels than existing devices will present a major technical challenge.

Billot et al. [158] have developed a semiconductor-based detector array which is being used in the PACS instrument on Herschel. This device uses MOSFET switches for multiplexing, and MOSFET amplifiers operating at the detector temperature. While this is conceptually appealing, the relatively high noise of MOSFETs at low frequencies requires very high resistance detectors to avoid amplifier noise limitations. This results in long RC time constants, which significantly limit the response bandwidth of the detector. The $\frac{1}{f}$ noise of the MOSFET can be corrected by sampling a stability reference in the course of the readout. The greatest practical limitation of this design is that given the level of amplifier noise, only about 16 channels can be multiplexed using a single amplifier output. This has significant power and complexity implications for large format arrays, but this simple and compact design has advantages for some applications

3.6. Present State of Development

Semiconductor bolometers are well developed for space flight missions. They have been flown and operated successfully on many missions, including COBE, IRTS, and Astro-E2, and large focal plane arrays will soon be launched on Planck and Herschel. This history suggests that the semiconducting detectors have a high technical readiness level, say TRL 9. However, since the detailed detector designs which we would fly on a CMB polarization mission could

Journal of Physics: Conference Series **155** (2009) 012001 doi:10.1088/1742-6596/155/1/012001

differ significantly from those flown on earlier missions, significant qualification work is must be anticipated to demonstrate flight readiness. To be more specific, the technology is well qualified, but the specific designs must be developed and tested.

3.7. Prospects

Semiconducting bolometers can provide adequate sensitivity for strongly background limited operation in future CMB missions. They are technically well developed, have predictable performance, and are easy to adapt to a wide variety of optical systems. However, significant work is required to integrate these detectors with planar mm-wave detector circuits being developed for CMB polarization missions. Despite their general maturity, significant effort would be required to adapt these detectors to the present generations of instruments, but it is likely that this could be successfully accomplished. The major factor limiting the utility of these detectors for future missions remains the lack of availability of low noise amplifiers which can be closely integrated with the detectors. This results in a complex system design and integration process. The mechanical and thermal engineering challenge of interfacing between hundreds of 120 K JFET amplifiers and detectors below 0.1 K is one that can be attempted, but with thousands of detectors, this daunting task is a major risk to mission development. However, if high performance alternatives did not exist, these detectors could be developed to enable CMB polarization science.

3.8. Acknowledgments

I thank Dan McCammon, Kent Irwin, Dominic Benford, and David Chuss for help in preparing this review.

4. Superconducting Microresonator Detectors

by Jonas Zmuidzinas

4.1. Technical Description

4.1.1. Basic Physics The basic concept of operation is described in a number of publications [159–164] and is illustrated in Fig. 19. Photons are absorbed in a superconducting film, which causes Cooper electron pairs to break. The binding energy of Cooper pairs is proportional to the critical temperature, $2\Delta \approx 3.5 k_B T_c$, which corresponds to a minimum photon frequency $\nu = 2\Delta/h \approx 90$ GHz for aluminum ($T_c = 1.2$ K). A perturbation δP of the absorbed mm-wave power thereby produces a perturbation $\delta Z_s = \delta R_s + j\delta X_s$ of the complex microwave surface impedance of the film. Both the real and imaginary parts of the impedance are affected in this process: Cooper pair breaking causes the surface reactance to increase (this is the "kinetic inductance" effect), and meanwhile the single electrons (or "quasiparticles") that are produced are capable of scattering, and are therefore dissipative, and are responsible for an increase in the surface resistance. Both of these effects can be sensed by incorporating the superconducting film into a resonant circuit and measuring the resulting changes in the resonance frequency and quality factor, as shown in Fig. 19.

The readout of the resonator can be explained more precisely by considering the trajectory of the complex microwave transmission $S_{21}(\omega)$ in the complex plane as the angular (microwave) frequency ω is varied. The linewidth of the resonance is $\Delta\omega_r = \omega_r/Q_r$, where the resonator's quality factor is determined by coupling and internal losses according to the usual formula, $Q_r^{-1} = Q_c^{-1} + Q_i^{-1}$. Away from resonance, the transmission is essentially unity $|S_{21}| = 1$, and near resonance describes a circle with the resonance frequency being the point of closest approach to the origin, as shown in Fig. 19. If the readout generator's frequency ω_g is fixed,

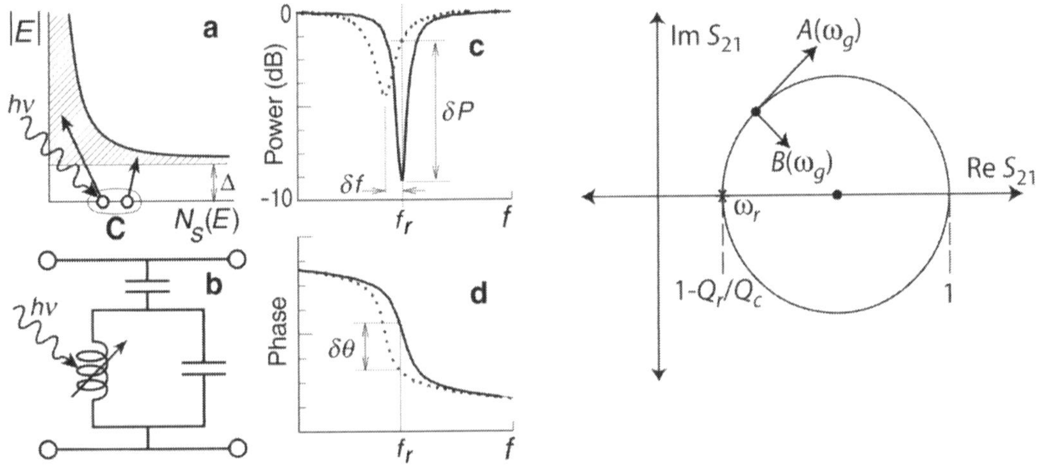

Figure 19. Left: Principle of operation of superconducting microresonator detectors, following [159]. The absorbed photon energy breaks Cooper electron pairs in a thin superconducting film (a) which is part of a lithographed microresonator circuit (b). The breaking of Cooper pairs causes a change in the surface reactance and resistance of the film, leading to a shift in the resonance frequency f_r and a reduction of the resonator's quality factor Q_r (c), which can be sensed by measuring the phase (d) and amplitude (c) of the microwave output signal. **Right**: This figure illustrates the trajectory of the resonator's transmission $S_{21}(\omega)$ as a function of angular frequency ω in the complex plane. The transmission is unity ($S_{21} = 1$) at frequencies away from resonance. Starting at a frequency below resonance, S_{21} traces out the circle in a clockwise direction as the frequency is increased, crossing the real axis at the resonance frequency ω_r indicated by the star, and corresponding to a transmission $S_{21}(\omega_r) = 1 - Q_r/Q_c$ as indicated, where Q_c is the coupling-limited quality factor. The adiabatic response coefficients for changes in the resonator frequency $A(\omega_g)$ and dissipation $B(\omega_g)$ are shown for the case that the generator frequency ω_g is tuned above the resonance. These coefficients are tangent to and perpendicular to the resonance circle, respectively.

and a change in optical power δP is introduced, the resulting change in the surface reactance δX_s will cause a perturbation in the complex transmission δS_{21} that is tangent to the resonance circle; meanwhile, δR_s causes δS_{21} to move in a perpendicular direction. The overall response is the vector sum of these two components. Therefore, by using readout electronics that are capable of real-time vector measurements of S_{21}, one can simultaneously measure and separate these two effects. However, the perturbation in the reactance is a stronger effect: the ratio is $\delta R_s/\delta X_s \equiv \tan\psi \approx 0.3$ [164–166]; this ratio also describes the relative strengths of the transmission perturbations δS_{21} for the tangent and perpendicular directions. This is of practical importance: the factor of $\sim 3-4$ lower responsivity for the dissipation readout places more stringent demands on the noise performance of the cryogenic microwave amplifier that follows the resonator (see section 4.1.4).

4.1.2. A millimeter-wave implementation A practical implementation of this concept for mm-wave detection is shown in Fig. 20, using a quarter-wavelength transmission-line resonator rather than a lumped-element circuit. A coplanar waveguide (CPW) transmission line allows the

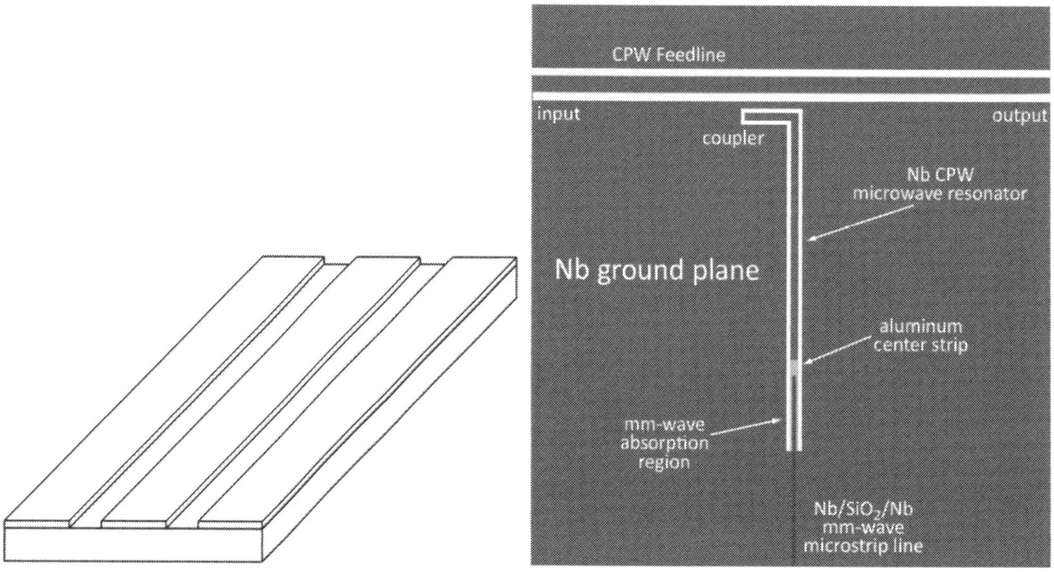

Figure 20. Left: A coplanar waveguide (CPW) transmission line is fabricated by etching two parallel slots in a metal film, which defines a center conductor and two ground planes. **Right**: A typical microstrip-coupled mm-wave detector made using a hybrid niobium/aluminum (Nb/Al) CPW quarter-wave resonator. The resonator is short-circuited at the bottom, where the mm-wave radiation is absorbed, and is open-circuited at the top, where the resonator is coupled to the feedline. The microwave readout signal propagates along the CPW feedline at the top, fed from the signal generator on the left and coupled to the cryogenic low-noise amplifier on the right. The readout signal excites the resonator via the coupler section; the length of the coupler and its separation from the feedline dictate the value of the coupling quality factor Q_c. At the short-circuit end, a strip of aluminum is used for the CPW resonator center strip in place of niobium, because the superconducting gap energy of aluminum allows it to absorb mm-wave radiation with frequencies above 90 GHz; the corresponding gap frequency of niobium is 700 GHz. The mm-wave radiation is coupled to the aluminum strip using a low-loss $Nb/SiO_2/Nb$ microstrip (parallel-plate) transmission line, as shown, and is absorbed over the ~ 1 mm length of the aluminum strip. The typical dimensions are $6\,\mu$m for the CPW center strip width, $2\,\mu$m for the CPW slot, and ~ 5 mm for the resonator length, giving a resonance frequency around 6 GHz.

resonator to be made without the use of deposited dielectric films (which are amorphous and generally lossy [167]), thereby allowing extremely high quality factors ($Q_r \sim 10^6$) to be achieved. The use of an aluminum section near the shorted end of the resonator allows absorption of mm-wave radiation at freuencies above 90 GHz (a lower T_c material would be needed for frequencies below 90 GHz). The mm-wave radiation is brought to the resonator using a low-loss $Nb/SiO_2/Nb$ microstrip line which runs across the aluminum center strip of the CPW resonator, at which point it becomes a $Nb/SiO_2/Al$ microstrip, causing mm-wave dissipation and pair-breaking in the Al strip.

The use of microstrip coupling allows this type of detector to be used with a variety of feeds, including narrow-beam phased-array planar antennas [168–174] with either single or dual-polarization response, broad-beam planar antennas [175–180], and the more traditional horn-waveguide-probe coupling approaches [181–185].

Other coupling techniques are possible, and indeed necessary at frequencies above 700 GHz where niobium microstrip lines become exceedingly lossy. Coupling of far-infrared radiation into an aluminum CPW resonator directly attached to a twin-slot antenna has been demonstrated [186]. Another possibility is to design the inductive portion of the resonator to simultaneously be a good far-infrared absorber. This approach has also been demonstrated [187, 188] and leads to two-dimensional pixel arrays similar to absorber-coupled transition-edge sensor (TES) bolometer arrays. Other approaches for far-infrared microresonator detectors are also being actively pursued (E. Wollack, priv. comm.).

4.1.3. Frequency multiplexing Because the resonators have high quality factors ($Q_r = 10^4 - 10^6$), a large number of resonators may be multiplexed in the frequency domain and read out with a single cryogenic microwave amplifier, as shown in Fig. 21. In this scheme, the resonators are coupled to a common feedline and therefore the detector chip is fully multiplexed, as opposed to TES designs which require multi-lead superconducting connections between the detector and SQUID multiplexer for each pixel (this is true for all types of SQUID multiplexing - time domain as well as MHz/GHz frequency-domain). The room-temperature electronics then has the job of generating a set of excitation frequencies - one per resonator - and then measuring the amplitudes and phases of these signals after they have been sent into the cryostat, transmitted past the resonator array, and amplified by the cryogenic amplifier and any additional room-temperature amplifiers. Fortunately, this task has become quite feasible using modern digital signal processing techniques.

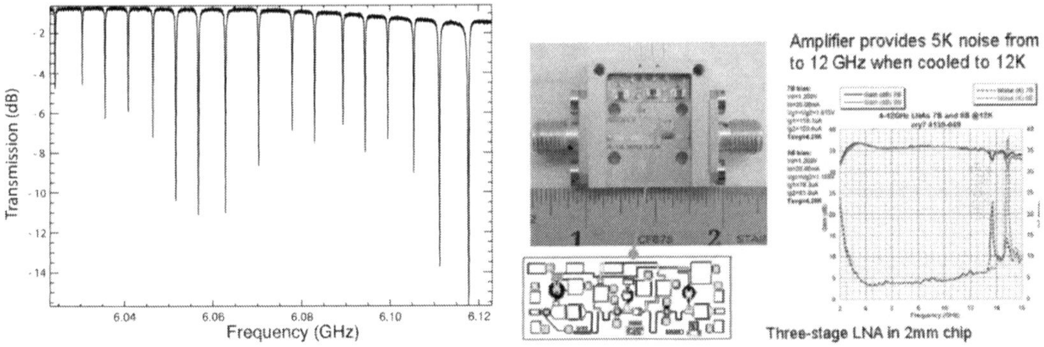

Figure 21. Left: an example of an array of resonators coupled to a single feedline. The depths of the resonances depend on the coupler design and the value of the coupling quality factor Q_c (credit: B. Mazin). **Right**: A single wide-band cryogenic HEMT amplifier (credit: S. Weinreb). can be used to simultaneously measure a large number ($\sim 10^3$) of resonators.

4.1.4. Sensitivity

Fundamental Limits: The fundamental sensitivity limit for this type of detector is set by the random generation of quasiparticles by pair-breaking thermal phonons, and their subsequent recombination. However, this mechanism is exponentially suppressed at low temperatures by the Boltzmann factor $\exp(-\Delta/k_BT)$ due to the superconducting energy gap, and therefore in practice the sensitivity is determined by other factors. Ideally, the sensitivity would be set by photon arrival statistics - the BLIP limit. One interesting subtlety is whether the BLIP limit for these detectors corresponds to a "photodiode" or "photoconductor" device – the latter being less sensitive by a factor of $\sqrt{2}$ due to the additional noise introduced by the random carrier lifetimes (recombination noise). The answer is "both"! For photon energies just above the gap, $h\nu = 2\Delta + \delta E$, only a single Cooper pair is broken, and an equal amount of noise is generated in quasiparticle recombination. Far above the gap, $h\nu >> 2\Delta$, a large number of quasiparticles are produced, and each quasiparticle recombines randomly, so the recombination noise is averaged down to an insignificant level and "photodiode" performance is obtained.

Conditions for Achieving BLIP Operation: Achieving the BLIP limit requires that the amplifier noise contribution be brought below the photon noise. This requires optimizing the responsivity of the device, which in practice amounts to choosing the correct value of the coupling quality factor Q_c. For low photon backgrounds, the number of photo-produced quasiparticles will be small, and so the resonator's internal dissipation Q_i^{-1} due to the quasiparticles will also be small; conversely, for large backgrounds the dissipation will be larger, and the quality factor will be lower. The coupling strength which maximizes the responsivity is given by $Q_c = Q_i$ [164]. This design optimization is analogous to choosing the desired value of thermal conductance G for a bolometer given the expected optical loading. The second key issue for achieving BLIP is the noise performance of the cryogenic amplifier that is required for an optimized detetector ($Q_c = Q_i$). This question has been studied in some detail theoretically; the answer [164] depends on a number of factors including the gap frequency, the photon frequency, the background loading, the quasiparticle lifetime, the maximum microwave readout power, and crucially, on whether one is using frequency or dissipation readout (see the discussion in section 4.1.1). Calculations [164] indicate that readily achievable amplifier noise temperatures in the range $T_n = 10-20$ K would allow BLIP operation using frequency readout, whereas much better performance ($T_n = 1-2$ K) is needed for dissipation readout. Because the amplifier contribution to the noise equivalent power (NEP) scales as $\sqrt{T_n}$, the use of a typical HEMT amplifier (see Fig. 21) with $T_n = 5$ K for dissipation readout will result in an NEP value that is roughly twice the BLIP limit. Finally, although the detailed physics of quasiparticle recombination is not fully understood yet [189], the measured lifetimes are sufficiently long and do not impose a serious limitation on CMB observations.

Excess Frequency Noise: Although amplifier noise considerations lead one to focus on using frequency readout, it turns out that CPW microresonators suffer from excess frequency noise. This excess noise prevents the detectors from achieving BLIP performance by a factor of around $5-6$ [164] when using frequency readout. The excess noise problem was appreciated early on [159] and stimulated efforts to determine the source of the noise and its general properties. The excess frequency noise rises at low frequencies but has a relatively shallow spectrum, $\sim 1/f^{0.5}$ (see Fig. 22), which implies that the NEP varies very slowly, as $1/f^{0.25}$. The early evidence indicated that the noise did not originate in the superconducting films but rather in the substrate or its surface [162]. Therefore, by using very thin superconducting aluminum films ($t \sim 40$ nm), the frequency responsivity of the device was increased dramatically, and allowed the NEP at 1 Hz to be reduced from the $\sim 10^{-15}\,\mathrm{W}/\sqrt{\mathrm{Hz}}$ level in 2003 to $\sim 10^{-17}\,\mathrm{W}/\sqrt{\mathrm{Hz}}$ in 2005.

Pushing Toward BLIP Performance: Further work [165] showed that the noise really affects just the resonator frequency; no excess noise was seen in the dissipation signal above the level of the amplifier noise (see Fig. 22). This of course opens up the possibility of using the dissipation readout with very high-Q resonators in order to achieve very low NEP values, and this idea has been actively pursued by the SRON group [191, 192]. Their lowest achieved NEP to date is $7 \times 10^{-19}\,\mathrm{W}/\sqrt{\mathrm{Hz}}$ and is steadily decreasing as the quality of the devices and measurement setup are improved. Although this NEP value is actually better than required for broadband CMB measurements, the NEP will degrade as optical loading is applied because the resonator

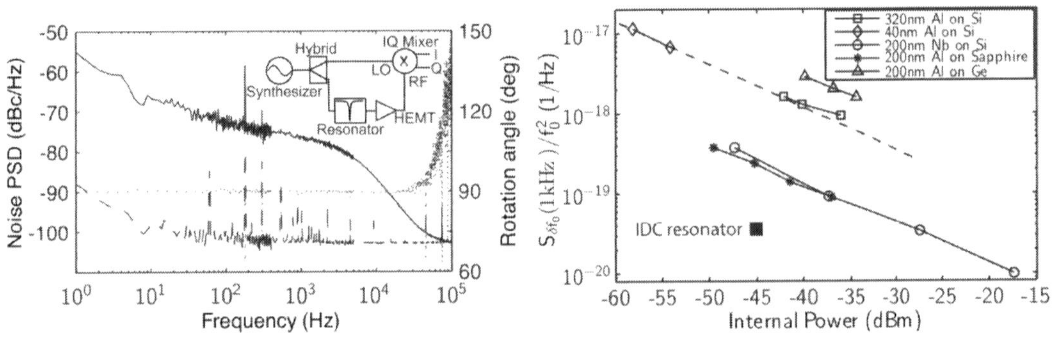

Figure 22. Left: An example [165] of the measured noise power spectra for frequency readout (solid line) and dissipation readout (dashed line). The dotted line indicates the rotation angle between the direction of minimum noise and the tangent to the resonance circle; a value of 90 degrees indicates that the noise is essentially entirely due to frequency noise. Significant excess frequency noise is seen, whereas the noise floor for the dissipation readout is set by the cryogenic HEMT amplifier. The quality of the noise spectra below 10 Hz are limited by insufficient integration time but may be showing $1/f$ amplifier fluctuations; if necessary, such fluctuations can be removed using a variety of techniques, or possibly by using SiGe bipolar amplifiers with low $1/f$ noise. The frequency noise shows a $1/f^{0.5}$ slope in the 10 Hz to 10 kHz range; the corresponding NEP varies as $1/f^{0.25}$. The inset shows the usual homodyne readout scheme. **Right**: This plot [165] shows the typical range of fractional frequency noise measured for CPW resonators. The frequency noise exhibits a $P^{-1/2}$ dependence on the microwave readout power. The square point represents very recent results using a new resonator design [190] that incorporates an interdigitated capacitor (IDC).

quality factor will be reduced. The conclusion is that even for CMB measurements from space, microresonator-based detectors used in dissipation readout mode can already achieve sensitivities within a factor of 2 or so from BLIP.

One approach to push closer towards the BLIP limit is to use dissipation readout with a lower-noise amplifier. Good progress is being made in this direction through the work by Weinreb and colleagues [193, 194] on cryogenic SiGe bipolar amplifiers; noise temperatures of order 2 K have been achieved at frequencies below 2 GHz. Note also that these amplifiers may be expected to have very low $1/f$ gain noise, which should reduce the need to apply other techniques for measuring and compensating the amplifier's gain fluctuations.

Another route is to reduce the resonator frequency noise. It has recently been demonstrated [164, 195, 196] that the noise almost certainly originates primarily from a layer on the surface of the device. This surface layer contains two-level defects such as are commonly found in amorphous dielectric materials, and the fluctuation of these two-level systems (TLS) introduces noise in the resonator due to coupling of the TLS electric dipole moments to the resonator's electric field. A semi-empirical, quantitative theory of this noise mechanism has been developed [196] in which the TLS noise contribution scales as the cube of the electric field, $|E|^3$, and therefore those TLS located near the open-circuit end of the resonator are the most harmful. This insight has been used to develop a modified resonator geometry, in which the high-field "capacitive" portion of the CPW resonator is replaced by an interdigitated capacitor (IDC) structure with $10-20$ μm electrode spacing, as compared to the 2 μm spacing used for the CPW line. Recent measurements [190] show that this new IDC design has dramatically lower noise (see Fig. 22), currently by about a factor of 10 in terms of the frequency noise power spectrum, corresponding to a factor of $\sqrt{10}$ in NEP. These results confirm essential aspects of the noise model and again show that the noise really is not related to the superconductor. Further significant noise reductions may be expected in the future as the insights offered by the new noise model are more fully exploited.

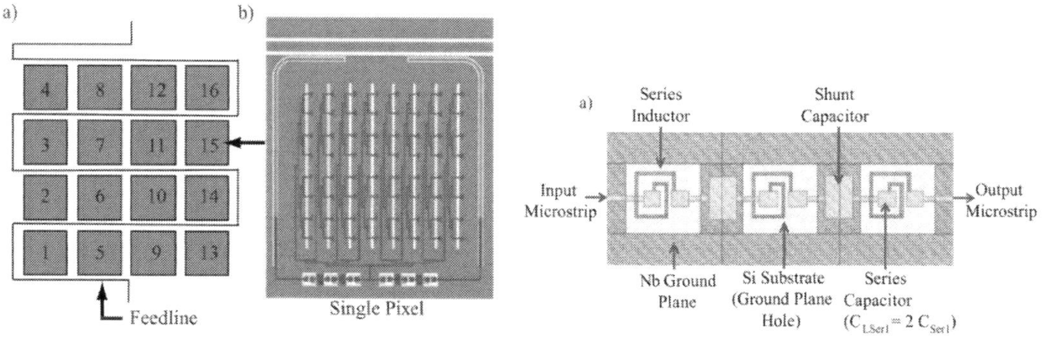

Figure 23. Left: The layout of a 4 × 4 array of dual-band pixels. Each pixel consists of a multislot antenna and an in-phase binary combining network; two on-chip filters that define the two bands centered at 230 GHz and 340 GHz; and two Nb/Al hybrid CPW resonators that serve as the detectors for the two bands. A common feedline is meandered past each pixel in the 4 × 4 array and simultaneously reads out all 32 detectors. **Right**: A close-up of the mm-wave bandpass filter design. The filter is a lumped-element design using spiral series inductors, and both series and shunt parallel-plate capacitors

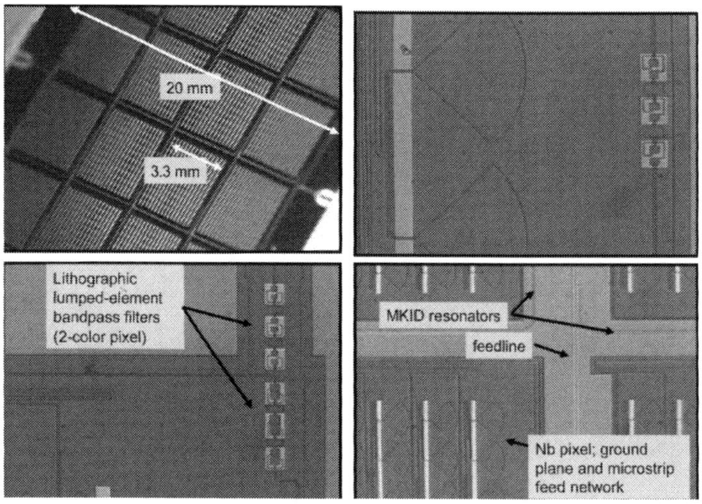

Figure 24. Microscope photographs of the 4×4 dual-band array. Each pixel is ∼ 3.3 mm square. Microstrip radial stubs acting as mm-wave short circuits are used to couple the microstrip in-phase combining network to the slot antennas.

4.2. State of Maturity

A small camera was built and taken to the Caltech Submillimeter Observatory (CSO) in March 2007 in order to demonstrate the technology [163, 197]. The camera was based on a 4 × 4, dual-

Journal of Physics: Conference Series **155** (2009) 012001 doi:10.1088/1742-6596/155/1/012001

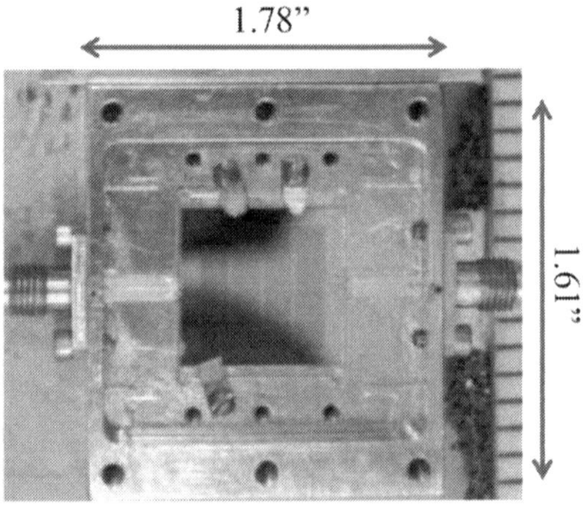

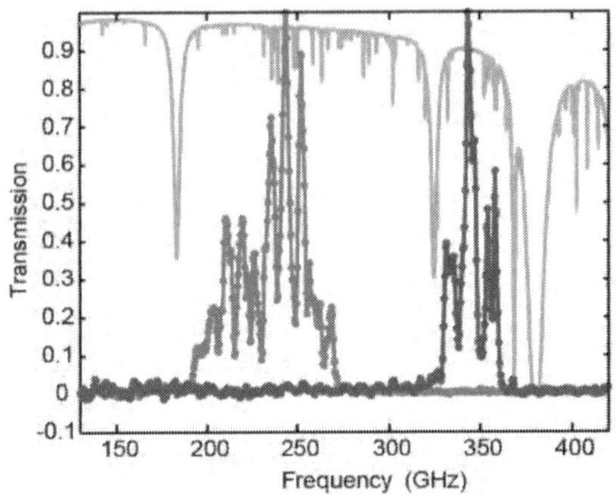

Figure 25. Left: The 4×4 array chip is mounted in a simple metal box. The input and output connections are wire-bonded to CPW to microstrip transitions fabricated on circuit boards, which are soldered to the center pins of the SMA/K coaxial connectors. **Right**: The frequency responses measured for a single dual-band pixel, indicating good agreement with the design values of the turn-on and turn-off frequencies of the on-chip filters. The lack of anti-reflection coatings on some of the optics may be causing the observed fringes; additional measurements are in progress.

band detector array as illustrated in Figures 23 and 24. The array chip was mounted into a simple housing (Fig. 25) and installed into a cryostat containing the optics and a 250 mK $^3He/^3He/^4He$ refrigerator. The frequency response was measured in the laboratory using a Fourier-transform spectrometer, which confirmed dual-band operation with the desired bandpasses (Fig. 25). The beam patterns were also measured in the laboratory and were found to generally agree with expectations. Using a digital readout system based on commercially available hardware [198], successful astronomical observations of bright objects (planets, HII regions) were made at the CSO, which again demonstrated the basic functionality of the system.

This quick ($\sim$ 6 month) exercise proved highly useful and revealed a number of significant issues. The resonators were found to be highly sensitive to the Earth's magnetic field, and this had a serious impact on the sensitivity achieved. A proper magnetic shield was subsequently designed and manufactured, and will be implemented for the next telescope demonstration. The physics of this effect is currently being investigated [199]; simple modifications to the resonator design may substantially reduce their magnetic field sensitivity. In addition, we have found that small field variations only affect the resonance frequency, so the problem could be circumvented using dissipation readout. A second issue was that the devices were found to be substantially overcoupled ($Q_c >> Q_i$), which again adversely impacts the sensitivity. This situation is now being investigated in detail and the lessons learned will be applied to the next version of the array design.

In order to verify the basic optical efficiency of the devices, measurements were performed in a laboratory dilution refrigerator cryostat containing a cryogenic blackbody. This setup has the advantage of having minimal optics between the detector and the source, and also has minimum levels of stray light. These results indicate that the overall efficiency is rather high, of order $\sim 50\%$.

The present situation can be summarized by saying that essentially all aspects of the basic physics of the detector operation have been confirmed and match expectations. In addition, a fully-functional detector array incorporating antennas and on-chip filters has been demonstrated. However, the operation of a fully optimized array at the expected sensitivity in a complete system capable of astronomical observations is still a work in progress - we expect to reach this milestone in about a years' time

A much larger camera consisting of a 24 × 24 array of four-band pixels is now under construction (NSF ATI) for the CSO and is expected to be completed in mid-2010. One of the major tasks is the implementation of the 2304-channel readout electronics. We have studied the technical requirements in detail and have produced a specifications document. Our requirements are well within the state of the art, and we are currently in discussions with potential vendors for the development of a turnkey system.

4.3. Advantages and Disadvantages

The primary advantage of superconducting microresonator detectors is the relatively low cost for implementing a full system. This is due to several factors:

- Detector fabrication is very straightforward and fast, requiring only 3-6 levels of lithography depending on the design, and yields are high. Critical temperature (T_c) uniformity is not a significant issue. Film thickness uniformity is necessary at the $\sim$ 10% level.
- The devices are already fully multiplexed and a separate complex multiplexer chip is not needed. In addition, procedures for attaching the detector array to a multiplexer array (using wire bonding, indium bump bonding, etc.) are not required.
- Compared to first-generation TDM/FDM SQUID multiplexing, much larger multiplexing factors ($\sim 10^3$ instead of $\sim 10^1$) are possible. This reduces the wire count dramatically.
- Microresonator detectors can be used as "drop-in" replacements for microstrip-coupled TES bolometers. This allows existing antenna designs, filter designs, etc. to be reused.

There are also several technical advantages. Perhaps the most useful is that the saturation characteristics are very benign. As the optical load is increased, the internal quality factor Q_i drops, and eventually no longer obeys the optimal coupling condition $Q_c = Q_i$. This just causes the amplifier noise contribution to increase - the detector remains functional with a graceful degradation of its performance. Another advantage is speed: leg-isolated bolometers will be several orders of magnitude slower for the same NEP. The low cost and ease of implementation also facilitates rapid development of the technology. Finally, superconducting microresonators are finding numerous applications in other areas of physics, and the research community in this area is growing rapidly.

The primary disadvantages are:

- Lower level of maturity. TES bolometer development has a $\sim$ 5-year head start.
- Sensitivity. Using dissipation readout and commonly available HEMT amplifiers, sensitivities around 2× BLIP are achievable today, and somewhat better using state-of-the-art SiGe amplifiers. Comparable performance levels should also be possible with frequency readout along with emerging low-noise resonator designs. Reaching BLIP-level performance should eventually be possible but will likely take several more years of effort.
- Readout electronics are not yet available. A first-generation, 2k-channel FPGA-based readout electronics system should be ready by mid-2010. A low power dissipation readout system suitable for space will likely require eventual development of a custom ASIC; however this is best done after the signal processing schemes have been developed and demonstrated using FPGAs.

Journal of Physics: Conference Series **155** (2009) 012001 doi:10.1088/1742-6596/155/1/012001

4.4. Technology Readiness Level

I judge the level to be around TRL4: a "low-fidelity" demonstration of a working system has been achieved (the demonstration camera on the CSO). I expect the TRL level to jump to 5-6 over the next several years; the milestone will be having the new large CSO camera working at its target performance. Prior to inclusion in a CMBpol mission proposal, one would also like to demonstrate dual-polarization versions working at the desired sensitivity levels in the laboratory.

References

[1] M. Ukibe, B. Belier, P. Chamus, C. Cobrea, L. Dumoulin, B. Fernandez, T. Fournier, O. Guillaudin, S. Marnieros, and S. Yates, "Fabrication of large NbSi bolometer arrays for CMB applications," *Nuc. Instr. and Meth. Phys. A*, vol. 559, p. 534, 2006.

[2] A. Orlando, M. Bruijn, H. Hoevers, P. Mauskopf, P. de Korte, E. Krouwer, and M. Ridder, "A waveguide-coupled millimetre-wave TES bolometer suitable for 2-D arrays," *Nucl. Instr. and Meth. A*, vol. 559, p. 534, 2006.

[3] M. D. Audley, D. Glowacka, D. J. Goldie, V. Tsaneva, S. Withington, P. K. Grimes, C. E. North, G. Yassin, L. Piccirillo, P. Ade, and R. V. Sudiwala, "Performance of microstrip-coupled TES bolometers with finline transitions," in *Society of Photo-Optical Instrumentation Engineers (SPIE) Conference Series*, ser. Society of Photo-Optical Instrumentation Engineers (SPIE) Conference Series, vol. 7020, Aug. 2008.

[4] M. Myers, P. Ade, K. Arnold, G. Engargiola, B. Holzapfel, A. Lee, R. O'Brient, P. Richards, A. Smith, H. Spieler, and H. Tran, "Antenna-coupled bolometer arrays using transition-edge sensors," *Nucl. Instr. and Meth. A*, vol. 559, p. 531, 2006.

[5] H. LeDuc, M. Kenyon, P. Day, M. Yun, and J. Bock, "Fabrication of antenna-coupled transition edge sensors for polarimeter applications," *Nucl. Instr. and Meth. A*, vol. 559, p. 459, 2006.

[6] C. Kuo, J. Bock, J. Bonetti, J. Brevik, G. Chattopadhyay, P. Day, S. Golwala, M. Kenyon, A. Lange, H. LeDuc, H. Nguyen, R. Ogburn, A. Orlando, A. Transgrud, A. Turner, G. Wang, and J. Zmuidzinas, "Antenna-Coupled TES Bolometer Arrays for CMB Polarimetry," *Proc. SPIE*, vol. 7020, p. 702011, 2008.

[7] T. Stevenson, W.-T. Hsieh, G. Schneider, D. Travers, N. Cao, E. Wollack, M. Limon, and A. Kogut, "Building blocks for a polarimeter-on-a-chip," *Nucl. Instr. and Meth. A*, vol. 559, p. 611, 2006.

[8] T. Stevenson, D. Benford, C. Bennett, N. Cao, D. Chuss, K. Denis, W. Hsieh, A. Kogut, S. Moseley, J. Panek, G. Schneider, D. Travers, K. U-Yen, G. Voellmer, and E. Wollack, "Cosmic Microwave Background Polarization Detector with High Efficiency, Broad Bandwidth, and Highly Symmetric Coupling to Transition Edge Sensor Bolometers," *Journal of Low Temperature Physics*, vol. 151, p. 471, 2008.

[9] F. Herschel, "Experiments on the Refrangibility of the Invisible Rays of the Sun," *Philosophical Transactions*, p. 294, 1800.

[10] S. Langley, "The Bolometer," *Nature*, vol. 25, p. 14, 1881.

[11] A. Goetz, "The Possible Use of Superconductivity for Radiometric Purposes," *Phys. Rev.*, vol. 55, p. 1270, 1939.

[12] D. Andrews, J. Bruksch, W.F., W. Ziegler, and E. . Blanchard, "Superconducting Films as Radiometric Receivers," *Phys. Rev.*, vol. 59, p. 1045, 1941.

[13] D. Andrews, R. Milton, and W. DeSorro, "A Fast Superconducting Bolometer," *JOSA*, vol. 36, no. 9, p. 518, 1946.

[14] N. Fuson, "The Infra-Red Sensitivity of Superconducting Bolometers," *JOSA*, vol. 38, no. 10, p. 845, 1948.

[15] W. Boyle and J. Rodgers, K.F., "Performance Characteristics of a New Low-Temperature Bolometer," *JOSA*, vol. JOSA, no. 1, p. 66, 1959.

[16] F. Low, "Low-Temperature Germanium Bolometer," *JOSA*, vol. 51, no. 11, p. 1300, 1961.

[17] J. Clarke, P. Richards, and N.-H. Yeh, "Composite superconducting transition edge bolometer," *Appl. Phys. Lett.*, vol. 30, p. 664, 1977.

[18] J. Clarke, G. Hoffer, P. Richards, and N.-H. Yeh, "Superconductive Bolometers for Submillimeter Wavelengths," *J. Appl. Phys.*, vol. 48, no. 2, p. 4865, 1977.

[19] S. Wentworth and D. Neikirk, "The Transition-Edge Microbolometer (TREMBOL)," *Proc. SPIE*, vol. 1292, p. 148, 1990.

[20] A. Lee, P. Richards, S. Nam, B. Cabrera, and K. Irwin, "A superconducting bolometer with strong electrothermal feedback," *Appl. Phys. Lett.*, vol. 69, p. 1801, 1996.

[21] K. Irwin, "An application of electrothermal feedback for high resolution cryogenic particle detection," *Appl. Phys. Lett.*, vol. 66, p. 1998, 1995.

[22] H. Dewhurst, "A Rapid Bolometer Made By Sputtering On Thin Films," *Proc. Phys. Soc.*, vol. 39, no. 1, p. 39, 1926.

[23] W. Cannon and M. Chester, "Fast Carbon Bolometer for Observing Thermal Pulses," *Rev. Sci. Inst.*, vol. 38, no. 3, p. 318, 1967.

[24] R. De Waard and S. Weiner, "Miniature Optically Immersed Thermistor Bolometer Arrays," *Ap. Opt.*, vol. 6, no. 8, p. 1327, 1967.

[25] H. Drew and A. Sievers, "A ^{3}He-Cooled Bolometer for the Far Infrared," *Ap. Opt.*, vol. 8, no. 10, p. 2067, 1969.

[26] P. Downey, A. Jeffries, S. Meyer, R. Weiss, F. Cahner, J. Donnelly, W. Lindley, R. Mountain, and D. Silversmith, "Monolithic Silicon Bolometers," *Ap. Opt.*, vol. 23, no. 6, p. 910, 1984.

[27] R. Bachmann, H. Kirsch, and T. Geballe, "Low Temperature Silicon Thermometer and Bolometer," *Rev. Sci. Inst.*, vol. 41, no. 4, p. 547, 1970.

[28] R. Britt and P. Richards, "An Adiabatic Demagnetization Refrigerator For Infrared Bolometers," *IJIMW*, vol. 2, no. 6, p. 1083, 1981.

[29] L. Lesyna, T. Roellig, and P. Kittel, "Bolometers Operated At 0.1 K And 0.2 K Cooled By Adiabatic Demagnetization," *IJIMW*, vol. 5, no. 6, p. 755, 1984.

[30] J. Mees, M. Nahum, and P. Richard, "New Designs for Antenna-coupled Superconducting Bolometers," *Appl. Phys. Lett.*, vol. 59, no. 18, p. 2329, 1991.

[31] J. Milatz and H. Van Der Velden, "Natural Limit of Measuring Radiation with a Bolometer," *Physica*, vol. X, no. 6, p. 24, 1943.

[32] B. H. Billings, W. L. Hyde, and E. E. Barr, "An investigation of the properties of evaporated metal bolometers," *Journal of the Optical Society of America (1917-1983)*, vol. 37, pp. 123–+, Mar. 1947.

[33] R. C. Jones, "Steady-state load curves for semi-conductor bolometers," *Journal of the Optical Society of America (1917-1983)*, vol. 36, pp. 448–+, Aug. 1946.

[34] ——, "The ultimate sensitivity of radiation detectors," *Journal of the Optical Society of America (1917-1983)*, vol. 37, pp. 879–+, Nov. 1947.

[35] R. Jones, "Factors of Merit for Radiation Detectors," *JOSA*, vol. 39, no. 8, p. 344, 1949.

[36] ——, "General Theory of Bolometer Performance," *JOSA*, vol. 43, no. 1, p. 1, 1953.

[37] J. Mather, "Bolometer noise: nonequilibrium theory," *Ap. Opt.*, vol. 21, no. 6, p. 1125, 1982.

[38] Y. Urano, "An Equivalent Circuit of a Bolometer," *J. Phys. Soc. Japan*, vol. 10, no. 10, p. 864, 1966.

[39] N. Fuson, "The Effect of Current Magnitude upon the Behavior of a Superconducting Bolometer n Its Transition Region," *J. App. Phys.*, vol. 30, p. 59, 1949.

[40] B. Lalevic, "Criteria for the Choice of a Superconducting Bolometer," *J. App. Phys.*, vol. 31, no. 7, p. 1234, 1960.

[41] M. Maul and M. Strandberg, "Equivalent Circuit of a Superconducting Bolometer," *J. Appl. Phys.*, vol. 40, no. 7, p. 2822, 1969.

[42] M. K. Maul, M. W. Strandberg, and R. L. Kyhl, "Excess Noise in Superconducting Bolometers," *Physical Review*, vol. 182, pp. 522–525, Jun. 1969.

[43] C. Bertin and K. Rose, "Comparison of Superconducting and Semiconducting Bolometers," *J. App. Phys.*, vol. 42, no. 1, p. 163, 1971.

[44] S. Tinchev, L. Bliek, and V. Kose, "A Consistent Model For 1/f-Noise In Thin-Film Devices Such As Bolometers, Josephson Junctions And Squids," *Phys. Lett.*, vol. 94A, no. 8, p. 381, 1983.

[45] C. Knoedler, "Phase-Slip Shot Noise Contribution to Excess Noise of Superconducting Bolometers," *J. Appl. Phys.*, vol. 54, no. 5, p. 2773, 1983.

[46] R. Sherlock and A. Wyatt, "The effect of self-heating on the dynamical response of bolometric detectors," *J. Phys. E: Sci. Instrum.*, vol. 16, p. 669, 1983.

[47] G. C. Irwin, K. D. & Hilton, *"Transition-Edge Sensors"*, C. Enss, Ed. Springer-Verlag, 2005, vol. Cryogenic Particle Detecion.

[48] B. Cabrera, "Introduction to TES Physics," *J. Low Temp. Phys.*, vol. 151, p. 82, 2008.

[49] M. Lindeman, K. Barger, D. Brandl, S. Crowder, L. Rocks, D. McCammon, and H. Hoevers, "The Superconducting Transition in 4-D: Temperature, Current, Resistance and Heat Capacity," *J. Low Temp. Phys.*, vol. 151, p. 190, 2008.

[50] K. Irwin, "Thermodynamics of nonlinear bolometers near equilibrium," *Nuclear Instruments and Methods in Physics Research A*, vol. 559, p. 718, 2006.

[51] D. McCammon, *"Thermal Equilibrium Calorimeters – An Introduction", in Cryogenic Particle Detection*, C. Enss, Ed. Springer-Verlag, 2005.

[52] S.-F. Lee, J. M. Gildemeister, W. Holmes, A. T. Lee, and P. L. Richards, "Voltage-Biased Superconducting Transition-Edge Bolometer with Strong Electrothermal Feedback Operated at 370 mK," *Applied Optics*, vol. 37, pp. 3391–3397, Jun. 1998.

[53] J. Gilemeister, A. Lee, and P. Richards, "A fully lithographed voltage-biased superconducting spiderweb bolometer," *APL*, vol. 74, no. 6, p. 868, 1999.

[54] ——, "Monolithic arrays of absorber-coupled voltage-biased superconducting bolometers," *APL*, vol. 77, no. 24, p. 4040, 2000.

[55] ——, "Model for excess noise in voltage-biased superconducting bolometers," *Ap. Opt*, vol. 40, no. 34, p. 6229, 2001.

[56] R. Romani, A. Miller, B. Cabrera, E. Figueroa-Feliciano, and S. Nam, "First Astronomical Application Of A Cryogenic Transition Edge Sensor Spectrophotometer," *Ap.J.*, vol. 521, p. L153, 1999.

[57] D. Benford, C. Allen, J. Chervenak, E. Grossman, K. Irwin, A. Kutyrev, M. J.M., S. Moseley, R. Shafer, and C. Reintsema, "Superconducting Bolometer Arrays for Submillimeter Astronomy," in *APS Conference*, vol. 217, 2000, p. 134.

[58] D. Benford, "Multiplexed Readout of Superconducting Bolometers," *IJIMW*, vol. 21, no. 12, p. 1909, 2000.

[59] J. Staguhn, C. Allen, D. Benford, J. Chervenak, M. Freund, S. Khan, A. Kutyrev, S. Moseley, R. Shafer, S. Deiker, E. Grossman, G. Hilton, K. Irwin, J. Martinis, S. Nam, D. Rudman, and D. Wollman, "TES Detector Noise Limited Readout Using SQUID Multiplexers," in *AIP Conference proceedings*, vol. 605, 2001, p. 321.

[60] J. Staguhn, S. Moseley, D. Benford, C. Allen, J. Chervenak, T. Stevenson, and W.-T. Hsieh, "Approaching the fundamental noise limit in Mo/Au TES bolometers with

transverse normal metal bars," *Nuclear Instruments and Methods in Physics Research A*, vol. 520, p. 336, 2004.

[61] D. Benford, S. Moseley, J. Staguhn, C. Allen, J. Chervenak, T. Stevenson, and W.-T. Hsieh, "Parameter Comparison for Low-Noise Mo/Au TES Bolometers," *Nuclear Instruments and Methods in Physics Research A*, vol. 520, p. 270, 2004.

[62] D. Benford, C. Ames, T.A., E. J.A., Grossman, K. Irwin, S. Khan, B. Maffei, S. Moseley, F. Pajot, T. Phillips, J.-C. Renault, C. Reintsema, C. Rioux, R. Shafer, J. Staguhn, C. Vastel, and G. Voellmer, "First Astronomical Use of Multiplexed Transition Edge Bolometers," in *AIP Conference proceedings*, vol. 605, 2001, p. 589.

[63] D. Benford, J. Staguhn, T. Ames, C. Allen, J. Chervenak, C. Kennedy, S. Lefranc, S. Maher, S. Moseley, F. Pajot, C. Rioux, R. Shafer, and G. Voellmer, "First astronomical images with a multiplexed superconducting bolometer array," *Proc. SPIE*, vol. 6275, p. 62751C, 2006.

[64] T. Saab, E. Apodacas, S. Bandler, K. Boyce, J. Chervenak, E. Figueroa-Feliciano, F. Finkbeiner, C. Hammock, R. Kelley, M. Lindeman, F. Porter, and C. Stahle, "Characterization and Modeling of Transition Edge Sensors for High Resolution X-ray Calorimeter Arrays," *Nucl. Instr. and Meth. A*, vol. 520, p. 281, 2003.

[65] M. Lindeman, S. Bandler, S. Brekosky, J. Chervenak, E. Figueroa-Feliciano, F. Finkbeiner, T. Saab, and C. Stahle, "Characterization and Reduction of Noise in Mo/Au Transition Edge Sensors," *Nucl. Instr. and Meth. A*, vol. 520, p. 348, 2003.

[66] S. Bandler, E. Figueroa-Feliciano, C. Stahle, K. Boyce, R. Brekosky, J. Chervenak, F. Finkbeiner, R. Kelley, M. Lindeman, F. Porter, and T. Saab, "Design of Transition Edge Sensor Microcalorimeters for Optimal Performance," *Nucl. Instr. and Meth. A*, vol. 520, p. 285, 2003.

[67] J. Chervenak, K. Irwin, E. Grossman, J. Martinis, C. Reintsema, and M. Huber, "Superconducting multiplexer for arrays of transition edge sensors," *App. Phys. Lett.*, vol. 74, no. 26, p. 4043, 1999.

[68] K. Irwin, "SQUID Multiplexers for Transition-Edge Sensors," *Physica C*, vol. 368, p. 203, 2002.

[69] K. Irwin, M. Audley, J. Beall, J. Beyer, S. Deiker, W. Doriese, W. Duncan, G. Hilton, W. Holland, C. Reintsema, J. Ullom, L. Vale, and Y. Xu, "In-Focal-Plane SQUID Multiplexer," *Nucl. Instr. and Meth. A*, vol. 520, p. 544, 2004.

[70] J. Staguhn, D. Benford, J. Chervenak, J. Moseley, S.H., C. Allen, T. Stevenson, and W.-T. Hsieh, "Design techniques for improved noise performance of superconducting Transition Edge Sensor bolometers," *Proc. SPIE*, vol. 5498, p. 390, 2004.

[71] J. Ullom, W. Doriese, G. Hilton, J. Beall, S. Deiker, W. Duncan, L. Ferreira, K. Irwin, C. Reintsema, and L. Vale, "Characterization and reduction of unexplained noise in superconducting transition-edge sensors," *App. Phys. Lett.*, vol. 84, no. 21, p. 4206, 2004.

[72] S. R. Dicker, P. A. Ade, D. J. Benford, M. J. Devlin, K. D. Irwin, P. R. Jewell, B. S. Mason, S. H. Moseley, M. P. Supanich, and C. Tucker, "A 90-GHz array for use on the Green Bank Telescope," in *Society of Photo-Optical Instrumentation Engineers (SPIE) Conference Series*, ser. Society of Photo-Optical Instrumentation Engineers (SPIE) Conference Series, J. M. Oschmann, Jr., Ed., vol. 5489, Oct. 2004, pp. 1221–1229.

[73] D. Benford, S. Dicker, E. Wollack, M. Supanich, J. Staguhn, S. Moseley, K. Irwin, M. Devlin, J. Chervenak, and T. Chen, "A planar two-dimensional superconducting bolometer array for the Green Bank Telescope," *Proc. SPIE*, vol. 5498, p. 208, 2004.

[74] A. Kosowsky, "The Atacama Cosmology Telescope," *New Astronomy Reviews*, vol. 47, p. 939, 2003.

[75] M. Niemack and the ACT Collaboration, "Measuring two-millimeter radiation with a prototype multiplexed TES receiver for ACT," *Proc. SPIE*, vol. 6275, p. 62750C, 2006.

[76] D. Benford, J. Staguhn, G. Stacey, L. Page, S. Moseley, K. Irwin, J. Chervenak, and C. Allen, "Design and Fabrication of Two-Dimensional Superconducting Bolometer Arrays," *Proc. SPIE*, vol. 5498, p. 647, 2004.

[77] C. Hunt, J. Bock, P. Day, A. Goldin, A. Lange, H. LeDuc, A. Vayonakis, and J. Zmuidzinas, "Transition-edge superconducting antenna-coupled bolometer," *Proc. SPIE*, vol. 4855, p. 318, 2003.

[78] A. Goldin, J. J. Bock, A. E. Lange, H. Leduc, A. Vayonakis, and J. Zmuidzinas, "Antennas for bolometric focal plane," *Nuclear Instruments and Methods in Physics Research A*, vol. 520, pp. 390–392, Mar. 2004.

[79] P. Day, H. LeDuc, A. Golding, C. Dowell, and J. Zmuidzinas, "Far Infrared/ Submillimeter Imager-Polarimeter Using Distributed Antenna-Coupled Transition Edge Sensors," *Proc. SPIE*, vol. 5498, p. 857, 2004.

[80] M. Myers, W. Holzapfel, A. Lee, R. O'Brient, P. Richards, D. Schwan, A. Smith, H. Spieler, and H. Tran, "Arrays of Antenna-Coupled Bolometers using Transition Edge Sensors," *Nucl. Instr. and Meth. A*, vol. 520, p. 424, 2003.

[81] M. Dobbs, N. Halverson, P. Ade, K. Basu, A. Beelen, F. Bertoldi, C. Cohalan, H. Cho, R. Güsten, W. Holzapfel, Z. Kermish, R. Kneissl, A. Kovács, E. Kreysa, T. Lanting, A. Lee, M. Lucker, J. Mehl, K. Menten, D. Muders, M. Nord, T. Plagge, P. Richards, P. Schilke, D. Schwan, H. Spieler, A. Weiss, and M. White, "APEX-SZ first light and instrument status," *New Astronomy Reviews*, vol. 50, no. 11-12, p. 960, 2006.

[82] K. D. Irwin, "The Thermodynamics of Nonlinear Bolometers Near Equilibrium," in *TES III, Gainesville, FL*, August 17-18, 2006.

[83] H. Hoevers, "Thermal physics of transition edge sensor arrays," *Nucl. Instr. and Meth. A*, vol. 559, p. 702, 2006.

[84] M. Bruijn, A. Germeau, H. Hoevers, P. de Korte, J. van der Kuur, and M. Ridder, "Steepness, noise and instabilities of Ti/Au transition edge thermometers," *Nucl. Instr. and Meth. A*, vol. 559, p. 709, 2006.

[85] H. Hoevers, M. Bruijn, L. Dirks, B.P.F. adn Gottardi, P. de Korte, J. van der Kuur, A. Popescu, M. Ridder, Y. Takei, and D. Takken, "Comparative Study of TiAu-Based TES Microcalorimeters with Different Geometries," *Journal of Low Temperature Physics*, vol. 151, p. 94, 2008.

[86] G. W. Fraser, "On the nature of the superconducting-to-normal transition in transition edge sensors," *Nuclear Instruments and Methods in Physics Research A*, vol. 523, p. 234, 2004.

[87] M. Galeazzi, "Flux Flow Noise in Transition Edge Sensors," in *TES III, Gainesville, FL*, August 17-18, 2006.

[88] M. Lindeman, M. Anderson, S. Bandler, N. Bilgri, J. Chervenak, S. Gwynne Crowder, S. Fallows, E. Figueroa-Feliciano, F. Finkbeiner, N. Iyomoto, R. Kelley, C. Kilbourne, T. Lai, J. Man, D. McCammon, K. Nelms, F. Porter, L. Rocks, T. Saab, J. Sadleir, and G. Vidugiris, "Percolation model of excess electrical noise in transition-edge sensors," *Nuclear Instruments and Methods in Physics Research A*, vol. 559, p. 715, 2006.

[89] H. Hoevers, A. Bento, M. Bruijn, L. Gottardi, M. Korevaar, W. Mels, and P. de Korte, "Thermal fluctuation noise in a voltage biased superconducting transition edge thermometer," *Applied Physics Letters*, vol. 77, p. 4422, 2000.

[90] A. Luukanen, K. M. Kinnunen, A. K. Nuottajaumlrvi, H. F. Hoevers, W. M. Bergmann Tiest, and J. P. Pekola, "Fluctuation-Limited Noise in a Superconducting Transition-Edge Sensor," *Physical Review Letters*, vol. 90, p. 238306, 2003.

[91] S. Crowder, M. Lindeman, M. Anderson, S. Bandler, N. Bilgri, M. Bruijn, J. Chervenak, E. Figueroa-Feliciano, F. Finkbeiner, A. Germeau, H. Hoevers, N. Iyomoto, R. Kelly, C. Kilbourne, T. Lai, J. Man, D. McCammon, K. Nelms, F. Porter, L. Rocks, T. Saab, J. Sadleir, and G. Vidugiris, "An investigation of excess noise in transition-edge sensors on a solid silicon substrate," *Nucl.Instr.and Meth.A*, vol. 559, p. 721, 2006.

[92] D. Brandt, G. Fraser, and S. Smith, "Excess Noise in TESs – A Comparison Between Theories," in *TES III, Gainesville, FL*, August 17-18, 2006.

[93] K. Kinnunen, A. Nuottajärvi, and I. Maasilta, "A Transition-Edge Sensor with Two Excess Noise Mechanisms," *Journal of Low Temperature Physics*, vol. 151, p. 144, 2008.

[94] E. Barrentine, P. Timbie, T. Stevenson, S. Ali, J. Chervenak, E. Wollack, S. Moseley, C. Allen, W. Hseih, T. Miller, D. Benford, and A. Brown, "Sensitivity Measurements of a Transition-Edge Hot-Electron Microbolometer for Millimeter-Wave Astrophysical Observations," *Journal of Low Temperature Physics*, vol. 151, p. 173, 2008.

[95] K. M. Kinnunen, A. K. Nuottajärvi, J. Leppäniemi, and I. J. Maasilta, "Reducing Excess Noise in Au/Ti Transition-Edge Sensors," *Journal of Low Temperature Physics*, vol. 151, p. 119, 2008.

[96] M. Griffin, J. Bock, and W. Gear, "Relative performance of filled and feedhorn-coupled focal-plane architectures," *Ap. Opt.*, vol. 41, no. 31, p. 6543, 2002.

[97] D. J. Benford, G. M. Voellmer, J. A. Chervenak, K. D. Irwin, S. H. Moseley, R. A. Shafer, and J. G. Staguhn, "Design and fabrication of a 2D superconducting bolometer array for SAFIRE," *Proc. SPIE*, vol. 4857, p. 125, 2003.

[98] C. Bradford, M. Kenyon, W. Holmes, J. Bock, T. Koch, and the BLISS study team, "Sensitive far-IR survey spectroscopy: BLISS for SPICA," *Proc. SPIE*, vol. 7020, p. 70201O, 2008.

[99] M. Kenyon, P. Day, C. Bradford, J. Bock, and H. LeDuc, "Electrical Properties of Background-Limited Membrane-Isolation Transition-Edge Sensing Bolometers for Far-IR/Submillimeter Direct-Detection Spectroscopy," *J. Low Temp. Phys.*, vol. 151, p. 112, 2008.

[100] P. Mauskopf, D. Morozov, D. Glowacka, D. Goldie, S. Withington, M. Bruijn, P. DeKorte, H. Hoevers, M. Ridder, J. Van Der Kuur, and J.-R. Gao, "Development of transition edge superconducting bolometers for the SAFARI Far-Infrared spectrometer on the SPICA space-borne telescope," *Proc. SPIE*, vol. 7020, p. 70200N, 2008.

[101] B. Karasik, D. Olaya, J. Wei, S. Pereverzev, M. Gershenson, J. Kawamura, W. McGrath, and A. Sergeev, "Record-Low NEP in Hot-Electron Titanium Nanobolometers," *IEEE Trans. App. Superconductivity*, vol. 17, no. 2, p. 293, 2007.

[102] J. Wei, D. Olaya, B. Karasik, S. Pereverzev, A. Sergeev, and M. Gershenson, "Ultra-Sensitive Hot-Electron Nanobolometers for Terahertz Astrophysics," *astro-ph*, p. 0710.5474, 2007.

[103] T. Marriage, J. Chervenak, and W. Doriese, "Testing and assembly of the detectors for the Millimeter Bolometer Array Camera on ACT," *Nucl. Instr. and Meth. A*, vol. 559, p. 551, 2006.

[104] M. Niemack, Y. Zhao, E. Wollack, R. Thornton, E. Switzer, D. Swetz, S. Staggs, L. Page, O. Sytrzak, S. Moseley, T. Marriage, M. Limon, J. Lau, J. Klein, M. Kaul, N. Jarosik, K. Irwin, A. Hincks, G. Hilton, M. Halpern, R. Fowler, J.W.and Fisher, R. Dünner,

W. Doriese, S. Dicker, M. Devlin, J. Chervenak, B. Burger, E. Battistelli, J. Appel, M. Amiri, C. Allen, and A. Aboobaker, "A Kilopixel Array of TES Bolometers for ACT: Development, Testing, and First Light," *J. Low Temp. Phys.*, vol. 151, p. 690, 2008.

[105] J. Fowler, "The Atacama Cosmology Telescope: 2007 Initial Observations," in *Aspen Conference on the Cosmic Microwave Background Radiation*, January 28, 2008.

[106] P. Korngut, S. Dicker, B. Mason, P. Ade, J. Aguirre, T. Ames, D. Benford, J. Chervenak, T. Chen, M. Compiegne, W. Cotton, M. Devlin, E. Figueroa-Feliciano, K. Irwin, S. Maher, P. Martin, M. Mello, S. Moseley, C. Tucker, J. Staguhn, B. Werner, and S. D. White, "High Resolution Continuum Observations Of The Orion Nebula At 3.3mm With MUSTANG And The GBT," *BAAS*, vol. 53.01, no. 40, p. 249, 2008.

[107] J. Staguhn, C. Allen, D. Benford, E. Sharp, T. Ames, R. Arendt, D. Chuss, E. Dwek, A. Kovacs, S. Maher, C. Marx, T. Miller, S. Moseley, S. Navaroo, A. Sievers, G. Voellmer, and E. Wollack, "GISMO, a 2 mm Bolometer Camera Optimized for the Study of High Redshift Galaxies," *J. Low Temp. Phys.*, vol. 151, p. 709, 2008.

[108] J. Staguhn, D. Benford, C. Allen, S. Moseley, E. Sharp, T. Ames, W. Brunswig, D. Chuss, E. Dwek, S. Maher, C. Marx, T. Miller, S. Navarro, and E. Wollack, "GISMO: A 2-Millimeter Bolometer Camera for the IRAM 30 m Telescope," *Proc. SPIE*, vol. 6275, p. 62751D, 2006.

[109] C. Allen, D. Benford, T. Miller, S. Moseley, J. Staguhn, and E. Wollack, "Design and Fabrication Highlights Enabling a 2 mm,128 Element Bolometer Array for GISMO," *J. Low Temp. Phys.*, vol. 151, p. 266, 2008.

[110] G. Bernstein, "Advanced Exposure-Time Calculations: Undersampling, Dithering, Cosmic Rays, Astrometry, and Ellipticities," *PASP*, vol. 114, p. 98, 2002.

[111] C. Allen, D. Benford, J. Chervenak, D. Chuss, T. Miller, S. Moseley, J. Staguhn, and E. Wollack, "Backshort-Under-Grid arrays for infrared astronomy," *Nucl. Instr. and Meth. A*,, vol. 559, p. 522, 2008.

[112] J. G. Staguhn, C. A. Allen, D. J. Benford, J. A. Chervenak, D. T. Chuss, T. M. Miller, S. H. Moseley, and E. J. Wollack, "Characterization of TES bolometers used in 2-dimensional Backshort-Under-Grid (BUG) arrays for far-infrared astronomy," *Nuclear Instruments and Methods in Physics Research A*, vol. 559, pp. 545–547, Apr. 2006.

[113] C. Allen, J. Abrahams, D. Benford, J. Chervenak, D. Chuss, J. Staguhn, T. Miller, S. Moseley, and E. Wollack, "Far infrared through millimeter Backshort-Under-Grid arrays," *Proc. SPIE*, vol. 6275, p. 62750B, 2006.

[114] T. Miller, J. Abrahams, and C. Allen, "Fabricating interlocking support walls, with an adjustable backshort, in a TES bolometer array for far-infrared astronomy," *Nucl. Instr. and Meth. A*, vol. 559, p. 548, 2006.

[115] J. Vieira, "The South Pole Telescope," in *Aspen Conference on the Cosmic Microwave Background Radiation*, 2008.

[116] A. Taylor, "Clover – A B-mode polarization experiment," *NAR*, vol. 50, p. 993, 2006.

[117] M. Audley, R. Barker, M. Crane, R. Dace, D. Glowacka, D. Goldie, A. Lasenby, H. Stevenson, V. Tsaneva, S. Withington, P. Grimes, B. Johnson, G. Yassin, L. Piccirillo, W. Pisano, G.and Duncan, G. Hilton, K. Irwin, C. Reintsema, and M. Halpern, "TES imaging array technology for CLOVER," *Proc. SPIE*, vol. 6275, p. 627524, 2006.

[118] G. Siringo, A. Weiß, E. Kreysa, F. Schuller, and A. Kovacs, "SABOCA Technical Verification Report," APEX memo., Tech. Rep., 2008.

[119] W. Holland, M. MacIntosh, A. Fairley, D. Kelly, D. Montgomery, D. Gostick, E. Atad-Ettedgui, M. Ellis, I. Robson, M. Hollister, A. Woodcraft, P. Ade, I. Walker, K. Irwin,

G. Hilton, W. Duncan, C. Reintsema, A. Walton, W. Parkes, C. Dunare, M. Fich, J. Kycia, M. Halpern, D. Scott, A. Gibb, J. Molnar, E. Chapin, D. Bintley, S. Craig, T. Chylek, T. Jenness, F. Economou, and G. Davis, "SCUBA-2: a 10,000-pixel submillimeter camera for the James Clerk Maxwell Telescope," *Proc. SPIE*, vol. 6275, p. 62751E, 2006.

[120] (2008) JCMT Newsletter #28 (Spring '08). [Online]. Available: `http://www.jach.hawaii.edu/JCMT/publications/newsletter/n28/jcmt-n28.pdf`

[121] P. Oxley, P. Ade, C. Baccigalupi, P. deBernardis, H.-M. Cho, M. Devlin, S. Hanany, B. Johnson, T. Jones, A. Lee, T. Matsumura, A. Miller, M. Milligan, T. Renbarger, H. Spieler, R. Stompor, G. Tucker, and M. Zaldarriaga, "The EBEX experiment," *Proc.SPIE*, vol. 5543, p. 320, 2004.

[122] B. Reichborn-Kjennerud, "EBEX: E and B EXperiment: A CMB Polarization Experiment," in *Aspen Conference on the Cosmic Microwave Background Radiation*, 2008.

[123] T. E. Montroy, P. A. R. Ade, R. Bihary, J. J. Bock, J. R. Bond, J. Brevik, C. R. Contaldi, B. P. Crill, A. Crites, O. Doreacute, L. Duband, S. R. Golwala, M. Halpern, G. Hilton, W. Holmes, V. V. Hristov, K. Irwin, W. C. Jones, C. L. Kuo, A. E. Lange, C. J. MacTavish, P. Mason, J. Mulder, C. B. Netterfield, E. Pascale, J. E. Ruhl, A. Trangsrud, C. Tucker, A. Turner, and M. Viero, "SPIDER: a new balloon-borne experiment to measure CMB polarization on large angular scales," *Proc. SPIE*, vol. 6267, p. 62670R, 2006.

[124] C. Hunt, J. Bock, P. Day, A. Goldin, A. Lange, H. LeDuc, A. Vayonakis, and J. Zmuidzinas, "Transition-edge superconducting antenna-coupled bolometer," *Proc. SPIE*, vol. 4855, p. 318, 2003.

[125] [Online]. Available: `http://en.wikipedia.org/wiki/Spider_CMB`

[126] C. MacTavish, "Spider Optimization: Probing the Systematics of a B-Mode Experiment," January 2008, Aspen Conference on the Cosmic Microwave Background Radiation, January 29, 2008.

[127] D. Benford, S. Moseley, G. Stacey, R. Shafer, and J. Staguhn, "Far-infrared imaging spectroscopy with SAFIRE on SOFIA," *Proc.SPIE*, vol. 4857, p. 105, 2003.

[128] C. Allen, D. Benford, T. Miller, J. Moseley, S.H.and Staguhn, and E. Wollack, "Technology developments toward large format long wavelength bolometer arrays," *Proc. SPIE*, vol. 6678, p. 667806, 2007.

[129] D. J. Benford, J. G. Staguhn, S. H. Moseley, C. A. Allen, J. A. Chervenak, K. D. Irwin, S. R. Dicker, M. J. Devlin, T. Nikola, T. E. Oberst, and G. J. Stacey, "Superconducting Bolometers for Submillimeter Spectroscopy from Ground-Based, Airborne, and Space Platforms," in *ASPC:From Z-Machines to ALMA: (Sub)Millimeter Spectroscopy of Galaxies*, vol. 375, 2007, p. 217.

[130] J. Mankins, "Technology Readiness Levels," NASA Office of Space Access and Technology, Tech. Rep., 1995.

[131] [Online]. Available: `http://www.hq.nasa.gov/office/codeq/trl/trlchrt.pdf`

[132] F. Menanteau and J. P. Hughes, "Southern Cosmology Survey II: Physical Properties of SZE-Selected Galaxy Clusters," *ArXiv e-prints*, Nov. 2008.

[133] Z. Staniszewski, P. A. R. Ade, K. A. Aird, B. A. Benson, L. E. Bleem, J. E. Carlstrom, C. L. Chang, H. . Cho, T. M. Crawford, A. T. Crites, T. de Haan, M. A. Dobbs, N. W. Halverson, G. P. Holder, W. L. Holzapfel, J. D. Hrubes, M. Joy, R. Keisler, T. M. Lanting, A. T. Lee, E. M. Leitch, A. Loehr, M. Lueker, J. J. McMahon, J. Mehl, S. S. Meyer, J. J. Mohr, T. E. Montroy, C. . Ngeow, S. Padin, T. Plagge, C. Pryke, C. L. Reichardt, J. E. Ruhl, K. K. Schaffer, L. Shaw, E. Shirokoff, H. G. Spieler, B. Stalder, A. A. Stark, K. Vanderlinde, J. D. Vieira, O. Zahn, and A. Zenteno, "Galaxy clusters discovered with a Sunyaev-Zel'dovich effect survey," *ArXiv e-prints*, Oct. 2008.

[134] J. C. Mather, E. S. Cheng, D. A. Cottingham, R. E. Eplee, Jr., D. J. Fixsen, T. Hewagama, R. B. Isaacman, K. A. Jensen, S. S. Meyer, P. D. Noerdlinger, S. M. Read, L. P. Rosen, R. A. Shafer, E. L. Wright, C. L. Bennett, N. W. Boggess, M. G. Hauser, T. Kelsall, S. H. Moseley, Jr., R. F. Silverberg, G. F. Smoot, R. Weiss, and D. T. Wilkinson, "Measurement of the cosmic microwave background spectrum by the COBE FIRAS instrument," *Astrophysical Journal*, vol. 420, pp. 439–444, Jan. 1994.

[135] J. Mather, "Bolometers: ultimate sensitivity, optimization, and amplifier coupling," *Ap. Opt*, vol. 23, p. 584, 1984.

[136] J. J. Bock, D. Chen, P. D. Mauskopf, and A. E. Lange, "A Novel Bolometer for Infrared and Millimeter-Wave Astrophysics," *Space Science Reviews*, vol. 74, pp. 229–235, Oct. 1995.

[137] Y. D. Takahashi, D. Barkats, J. O. Battle, E. M. Bierman, J. J. Bock, H. C. Chiang, C. D. Dowell, E. F. Hivon, W. L. Holzapfel, V. V. Hristov, W. C. Jones, J. P. Kaufman, B. G. Keating, J. M. Kovac, C.-L. Kuo, A. E. Lange, E. M. Leitch, P. V. Mason, T. Matsumura, H. T. Nguyen, N. Ponthieu, G. M. Rocha, K. W. Yoon, P. Ade, and L. Duband, "CMB polarimetry with BICEP: instrument characterization, calibration, and performance," in *Society of Photo-Optical Instrumentation Engineers (SPIE) Conference Series*, ser. Society of Photo-Optical Instrumentation Engineers (SPIE) Conference Series, vol. 7020, Aug. 2008.

[138] C. D. Dowell, C. A. Allen, R. S. Babu, M. M. Freund, M. Gardner, J. Groseth, M. D. Jhabvala, A. Kovacs, D. C. Lis, S. H. Moseley, Jr., T. G. Phillips, R. F. Silverberg, G. M. Voellmer, and H. Yoshida, "SHARC II: a Caltech submillimeter observatory facility camera with 384 pixels," in *Society of Photo-Optical Instrumentation Engineers (SPIE) Conference Series*, ser. Society of Photo-Optical Instrumentation Engineers (SPIE) Conference Series, T. G. Phillips and J. Zmuidzinas, Eds., vol. 4855, Feb. 2003, pp. 73–87.

[139] J. Glenn, P. A. R. Ade, M. Amarie, J. J. Bock, S. F. Edgington, A. Goldin, S. Golwala, D. Haig, A. E. Lange, G. Laurent, P. D. Mauskopf, M. Yun, and H. Nguyen, "Current status of Bolocam: a large-format millimeter-wave bolometer camera," in *Society of Photo-Optical Instrumentation Engineers (SPIE) Conference Series*, ser. Society of Photo-Optical Instrumentation Engineers (SPIE) Conference Series, T. G. Phillips and J. Zmuidzinas, Eds., vol. 4855, Feb. 2003, pp. 30–40.

[140] M. J. Devlin, P. A. R. Ade, I. Aretxaga, J. J. Bock, J. Chung, E. Chapin, S. R. Dicker, M. Griffin, J. Gundersen, M. Halpern, P. Hargrave, D. Hughes, J. Klein, G. Marsden, P. Martin, P. D. Mauskopf, B. Netterfield, L. Olmi, E. Pascale, M. Rex, D. Scott, C. Semisch, M. Truch, C. Tucker, G. Tucker, A. D. Turner, and D. Weibe, "The Balloon-borne Large Aperture Submillimeter Telescope (BLAST)," in *Society of Photo-Optical Instrumentation Engineers (SPIE) Conference Series*, ser. Society of Photo-Optical Instrumentation Engineers (SPIE) Conference Series, C. M. Bradford, P. A. R. Ade, J. E. Aguirre, J. J. Bock, M. Dragovan, L. Duband, L. Earle, J. Glenn, H. Matsuhara, B. J. Naylor, H. T. Nguyen, M. Yun, and J. Zmuidzinas, Eds., vol. 5498, Oct. 2004, pp. 42–54.

[141] J. Dunkley, A. Amblard, C. Baccigalupi, M. Betoule, D. Chuss, A. Cooray, J. Delabrouille, C. Dickinson, G. Dobler, J. Dotson, H. K. Eriksen, D. Finkbeiner, D. Fixsen, P. Fosalba, A. Fraisse, C. Hirata, A. Kogut, J. Kristiansen, C. Lawrence, A. M. Magalhaes, M. A. Miville-Deschenes, S. Meyer, A. Miller, S. K. Naess, L. Page, H. V. Peiris, N. Phillips, E. Pierpaoli, G. Rocha, J. E. Vaillancourt, and L. Verde, "CMBPol Mission Concept Study: Prospects for polarized foreground removal," *ArXiv e-prints*, Nov. 2008.

[142] D. Samtleben and for the QUIET Collaboration, "Measuring the Cosmic Microwave Background Radiation (CMBR) polarization with QUIET," *ArXiv e-prints*, Feb. 2008.

[143] D. Benford and S. Moseley, "Cryogenic detectors for infrared astronomy: the Single Aperture Far-Infrared (SAFIR) Observatory," *Nucl. Instr. and Meth. A*, vol. 520, p. 379, 2004.

[144] D. Benford, S. Rinehart, D. Leisawitz, and T. Hyde, "Cryogenic Far-Infrared Detectors for the Space Infrared Interferometric Telescope (SPIRIT)," *Proc. SPIE*, vol. 6687, p. 66870E, 2007.

[145] J. C. Mather, "Cosmic background explorer (COBE) mission," in *Society of Photo-Optical Instrumentation Engineers (SPIE) Conference Series*, ser. Presented at the Society of Photo-Optical Instrumentation Engineers (SPIE) Conference, M. S. Scholl, Ed., vol. 2019, Oct. 1993, pp. 146–157.

[146] J. C. Mather, D. J. Fixsen, and R. A. Shafer, "Design for the COBE far-infrared absolute spectrophotometer (FIRAS)," in *Society of Photo-Optical Instrumentation Engineers (SPIE) Conference Series*, ser. Presented at the Society of Photo-Optical Instrumentation Engineers (SPIE) Conference, M. S. Scholl, Ed., vol. 2019, Oct. 1993, pp. 168–179.

[147] R. F. Silverberg, M. G. Hauser, N. W. Boggess, T. J. Kelsall, S. H. Moseley, and T. L. Murdock, "Design of the diffuse infrared background experiment (DIRBE) on COBE," in *Society of Photo-Optical Instrumentation Engineers (SPIE) Conference Series*, ser. Presented at the Society of Photo-Optical Instrumentation Engineers (SPIE) Conference, M. S. Scholl, Ed., vol. 2019, Oct. 1993, pp. 180–189.

[148] D. J. Fixsen, E. S. Cheng, D. A. Cottingham, W. C. Folz, C. A. Inman, M. S. Kowitt, S. S. Meyer, L. A. Page, J. L. Puchalla, J. E. Ruhl, and R. F. Silverberg, "A Balloon-borne Millimeter-Wave Telescope for Cosmic Microwave Background Anisotropy Measurements," *Ap.J.*, vol. 470, pp. 63–+, Oct. 1996.

[149] H. Murakami, J. Bock, M. M. Freund, H. Guo, T. Hirao, A. E. Lange, H. Matsuhara, T. Matsumoto, S. Matsuura, T. J. McMahon, M. Murakami, T. Nakagawa, M. Noda, K. Noguchi, H. Okuda, K. Okumura, T. Onaka, T. L. Roellig, S. Sato, H. Shibai, T. Tanabe, T. Watabe, T. Yagi, N. Yajima, and M. Yui, "The Infrared Telescope in Space (IRTS)," *Ap.J.*, vol. 428, pp. 354–362, Jun. 1994.

[150] B. P. Crill, P. A. R. Ade, D. R. Artusa, R. S. Bhatia, J. J. Bock, A. Boscaleri, P. Cardoni, S. E. Church, K. Coble, P. de Bernardis, G. de Troia, P. Farese, K. M. Ganga, M. Giacometti, C. V. Haynes, E. Hivon, V. V. Hristov, A. Iacoangeli, W. C. Jones, A. E. Lange, L. Martinis, S. Masi, P. V. Mason, P. D. Mauskopf, L. Miglio, T. Montroy, C. B. Netterfield, C. G. Paine, E. Pascale, F. Piacentini, G. Polenta, F. Pongetti, G. Romeo, J. E. Ruhl, F. Scaramuzzi, D. Sforna, and A. D. Turner, "BOOMERANG: A Balloon-borne Millimeter-Wave Telescope and Total Power Receiver for Mapping Anisotropy in the Cosmic Microwave Background," *Ap. J. Supp.*, vol. 148, pp. 527–541, Oct. 2003.

[151] S. Masi, P. A. R. Ade, J. J. Bock, J. R. Bond, J. Borrill, A. Boscaleri, P. Cabella, C. R. Contaldi, B. P. Crill, P. de Bernardis, G. de Gasperis, A. de Oliveira-Costa, G. de Troia, G. di Stefano, P. Ehlers, E. Hivon, V. Hristov, A. Iacoangeli, A. H. Jaffe, W. C. Jones, T. S. Kisner, A. E. Lange, C. J. MacTavish, C. Marini Bettolo, P. Mason, P. D. Mauskopf, T. E. Montroy, F. Nati, L. Nati, P. Natoli, C. B. Netterfield, E. Pascale, F. Piacentini, D. Pogosyan, G. Polenta, S. Prunet, S. Ricciardi, G. Romeo, J. E. Ruhl, P. Santini, M. Tegmark, E. Torbet, M. Veneziani, and N. Vittorio, "Instrument, method, brightness, and polarization maps from the 2003 flight of BOOMERanG," *Astron. Astrophy.*, vol. 458, pp. 687–716, Nov. 2006.

[152] P. L. Richards, M. Abroe, P. Ade, A. Balbi, J. Bock, J. Borrill, A. Boscaleri, P. de Bernardis, J. Collins, P. G. Ferreira, S. Hanany, V. V. Hristov, A. H. Jaffe, B. Johnson, A. T. Lee, T. Matsumu, P. D. Mauskopf, C. B. Netterfield, E. Pascale, B. Rabii, G. F.

Smoot, R. Stompor, C. D. Winant, and J. H. P. Wu, "The MAXIMA and MAXIPOL experiments," in *Experimental Cosmology at Millimetre Wavelengths*, ser. American Institute of Physics Conference Series, M. de Petris and M. Gervasi, Eds., vol. 616, May 2002, pp. 12–17.

[153] E. E. Haller, "Physics and design of advanced IR bolometers and photoconductors." *Infrared Physics*, vol. 25, pp. 257–266, 1985.

[154] J. Zhang, W. Cui, M. Juda, D. McCammon, R. L. Kelley, S. H. Moseley, C. K. Stahle, and A. E. Szymkowiak, "Hopping Conduction in Partially Compensated Silicon," *Phys. Rev. B*, vol. 48, p. 2312, 1993.

[155] ——, "Non-ohmic Effects in Hopping Conduction in Doped Silicon and Germanium between 0.05 - 1 Kelvin," *Phys. Rev. B*, vol. 57, p. 4472, 1998.

[156] B. Schulz, J. J. Bock, N. Lu, H. T. Nguyen, C. K. Xu, L. Zhang, C. D. Dowell, M. J. Griffin, G. T. Laurent, T. L. Lim, and B. M. Swinyard, "Noise performance of the Herschel-SPIRE bolometers during instrument ground tests," in *Society of Photo-Optical Instrumentation Engineers (SPIE) Conference Series*, ser. Presented at the Society of Photo-Optical Instrumentation Engineers (SPIE) Conference, vol. 7020, Aug. 2008.

[157] C. K. Stahle, M. D. Audley, K. R. Boyce, R. P. Brekosky, R. Fujimoto, K. C. Gendreau, J. D. Gygax, Y. Ishisaki, R. L. Kelley, R. A. McClanahan, T. Mihara, K. Mitsuda, S. H. Moseley, D. B. Mott, F. S. Porter, C. M. Stahle, and A. E. Szymkowiak, "Design and performance of the ASTRO-E/XRS microcalorimeter array and anticoincidence detector," in *Society of Photo-Optical Instrumentation Engineers (SPIE) Conference Series*, ser. Presented at the Society of Photo-Optical Instrumentation Engineers (SPIE) Conference, O. H. Siegmund and K. A. Flanagan, Eds., vol. 3765, Oct. 1999, pp. 128–136.

[158] N. Billot, P. Agnèse, J.-L. Auguères, A. Béguin, A. Bouère, O. Boulade, C. Cara, C. Cloué, E. Doumayrou, L. Duband, B. Horeau, I. le Mer, J. Lepennec, J. Martignac, K. Okumura, V. Revéret, M. Sauvage, F. Simoens, and L. Vigroux, "The Herschel/PACS 2560 bolometers imaging camera," in *Society of Photo-Optical Instrumentation Engineers (SPIE) Conference Series*, ser. Presented at the Society of Photo-Optical Instrumentation Engineers (SPIE) Conference, vol. 6265, Jul. 2006.

[159] P. K. Day, H. G. LeDuc, B. A. Mazin, A. Vayonakis, and J. Zmuidzinas, "A broadband superconducting detector suitable for use in large arrays," *Nature*, vol. 425, no. 6960, pp. 817–821, October 2003.

[160] B. A. Mazin, P. K. Day, J. Zmuidzinas, and H. G. LeDuc, "Multiplexable kinetic inductance detectors," *AIP Conference Proceedings*, no. 605, pp. 309–312, 2002.

[161] B. A. Mazin, P. K. Day, H. G. LeDuc, A. Vayonakis, and J. Zmuidzinas, "Superconducting kinetic inductance photon detectors," *Proceedings of the SPIE - The International Society for Optical Engineering*, vol. 4849, pp. 283–293, 2002.

[162] B. A. Mazin, "Microwave kinetic inductance detectors," Ph.D. dissertation, California Institute of Technology, Pasadena, CA 91125, August 2004.

[163] S. Kumar, "Submillimeter wave camera using a novel photon detector technology," Ph.D. dissertation, California Institute of Technology, Pasadena, CA 91125, 2008.

[164] J. Gao, "The physics of superconducting microwave resonators," Ph.D. dissertation, California Institute of Technology, Pasadena, CA 91125, 2008.

[165] J. S. Gao, J. Zmuidzinas, B. A. Mazin, H. G. LeDuc, and P. K. Day, "Noise properties of superconducting coplanar waveguide microwave resonators," *Applied Physics Letters*, vol. 90, no. 10, p. 102507, Mar 2007.

[166] J. Gao, J. Zmuidzinas, A. Vayonakis, P. Day, B. Mazin, and H. Leduc, "Equivalence of the effects on the complex conductivity of superconductor due to temperature change and external pair breaking," *Journal of Low Temperature Physics*, vol. 151, no. 1-2, pp. 557–563, April 2008.

[167] A. D. O'Connell, M. Ansmann, R. C. Bialczak, M. Hofheinz, N. Katz, E. Lucero, C. McKenney, M. Neeley, H. Wang, E. M. Weig, A. N. Cleland, and J. M. Martinis, "Microwave dielectric loss at single photon energies and millikelvin temperatures," *Applied Physics Letters*, vol. 92, no. 11, March 2008.

[168] J. J. Bock, A. Goldin, C. Hunt, A. E. Lange, H. G. LeDuc, P. K. Day, A. Vayonakis, and J. Zmuidzinas, "Integrated focal plane arrays for millimeter-wave astronomy," *AIP Conference Proceedings*, no. 605, pp. 243–246, 2002.

[169] A. Goldin, J. J. Bock, C. Hunt, A. E. Lange, H. LeDuc, A. Vayonakis, and J. Zmuidzinas, "SAMBA: superconducting antenna-coupled, multi-frequency, bolometric array," *AIP Conference Proceedings*, no. 605, pp. 251–254, 2002.

[170] A. Goldin, J. J. Bock, C. L. Hunt, A. E. Lange, H. G. LeDuc, A. Vayonakis, and J. Zmuidzinas, "Design of broadband filters and antennas for SAMBA," *Proceedings of the SPIE - The International Society for Optical Engineering*, vol. 4855, pp. 163–171, 2003.

[171] A. Goldin, J. J. Bock, A. E. Lange, H. LeDuc, A. Vayonakis, and J. Zmuidzinas, "Antennas for bolometric focal plane," *Nucl. Instrum. Meth. Phys. Res. A*, vol. 520, no. 1-3, pp. 390–392, March 2004.

[172] P. K. Day, H. G. Leduc, A. Goldin, T. Vayonakis, B. A. Mazin, S. Kumar, J. Gao, and J. Zmuidzinas, "Antenna-coupled microwave kinetic inductance detectors," *Nucl. Instrum. Meth. Phys. Res. A*, vol. 559, no. 2, pp. 561–563, Apr 2006.

[173] C. L. Kuo, P. Ade, J. J. Bock, P. Day, A. Goldin, S. Golwala, M. Halpern, G. Hilton, W. Holmes, V. Hristov, K. Irwin, W. C. Jones, M. Kenyon, A. E. Lange, H. G. LeDuc, C. MacTavish, T. Montroy, C. B. Netterfield, P. Rossinot, J. Ruhl, A. Vayonakis, G. Wang, M. Yun, and J. Zmuidzinas, "Antenna-coupled tes bolometers for the SPIDER experiment," *Nucl. Instrum. Meth. Phys. Res. A*, vol. 559, no. 2, pp. 608–610, Apr 2006.

[174] G. Chattopadhyay, C.-L. Kuo, P. Day, J. J. Bock, J. Zmuidzinas, and A. E. Lange, "Planar antenna arrays for CMB polarization detection," in *Proc. 2007 Joint 32nd International Conference on Infrared and Millimeter Waves and the 15th International Conference on Terahertz Electronics (IRMMW-THz)*, 2008, pp. 184–185.

[175] C. L. Hunt, J. J. Bock, P. K. Day, A. Goldin, A. E. Lange, H. G. LeDuc, A. Vayonakis, and J. Zmuidzinas, "Transition-edge superconducting antenna-coupled bolometer," *Proceedings of the SPIE - The International Society for Optical Engineering*, vol. 4855, pp. 318–321, 2003.

[176] M. Myers, W. Holzapfel, A. Lee, R. O'Brient, P. Richards, H. Tran, P. Ade, G. Engargiola, A. Smith, and H. Spieler, "An antenna-coupled bolometer with an integrated microstrip bandpass filter," *Applied Physics Letters*, vol. 86, no. 11, March 2005.

[177] M. Myers, P. Ade, G. Engargiola, W. Holzapfel, A. Lee, R. O'Brient, P. Richards, A. Smith, H. Spieler, and H. Tran, "Antenna-coupled bolometers for millimeter waves," *IEEE Transactions On Applied Superconductivity*, vol. 15, no. 2, pp. 564–566, June 2005.

[178] M. Myers, A. Peter, K. Arnold, G. Engargiola, B. Holzapfel, A. Lee, R. O'Brient, P. Richards, A. Smith, H. Spieler, and T. Huan, "Antenna-coupled bolometer arrays using transition-edge sensors," *Nucl. Instrum. Meth. Phys. Res. A*, vol. 559, no. 2, pp. 531–533, April 2006.

[179] M. J. Myers, K. Arnold, P. Ade, G. Engargiola, W. Holzapfel, A. T. Lee, X. Meng, R. O'Brient, P. L. Richards, H. Spieler, and H. T. Tran, "Antenna-coupled bolometer arrays for measurement of the cosmic microwave background polarization," *Journal of Low Temperature Physics*, vol. 151, no. 1-2, pp. 464–470, April 2008.

[180] R. O'Brient, P. A. R. Ade, K. Arnold, G. Engargiola, W. Holzapfel, A. T. Lee, M. J. Myers, X. F. Meng, E. Quealy, P. L. Richards, H. Spieler, and H. T. Tran, "A multi-band dual-polarized antenna-coupled tes bolometer," *Journal of Low Temperature Physics*, vol. 151, no. 1-2, pp. 459–463, April 2008.

[181] J. W. Kooi, G. Chattopadhyay, S. Withington, F. Rice, J. Zmuidzinas, C. Walker, and G. Yassin, "A full-height waveguide to thin-film microstrip transition with exceptional rf bandwidth and coupling efficiency," *Int. J. IR and MM Waves*, vol. 24, no. 3, pp. 261–284, March 2003.

[182] T. Stevenson, W. Hsieh, G. Schneider, D. Travers, N. Cao, E. Wollack, M. Limon, and A. Kogut, "Building blocks for a polarimeter-on-a-chip," *Nucl. Instrum. Meth. Phys. Res. A*, vol. 559, no. 2, pp. 611–613, April 2006.

[183] D. M. Glowacka, D. J. Goldie, S. Withington, M. Crane, V. Tsaneva, M. D. Audley, and A. Bunting, "A fabrication process for microstrip-coupled superconducting transition edge sensors giving highly reproducible device characteristics," *Journal of Low Temperature Physics*, vol. 151, no. 1-2, pp. 249–254, April 2008.

[184] D. J. Goldie, M. D. Audley, D. M. Glowacka, V. N. Tsaneva, and S. Withington, "Transition edge sensors for bolometric applications: responsivity and saturation," *Journal of Applied Physics*, vol. 103, no. 8, April 2008.

[185] T. Stevenson, D. Benford, C. Bennett, N. Cao, D. Chuss, K. Denis, W. Hsieh, A. Kogut, S. Moseley, J. Panek, G. Schneider, D. Travers, K. U-Yen, G. Voellmer, and E. Wollack, "Cosmic microwave background polarization detector with high efficiency, broad bandwidth, and highly symmetric coupling to transition edge sensor bolometers," *Journal of Low Temperature Physics*, vol. 151, no. 1-2, pp. 471–476, April 2008.

[186] S. J. C. Yates, J. J. A. Baselmans, R. Barends, Y. Lankwarden, H. F. Hoevers, J. Gao, T. Klapwijk, A. Neto, D. J. Bekers, G. Gerini, S. Doyle, P. D. Mauskopf, and P. Ade, "Antenna coupled kinetic inductance detectors for space based sub-mm astronomy," 2008, presented at the 2nd Workshop on the Physics and Applications of Superconducting Microresonators.

[187] S. Doyle, P. Mauskopf, J. Naylon, A. Porch, and C. Duncombe, "Lumped element kinetic inductance detectors," *Journal of Low Temperature Physics*, vol. 151, no. 1-2, pp. 530–536, April 2008.

[188] S. Doyle, "Kinetic inductance detectors," Ph.D. dissertation, Cardiff University, 2008.

[189] R. Barends, J. J. A. Baselmans, S. J. C. Yates, J. R. Gao, J. N. Hovenier, and T. M. Klapwijk, "Quasiparticle relaxation in optically excited high-Q superconducting resonators," *Physical Review Letters*, vol. 100, no. 25, June 2008.

[190] O. Noroozian, P. Day, J.-S. Gao, J. Glenn, S. Golwala, H. LeDuc, P.Maloney, B. Mazin, J. Sayers, J. Schlaerth, A. Vayonakis, and J.Zmuidzinas, "New resonator designs for reduced frequency noise in microwave kinetic inductance detectors (MKID)," 2008, presented at the 2nd Workshop on the Physics and Applications of Superconducting Microresonators.

[191] J. Baselmans, S. J. C. Yates, R. Barends, Y. J. Y. Lankwarden, J. R. Gao, H. Hoevers, and T. M. Klapwijk, "Noise and sensitivity of aluminum kinetic inductance detectors for sub-mm astronomy," *Journal of Low Temperature Physics*, vol. 151, no. 1-2, pp. 524–529, April 2008.

[192] J. Baselmans, S. Yates, R. Barends, J.-J. Lankwarden, J. Gao, H. F. Hoevers, and T. Klapwijk, "Optimizing kinetic inductance detectors for low background applications," June 19-20 2008, presented at the 2nd Workshop on the Physics and Applications of Superconducting Microresonators, SRON Utrecht.

[193] S. Weinreb, J. C. Bardin, and H. Mani, "Design of cryogenic SiGe low-noise amplifiers," *IEEE Transactions On Microwave Theory and Techniques*, vol. 55, no. 11, pp. 2306–2312, November 2007.

[194] T. K. Thrivikraman, J. Yuan, J. C. Bardin, H. Mani, S. D. Phillips, W.-M. L. Kuo, J. D. Cressler, and S. Weinreb, "SiGe HBT X-band LNAs for ultra-low-noise cryogenic receivers," *IEEE Microwave and Wireless Components Letters*, vol. 18, no. 7, pp. 476–478, July 2008.

[195] J. Gao, M. Daal, A. Vayonakis, S. Kumar, J. Zmuidzinas, B. Sadoulet, B. A. Mazin, P. K. Day, and H. G. Leduc, "Experimental evidence for a surface distribution of two-level systems in superconducting lithographed microwave resonators," *Applied Physics Letters*, vol. 92, no. 15, April 2008.

[196] J. Gao, M. Daal, J. M. Martinis, A. Vayonakis, J. Zmuidzinas, B. Sadoulet, B. A. Mazin, P. K. Day, and H. G. Leduc, "A semiempirical model for two-level system noise in superconducting microresonators," *Applied Physics Letters*, vol. 92, no. 21, May 2008.

[197] J. Schlaerth, A. Vayonakis, P. Day, J. Glenn, J. Gao, S. Golwala, S. Kumar, H. LeDuc, B. Mazin, J. Vaillancourt, and J. Zmuidzinas, "A millimeter and submillimeter kinetic inductance detector camera," *Journal of Low Temperature Physics*, vol. 151, no. 3-4, pp. 684–689, May 2008.

[198] B. A. Mazin, P. K. Day, K. D. Irwin, C. D. Reintsema, and J. Zmuidzinas, "Digital readouts for large microwave low-temperature detector arrays," *Nucl. Instrum. Meth. Phys. Res. A*, vol. 559, no. 2, pp. 799–801, April 2006.

[199] C. Muirhead, J. Healey, and M. Colclough, "Magnetic field tuning of coplanar waveguide resonators," 2008, presented at the 2nd Workshop on the Physics and Applications of Superconducting Microresonators.

Journal of Physics: Conference Series **155** (2009) 012002 doi:10.1088/1742-6596/155/1/012002

Coherent detectors

C R Lawrence[1], S Church[2], T Gaier[3], R Lai[4], C Ruf[5], E Wollack[6]

[1]M/C 169-327, Jet Propulsion Laboratory, 4800 Oak Grove Drive, Pasadena, CA 91109
[2]Room 324 Varian Physics Bldg, 382 Via Pueblo Mall, Stanford, CA 94305-4060
[3]M/C 168-314, Jet Propulsion Laboratory, 4800 Oak Grove Drive, Pasadena, CA 91109
[4]Northrop Grumman Corporation, Redondo Beach, CA 90278
[5]1533 Space Research Building, The University of Michigan, Ann Arbor, MI 48109-2143
[6]NASA/GSFC, Code 665, Observational Cosmology Laboratory, Greenbelt, MD 20771

E-mail: charles.lawrence@jpl.nasa.gov

Abstract. Coherent systems offer significant advantages in simplicity, testability, control of systematics, and cost. Although quantum noise sets the fundamental limit to their performance at high frequencies, recent breakthroughs suggest that near-quantum-limited noise up to 150 or even 200 GHz could be realized within a few years. If the demands of component separation can be met with frequencies below 200 GHz, coherent systems will be strong competitors for a space CMB polarization mission. The rapid development of digital correlator capability now makes space interferometers with many hundreds of elements possible. Given the advantages of coherent interferometers in suppressing systematic effects, such systems deserve serious study.

1. Executive Summary

Coherent detectors offer significant advantages in simplicity and control of systematic errors for CMB polarization measurements, due to fact that the full phase and amplitude information in the incoming signal is available for multiple uses in later signal processing. Operation at 20 K rather than ~ 0.1 K also results in simpler systems.

The cost of preserving the phase of the incoming signal is quantum noise, which is proportional to frequency. We assess the current state of coherent receivers, and discuss the technology developments required to improve performance. There are good reasons to believe that noise within a factor of 2 or 3 of the quantum limit up to 150–200 GHz can be achieved within a few years.

Whether detector noise is a serious and limiting problem for coherent detection of CMB polarization depends on the frequency range required for accurate foreground removal in CMB polarization observations and how close to the quantum limit amplifiers can get. The frequency range required is under consideration by a separate activity of the mission concept study; however, a definitive answer must await the better understanding of polarized foregrounds that will be gained by Planck and suborbital experiments over the next few years. If noise levels of 2–3 $\times$ quantum can be achieved, experiments based on amplifiers could reach the noise levels recommended by the TFCR in space. The active cooling requirements at 20 K for such experiments could be met with existing technology in space. If, on the other hand, observations at frequencies much higher than 150 or 200 GHz are required for control of foregrounds, adequate noise levels will be hard to achieve with amplifiers.

The ultimate control of many systematics is afforded by coherent interferometers. In the past, the power requirements of large correlators made coherent interferometers impossible to contemplate for space. No longer. Rapid development of correlator technology driven by commercial forces has reduced the power required for large digital correlators by orders of magnitude. It is now possible to correlate of hundreds of elements for an amount of power readily available in space. As a result, a serious investigation of the potential for B-mode measurements of coherent interferometers in space should now be undertaken.

The Task Force on Cosmic Microwave Background Research (TFCR 2005) recommended:

- "Technology development leading to receivers that contain a thousand or more polarization sensitive detectors, and adequate support for the facilities that produce these detectors.... It is important to keep open a variety of approaches until a clear technological winner has emerged."
- "A strategy that supports alternative technical approaches to detectors and instruments. Advances in CMB science have been based on a variety of technologies. Though we expect that bolometers will be the clear choice for CMBPOL, it is premature to shut down the development of alternatives. We recommend the continued development of HEMT-based detectors as they might lead to an alternative space mission and will certainly be used in ground-based measurements."

The noise projected for amplifier systems has changed dramatically even since the TFCR deliberated, when the expectation for amplifier noise at 140 GHz was too high to list in their table. This elevates the importance of the TFCR recommendation to support multiple approaches including HEMT-based detectors. Transistors now exist that far exceed the performance of anything that existed or was projected in 2005 at millimeter-wave frequencies.

A downselect of detector technology should not be made before foregrounds are better measured, or before all new technologies required for either bolometer or amplifier missions have been demonstrated in suborbital experiments and shown to be suitable for CMB work.

2. INTRODUCTION

In this paper we review the status and promise of phase-preserving or *coherent* detection systems for measurements of CMB polarization, and outline the scope and scale of the development program that would be required to realize the promise of the technologies involved. We consider both large arrays of independent continuum detectors looking at the sky through feeds or telescopes, as well as interferometers, in which signals from different elements are correlated. Low-noise amplifiers are common to both approaches. Inteferometers require in addition local oscillators (LOs), mixers, and correlators.

Transistor performance has improved dramatically over the last 20 years, driven primarily by military and civilian communication requirements at room temperature. High-electron-mobility transistors (HEMTs) offer low noise, low power dissipation, high reliability, inherently wide bandwidths, insensitivity to electromagnetic and charged particle radiation, the ability to handle high signal levels without damage, and operation over a wide temperature range. Transistors are not the only possible active element. SIS mixers have extremely low noise, and are still in widespread use for spectroscopy and interferometry at frequencies > 100 GHz or so. However, it is difficult to achieve the 20–30% fractional bandwidths needed for CMB work with SIS mixers. As a result, *we will restrict our attention to amplifiers only.*

The theoretical sensitivity of a "total power" radiometer array of N identical elements (to a signal filling the beam or in surveying the sky) is given by

$$\Delta T_{\rm min} = \frac{T_{\rm sys}}{\sqrt{N\beta\tau}} \tag{1}$$

where $T_{\rm sys}$ is the system noise temperature, β is the RF bandwidth of the radiometer, τ is the post-detection integration time, and $\Delta T_{\rm min}$ is the minimum detectable signal in that integration time. The sensitivity of other radiometer designs is of the same form, but with additional factors of order unity. Clearly there are three basic ways to make a more sensitive radiometer: 1) reduce the noise $T_{\rm sys}$; 2) increase the bandwidth β; or 3) increase the number of "pixels" N.

The path to improved sensitivity, therefore, is the same in the microwave and millimeterwave part of the spectrum as in others: push detector sensitivity to fundamental limits, and make large arrays of detectors.

3. ADVANTAGES & DISADVANTAGES OF COHERENT DETECTORS

There are both benefits and costs of preserving the full amplitude and phase information in the incoming signal. The cost is a noise floor set by quantum fluctuations. In terms of system noise equivalent temperature, this "quantum noise" or "quantum limit" q can be written as $q = h\nu/(k \log 2) \approx [\nu_{\rm GHz}/20]\,{\rm K}$.

The most important benefit stems from the availability of gain in the device defining the noise temperature. The "quantum tax" must be paid only once, but multiple copies of the incident signal can be produced at a high level, so that further processing adds essentially no noise. These muliple copies may be used in ways that allow significant control of systematic effects. One could imagine digitizing the raw signal after amplification and performing arbitrary phase and amplitude operations in software with essentially unlimited fidelity. Even without going that far, one one can design a system to be largely free of many important systematics at the outset (e.g., band leveling and band subdivision can reduce bandpass issues; multiplying interferometry solves many calibration issues).

Amplifiier-based detector systems have practical advantages as well, including: large dynamic range; operation over a broad temperature range, with lowest noise achieved at a relatively high 20 K; insensitivity to cosmic rays and microphonics; and simplified filtering, taking advantage of their inherent bandpass. These advantages provide a significant reduction in overall system complexity, both in construction and testing/validation, with concommitant decrease in cost.

We discuss these factors in more detail in the following sections, starting with noise because it is the factor that has dominated recent discussions about detectors for CMB polarization.

3.1. Noise

The characteristic of bolometers that has lifted them to prominence in the CMB field is the potential for extremely low noise. The linear dependence of quantum noise on frequency in coherent systems guarantees that above some frequency, bolometer systems can achieve lower raw noise than amplifier systems. What is that frequency?

It depends on many factors, both intrinsic to the detectors and as part of the receiver system and its radiation environment. For example, the small dynamic range of bolometers means that the same bolometers cannot be used on the ground as in space; the background radiation level from the atmosphere and ground would saturate space bolometers. Although it should be possible in principle to make bolometers that are close to photon noise limited, whatever the photon environment, manufacturing tolerances, the need for margins, and other practical matters mean that in practice bolometers built for use on the ground are significantly less sensitive than those built for space. Moreover, unlike amplifiers, bolometers must be the final element in the detector chain, so any optical inefficiency through the optical elements leads directly to a loss of sensitivity.

Figures 1 and 2 compare sensitivities and noise levels for amplifiers and bolometers for ground-based observations (Figure 1) and space (Figure 2). See the Appendix for details on the comparison. The sensitivities of a "perfect" quantum-noise-limited amplifier and a "perfect"

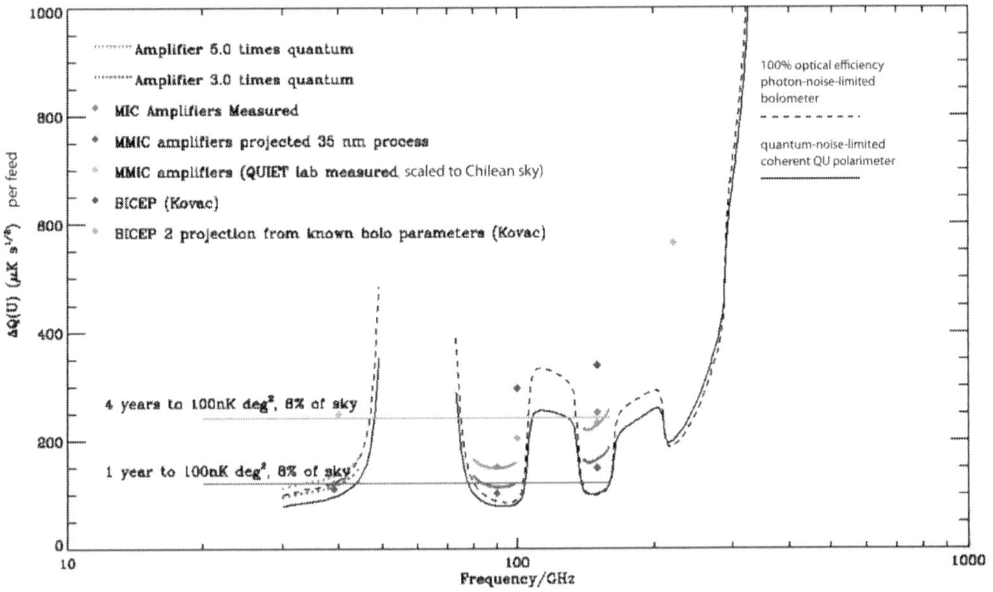

Figure 1. Sensitivity of a quantum-limited amplifier compared with the sensitivity of a photon-noise-limited bolometer (i.e., "perfect" amplifier and bolometer, respectively), from the ground at a high dry site. The dramatic effect of the atmospheric loading on bolometer sensitivity even in good atmospheric windows can be seen by comparison with Figure 2. We assume 8 K of thermal noise from the telescope and ground, independent of frequency (surely is an overestimate at low frequencies) and a 25% bandwidth for both detector systems. Amplifier sensitivities are divided by $\sqrt{2}$ because an amplifier polarimeter can measure Q and U simultaneously, while a bolometer measures Q or U at a given time. For a description of the various colored points and lines, see text.

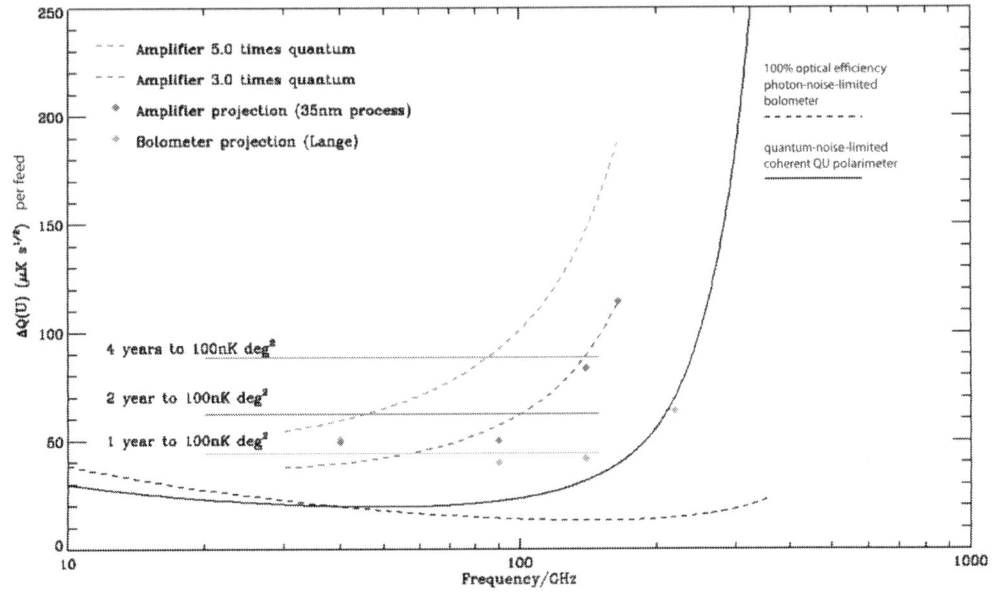

Figure 2. Same as Figure 1, but in space. Below 40 GHz photon noise would dominate quantum noise for the amplifier. We have not added the two together so that their relative strength can be seen. The amplifier sensitivities are divided by $\sqrt{2}$ because an amplifier polarimeter can measure Q and U simultaneously, while a bolometer measures Q or U at a given time. For a description of the various colored points and lines, see text.

photon-noise-limited bolometer are indicated by black solid and dashed curves, respectively. A 25% bandwidth is assumed for both. The amplifier sensitivity includes a factor of $1/\sqrt{2}$ because amplifiers can measure Q and U simultaneously through a single feed, whereas direct detectors can measure only Q *or* U at one time. The effect of atmospheric loading of the bolometers on sensitivity can be seen by comparing the bolometer curve in the atmospheric windows in Figure 1 with the curve in Figure 2.

In both Figure 1 and 2, sensitivity curves are shown also for amplifiers at 3 and 5 times the quantum limit. Discrete points show measured or projected sensitivity values for various current or planned experiments, as follows:

- BICEP bolometer (from John Kovac), measured
- BICEP2, projected from known bolometer parameters (John Kovac)
- MIC amplifiers, measured in lab
- MMIC amplifier modules for QUIET, measured in lab
- MMIC amplifiers, projected from high frequency, room temperature measurements of 35 nm gate process devices

The horizontal blue and green lines in Figure 1 show the sensitivity required for each of 512 feeds (or planar antennas) at each frequency, to reach 100 nK noise per square degree in the 4% of the sky with the lowest foreground levels, integrating for 1 year (blue) or 4 years (green) from the ground. This is the target noise level recommended by the Task Force on Cosmic Microwave Background Research. Such an array of amplifiers at the noise levels projected for 35-nm gate process HEMTs (see § 4.2) could reach this level in less than a year of integration time up to about 160 GHz. An array of projected bolometers could reach this level in four years up to same frequency.

With the performance expected from both technologies in the forseeable future, both can reach the recommended noise level in a reasonable experimental lifetime. As will be seen in later sections, the work required to bring amplifiers to the projected level of performance will take several years. But if an instrument with $3 \times q$ amplifiiers can be deployed in 2012, and if the performance expected of bolometers is realized on the projected schedule, then by 2013, both bolometer and HEMT experiments could reach $100\,\mathrm{nK\,deg}^2$ over the cleanest 4% of the sky. This would result in:

- Possibly a first detection of B-modes
- Certainly a great deal of essential information on foregrounds
- An assessment of systematics for two different detection systems operated in a real and stressful environment.

We believe that we should work towards this situation, and that a down-select between the two technologies should not take place before this comparison can be made.

In Figure 2, discrete points show:

- MMIC amplifiers, projected from high frequency, room temperature measurements of 35 nm gate process devices
- Bolometers, projected (Andrew Lange)

The horizontal green, blue, and magenta lines in Figure 2 show the sensitivity required for each of 512 feeds (or planar antennas) at each frequency, to reach 100 nK noise per square degree over the full sky, integrating for 1 year (green), 2 years (blue), and 4 years (magenta) in space. Projected bolometers could reach this level in 1 year up to about 140 GHz, and in 2 years up to about 220 GHz. Amplifiers could reach this level up to about 140 GHz in 4 years.

3.2. Foregrounds & Frequency range
For CMB polarization. foregrounds are a critical consideration, and may well be the limiting factor in measuring B modes. The June workshop has considered foregrounds at length, and work continues that will be reported in the mission concept study report and other papers. This section is not meant to substitute for that work, but rather to highlight the critical role that foreground separation will play in setting requirements for a CMB polarization mission.

Figure 3 shows foreground levels compared to CMB fluctuations for various sky areas, from full sky to one of the quietest (i.e., lowest foreground fluctuation) hundreds-of-square-degree patches on the sky, as calculated from angular power spectra of the various components based on the Planck Sky Model (see June workshop). Limited regions of the sky have much smaller fluctuations; however, the lowest multipoles, where the reionization bump makes the B-modes strongest compared to other sky signals, are the prime target for a space CMB polarization mission.

Simulations of component separation by Dickinson et al. (reported at the June workshop) suggest that experiments with SNR independent of frequency will give the best results (i.e., lowest error on the CMB itself for a given focal plane detector area). The signal of relevance is the *total* signal, CMB+foregrounds, not the CMB alone. If this preliminary result holds up, it is clear from Figure 3 that experimental noise can and should be much higher at low and high frequencies than near the foreground minimum. Neither a large number of detectors achieving low noise at high frequencies nor a large number of (large) detectors dominating the focal plane at low frequencies would be necessary or desirable.

If frequencies much higher than 150 or 200 GHz are required, amplifiers will not be the detector of choice because of their raw sensitivity disadvantage compared to bolometers. If frequencies up to this frequency range are enough, however, an amplifier mission will be very attractive.

3.3. Advantages
We can divide advantages of coherent systems into two categories: those that involve the signals and their processing, and therefore may have a direct impact on systematic errors and their control as well as on practical matters of complexity, reliability, risk, testability, and cost of the overall system; and those that are "merely" practical. § 3.3.1 discusses the former; §§ 3.3.2–3.3.7 the latter.

The systematic requirements on CMB polarization experiments are demanding. For both coherent and incoherent systems, satisfactory performance can be demonstrated only with observations of the sky. The importance of a strong suborbital experimental program in providing this testing cannot be overemphasized

3.3.1. Signal processing and design possibilities As mentioned before, the availability of gain in the device defining the noise temperature allows multiple copies of the full incident signal to be produced and used to measure the properties of the incident radiation without further noise penalty. This ability opens up a variety of signal processing and instrumentation approaches, with significant advantages.

- The "science data channel" can have a null signal. Baselines and offsets can be made small and stable.
- The desired set of measurements (e.g., $Q/U/I$, $Q/V/I$, etc.) can be made simultaneously by the receiver topology. Table 1 gives examples of phase switched pseudo-correlation signal detection schemes. (A Jones matrix formulism was used to compute the response of a selection of polarization sensitive receiver topologies [Jones 1941; 1942].) In selecting a receiver configuration for polarimetry, the residuals and baseline (i.e., the portion of the

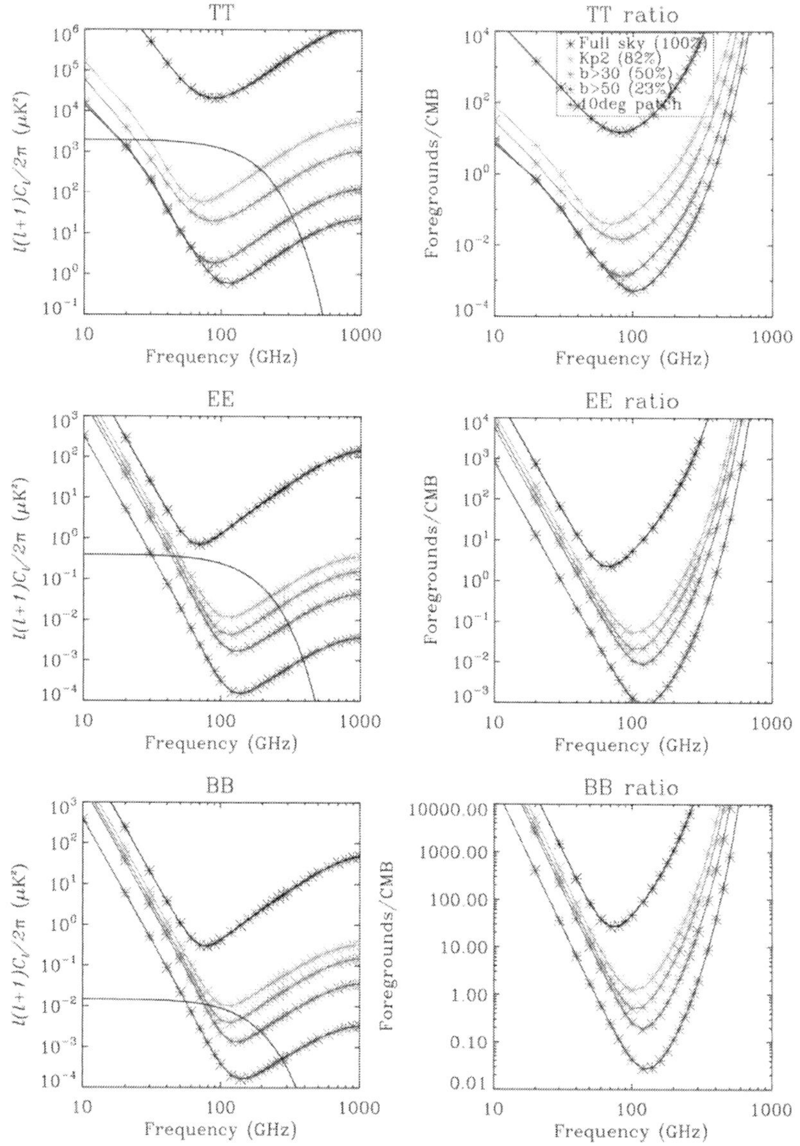

Figure 3. Level of synchrotron and dust temperature fluctuations in absolute units (*left*) and relative to the CMB (*right*), calculated over different regions of the sky (i.e., different Galactic "cuts", plus a $\sim 10^\circ$ patch with minimal foregrounds) as indicated in the upper right panel. The lowest multipoles can be observed only from space, and only by observing a large fraction of the sky. The lower curves representing small sky fractions are therefore relevant only for suborbital experiments. The foregrounds are based on the Planck Sky Model (see Theory and Foregrounds workshop for details). Too little is known about polarized foregrounds to support conclusive simulations, but there is no question that at high and low frequencies foreground fluctuations will dominate the measured signal. For example, in TT for a Kp2 sky cut, for which the foregrounds are known moderately well from WMAP up to 94 GHz, at 40 and 130 GHz dust and synchrotron fluctuations are up by a factor of five over their values at the minimum at 70 GHz. At 30 and 180 GHz the foregrounds are up by a factor of 25. By 300 GHz, dust fluctuations are up by three orders of magnitude compared to CMB fluctuations, which drop fast on the Wien side of the spectrum. If the foreground spectra are complicated, a very wide frequency range may be disadvantageous. [Figure from Clive Dickinson.] For EE the foreground data are much more uncertain, especially concerning dust, for which there is suspicion that the Planck Sky Model underestimates polarization fluctuations. The green curve in the EE ratio plot reaches the same values at about 50 and 240 GHz.

Table 1. Phase-switched pseudo-correlation signal detection schemes.

Example	Measurement Approach	Sensitivity/Residual (Baseline)
1.	$J_{180H} J_{\Delta\phi} J_G J_{OMT} \vert E_{\text{in}}\rangle$	$\pm U/V(I$ and $Q)$
2.	$J_{180H} J_{\Delta\phi} J_G J_{90H} J_{OMT}$ $J_{180H} J_{\Delta\phi} J_G J_{CPH} \vert E_{\text{in}}\rangle$ $J_{180H} J_{\Delta\phi} J_G J_{OMT} J_{QWP}(\pi/4) \vert E_{\text{in}}\rangle$	$\pm U/Q(I$ and $V)$
3.	$J_{180H} J_{\Delta\phi} J_G J_{180H} J_{OMT} \vert E_{\text{in}}\rangle$	$\pm Q/V(I$ and $U)$
4.	$J_{180H} J_{\Delta\phi} J_G J_{180H} J_{OMT}(0) \vert E_{\text{in}}\rangle$	$\pm V/U(I$ and $Q)$
5.	$J_{90H} J_{\Delta\phi} J_G J_{OMT} \vert E_{\text{in}}\rangle$	$\pm V/U(I$ and $Q)$
6.	$J_{90H} J_{\Delta\phi} J_G J_{OMT} \vert E_{\text{in}}\rangle$ $J_{90H} J_{\Delta\phi} J_G J_{CPM} \vert E_{\text{in}}\rangle$ $J_{90H} J_{\Delta\phi} J_G J_{OMT} J_{QWP}(\pi/4) \vert E_{\text{in}}\rangle$	$\pm Q/U(I$ and $V)$
7.	$J_{90H} J_{\Delta\phi} J_G J_{180H} J_{OMT} \vert E_{\text{in}}\rangle$	$\pm V/Q(I$ and $U)$
8.	$J_{90H} J_{\Delta\phi} J_G J_{OMT} J_{QWP}(0(\vert E_{\text{in}}\rangle$	$\pm U/V(I$ and $Q)$

signals which ideally would not be present) should be considered in addition to desired lock-in channel sensitivity.

- The desired signal (e.g., polarization state, incident angle, intensity, frequency, phase) can be rapidly and appropriately modulated, strongly supressing undesired and competing signals that may be present in the instrument's environment, giving excellent systematic control. The addition of beam waveguide polarization modulators (e.g., half wave plates, variable(?) polarization modulators, etc.), which add complexity and risk to the system, is not needed.
- The signal can be appropriately processed by subsequent circuitry (e.g., polarization diplexers, quadrature hybrids, magic-tees, in-phase power splitters, phase delay/modulators) before power detection. This ability to encode the desired properties of the signal fields and subsequently look for correlations is a powerful tool for obtaining precision control over systematic effects. More importantly, in a single-mode coherent system, each incident mode (i.e., in our case of interest polarization state) can be independently processed. In practice, examples from precision radiometry (Predmore et al. 1985), interferometry (Blum 1959; Faris et al. 1967), and polarimetry (Tinbergen 1996) have demonstrated the power and maturity of this overall approach.

Incoherent systems do not have these options. They typically rely upon synchronous modulation/demodulation of the incident power in front of the detection. In the absence of detector gain, this places stringent constraints upon the allowable loss, emission, temperature and temperature stability, as well as achievable switching speeds. As described in the other papers associated with the Boulder Workshop, a great deal of work is being devoted to the development of pre-detector components for incoherent systems. Success is nearer at hand in some areas than other, and ultimately success can only be demonstrated in on-sky observations.

In addressing thermal drifts, finite beams and sidelobes, differing frequency responses, astrophysical foregrounds, and other forms of radiation., merely encoding/decoding the incident signals is not sufficient. Calibration of the system response requires instrument stability on a time scale long compared to the modulation in order to achieve the ultimate detector sensitivity. The resultant linkages between detector speed/noise, scan, modulation, and calibration rates required to achieve the mission goals can be addressed in principle in either system approach; however, when considering a system's overall testability, the ability to minimize/retire risk before flight, and simplicity, a coherent system has strong advantages.

Ultimately, the most important advantage of coherent systems is likely to be that the required control of systematics can be achieved at lower complexity, risk, and cost.

We turn now to a number of practical considerations.

3.3.2. Dynamic range Amplifiers can handle a very large range of input signal levels. Non-linearities are much less than seen for bolometers. Even more importantly, the noise temperature depends very little on input signal level up to very high levels. Although in principle bolometers can be built that are photon noise limited at any level, in practice the limited dynamic range of bolometers coupled with the difficulty of knowing in advance the total photon load (CMB + foregrounds + sky + telescope + ...) at a given site for a given instrument, and the uncertainties in fabrication, mean that bolometers built for the ground are significantly less sensitive than those built for space. This is not the case with amplifiers, whose large dynamic range can easily accomodate the difference in signal between ground and space. This simplifies design and testing.

3.3.3. Operating temperature and testability Physical temperature requirements for minimum noise are modest, at around 20 K. Moreover, amplifiers work over a broad temperature range, from above room temperature to well below 4 K. The system noise properties degrade gracefully as detector physical temperature increases above 20 K. This has broad impact on system design:

- Thermal and cooler systems to achieve 20 K are substantially simpler than systems to achieve sub-kelvin temperatures, as well as faster and easier to test.
- An amplifier system designed to operate cryogenically can be checked out and tested very well at higher temperatures. Both for ground testing and for in-orbit checkout this can result in substantial simplifications and savings.

3.3.4. Insensitivity to cosmic rays and microphonics Cosmic rays produce no detectable disturbance in the output signal of amplifiers. Although "spider web" type bolometers have reduced the susceptibility of bolometers to cosmic rays dramatically, detection and removal of cosmic rays spikes is still a first and necessary step in the processing of bolometer data from Planck. Up to a few percent of Planck HFI data will be thrown away because of this.

3.3.5. Filtering As an added benefit, the presence of in-band gain lowers the overall filtering requirements to limit the influence of out-of-band response.

3.3.6. Straightforward arrayability As shown below in § 3.2, cost-effective techniques for building large arrays of amplifier detectors have been demonstrated. Signals can be digitized at the cryogenic stage, eliminating the need for complicated cryogenic readout/multiplexer circuits.

3.3.7. Industrial infrastructure Amplifiers are the detector of choice for communications and remote sensing applications in the commercial and military world up to frequencies of several hundred gigahertz, at least. Moreover, while optimization of amplifiers for cryogenic operation may well be different from optimization at room temperature, it is a general rule that the best devices (transistors, MMICs) at room temperature are the best devices at cryogenic temperatures. As a result, the commercial and military development of high performance amplifier systems is of direct benefit to the scientific world. The area of low noise amplifier technology is one that NASA does not have to develop on its own. Significant synergies exist. This substantially decreases the size of the investment in technology required to realize the best cryogenic performance.

Table 2. Operational and systematic comparison of HEMT and bolometer arrays

Requirement	MMIC HEMTs	Bolometers
Response time	Excellent, $\tau \approx 1/\beta$	Adequate, can be modeled
Linearity	Excellent	Adequate, can be modeled.
Dynamic range	Large	Small. Different devices required for space and ground.
Gain stability	Excellent with modulation	Excellent for Planck NTD Ge technology, not yet public for TES
Offset stability	Excellent with modulation	Excellent for Planck NTD Ge technology, not yet public for TES
Polarization systematics	Excellent; Q and U from same pixel. Modulation electronic and fast, after amplification. Effect of gain mismatch removed by radiometer design.	Good; Q and U from different pixels, or complicated and relatively slow modulation. Gain mismatch leads to leakage of temperature to polarization.
B-field, microphonics, EMI, RFI susceptibility, cosmic rays	Good	Adequate (?). B-fields a particular issue for TES.
Device uniformity	Not established	Good for Planck, not yet public for TES
	PRACTICALITIES FOR SPACE	
Cooling requirements	Passive + active (20 K). Heat lift required is large but achievable.	0.1 K operation requires multi-stage cooling chain. Heat lift required at 0.1 K depends on parasitics. Complicated.
System testability & risk reduction before flight	Easier & less expensive	Harder & more expensive
Cryogenic readout & multiplexer .	Not needed	Complicated

3.4. Disadvantages

3.4.1. Quantum noise As described in § 2, there is a lower limit to $T_{\rm sys}$ for coherent receivers set by quantum fluctuations. But as shown in § 2.1, even with quantum noise amplifiers can reach the noise levels required for CMB polarizaiton over a broad frequency range.

3.4.2. Power dissipation Amplifiers dissipate significant power. A W-band (75–110 GHz) module to measure Q and U simultaneously with present technology dissipates about 12 mW, from which predictions at other frequencies can be made: 4 mW and 30 GHz, 7 mW at 40 GHz, 10 mW at 70 GHz, and 15 mW at 150 GHz. Hundreds of detectors would require watts of heat lift at 20 K or so. A cooler with this kind of heat lift would have a major impact on flight system design, but is not a showstopper by any means. For example, the two hydrogen sorption coolers on Planck, combined together into a single cooler with a JT valve optimized for a 45 K radiative precool temperature, would have a heat lift of about 4 W with 1100 W of input power. If a radiative precool temperature of 35 K could be achieved, the heat lift for the same power input would be 10 W. 1100 W is serious but not scary power in space.

3.5. Summary of Operational and Systematic Comparisons

Table 2 summarizes the characteristics and practicalities of amplifier detectors, with comparison to bolometer systems.

Journal of Physics: Conference Series **155** (2009) 012002 doi:10.1088/1742-6596/155/1/012002

4. TECHNOLOGY STATUS AND PROSPECTS

Dramatic improvements in noise and power dissipation of transistors have been achieved over the last 20 years, first with high electron mobility transistors (HEMTs) fabricated on GaAs substrates, then on InP substrates. Developments over the last two years promise noise performance of a few times the quantum limit in the forseeable future over a large frequency range.

Similarly dramatic advances have been made over the last few years in the packaging of amplifier receivers, making large arrays of receivers possible and achievable for CMB polarization. We discuss the state of both areas below, and identify technology development required to realize the kind of performance required for B-mode measurements on the ground or in space.

4.1. Transistor and Amplifier Performance

Figures 4 and 5 show the state of the art for cryogenic performance of transistor amplifiers in 2007, as well as goals for the future. In 2006, Northrop Grumman Corporation (NGC) developed a new ultra-short-gate-length high electron mobility transistor (HEMT) process[1] (Deal et al. 2006), incorporating the following changes:

- gate length reduced from 0.1 μm to 0.035 μm $\Rightarrow$ parasitic gate-source capacitance is reduced and cutoff frequency is increased by a factor of 2
- ohmic metal and contact resistance reduced by a factor of 2
- InAs channel enabled through new epitaxy design $\Rightarrow$ improved electron mobility by 25%
- 2-mil thick wafer process $\Rightarrow$ reduction of via hole size and pads by as much as a factor of 4, to support higher frequency designs
- reduction of minimum line size and spacings by 30-40%

Amplifiers using this technology give breakthrough performance up to 340 GHz at room temperature, as reported by Dawson et al. (2005), Deal et al. (2006, 2007), Gaier et al. (2007), Pukala et al. (2008), Samoska et al. (2008), and Kangaslahti et al. 2008). Figure 6 shows a 3-stage amplifier with 15 dB of gain at room temperature, or 5 dB per stage. This design is the highest frequency amplifier reported to date, and shows excellent correspondence between modelled and measured results.

A transistor model based on measurements on-wafer predicts that the 35 nm devices have a maximum stable gain (MaxGain) of 5 dB up to 600 GHz (Fig. 7), with 10–17 dB gain per stage predicted from 30–200 GHz, under ideal circuit matching conditions.

The technology has demonstrated very low noise at these frequencies (Deal et al. 2006; Gaier et al. 2007), with a noise figure[2] of 7.5 dB at room temperature for a 270 GHz LNA (Fig. 8).

4.1.1. Cryogenic Performance The lowest noise is achieved with amplifiers cooled to cryogenic temperatures. Little change in noise is observed below 20 K physical temperature, so hereafter "cryogenic" will mean $\sim$ 20 K unless otherwise specified. Kangaslahti et al. (2008) characterized the noise performance of a three-stage MMIC amplifier for the 180 GHz band as a function of physical temperature (Fig. 9). At room temperature, the noise temperature at 160 GHz is less than 400 K, a factor of two smaller than the previous state of the art, with gain of $\sim$16 dB. At 30 K physical temperature, the noise temperature is < 100 K up to 170 GHz, the lowest cryogenic LNA noise temperature ever reported at these frequencies. These improved cryogenic results

[1] Funded in part by the DARPA Submillimeter Wave Imaging Focal-plane Technology (SWIFT) program, whose goal is room-temperature imaging arrays at submillimeter frequencies.

[2] *Noise figure* in dB is the standard measure of noise in the non-astronomical world. The correspondance between NF and noise temperature T, the astronomical standard, is given by $NF = 10\log\{T/290 + 1\}$, with T in kelvin.

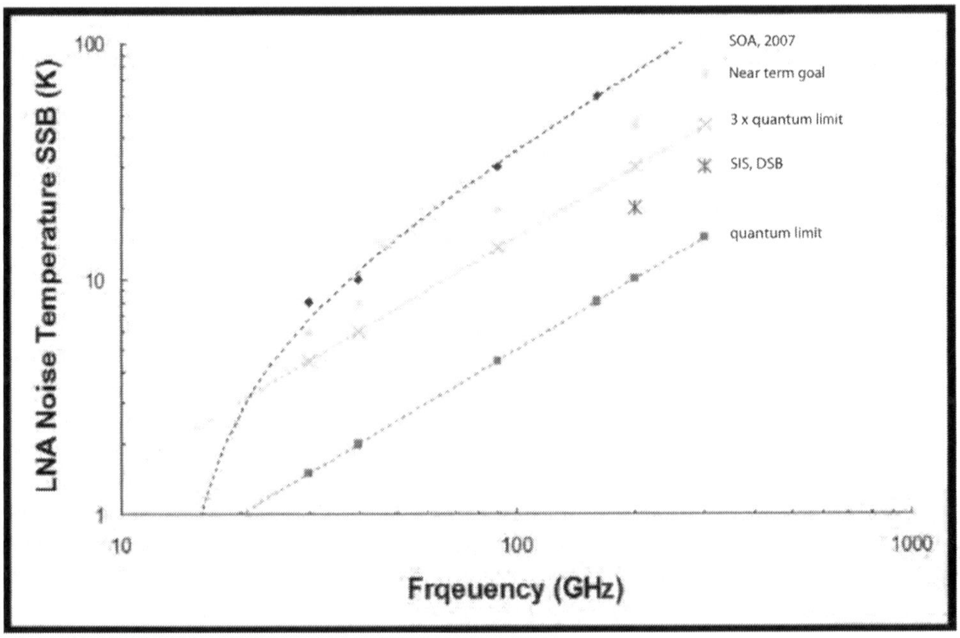

Figure 4. Amplifier noise temperature vs. frequency for the state-of-the-art in cryogenic noise performance in 2007, as well as near-term and $3q$ goals for the future. Double sideband SIS mixers are shown for comparison. Single sideband SIS mixers would have twice the temperature.

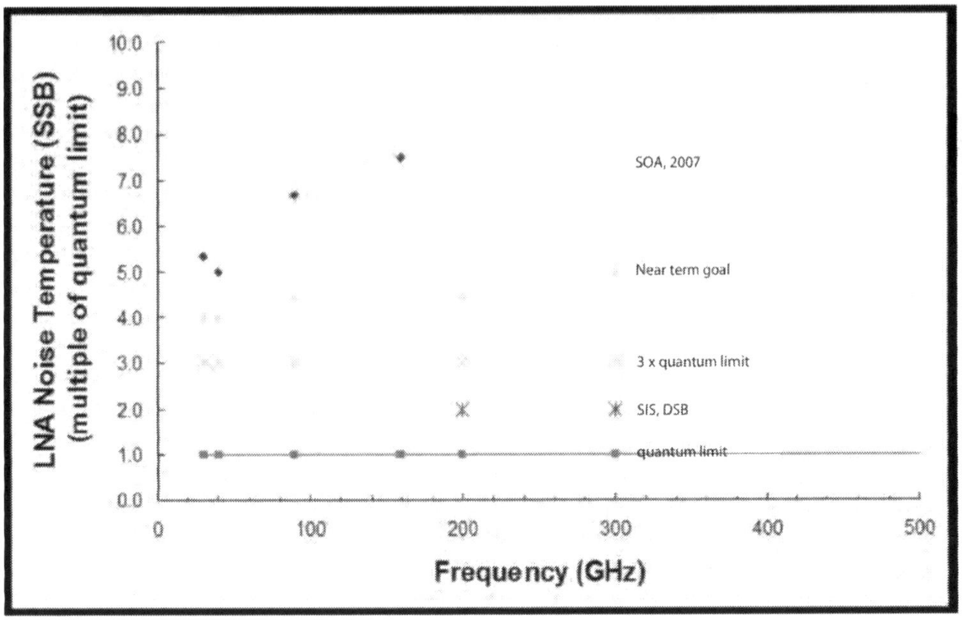

Figure 5. Same as Figure 4, but as a multiple of the quantum limit.

were obtained even though the amplifier showed a clear "kink effect" in its cryogenic DC I-V curves. This precluded proper biasing of the amplifier, increasing its noise.

In general, the noise of previous InP transistors (whether MIC of MMIC) decreased by roughly an order of magnitude from room temperature to 20 K. The factor of four measured for the 180 GHz amplifier is clearly less than a factor of 10, even allowing for the small additional

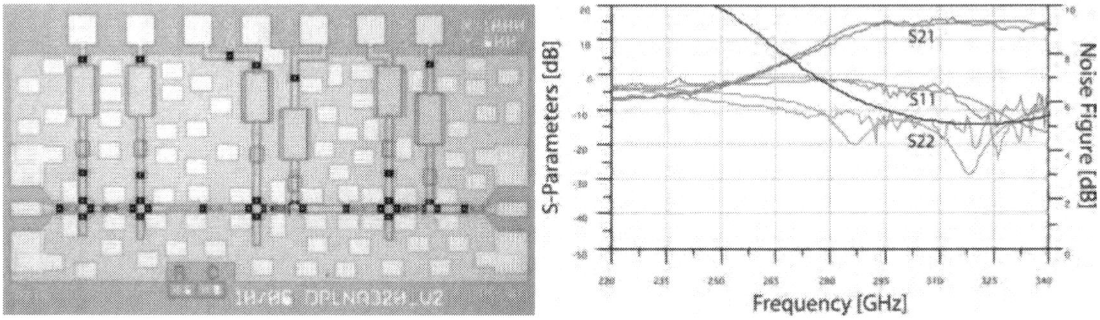

Figure 6. *Left*—Three-stage MMIC amplifier with 15 dB of gain from 285–340 GHz at room temperature. *Right*—Three-stage MMIC amplifier measurements and model of S21 (gain), S11 (input return loss), and S22 (output return loss). In each case, the smoother curve of the pair shows the model. Excellent agreement is seen between the measurements and the model. The black curve shows modelled noise figure (right-hand scale, see § 2.2.2). The high frequency is limited by the frequency range of the measurement, not of the amplifier.

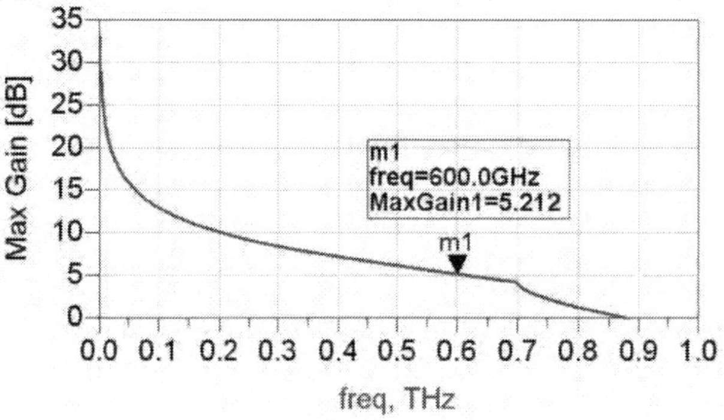

Figure 7. Predicted maximum available/maximum stable gain (MaxGain) of the 35 nm gate HEMTs. MaxGain is above 5 dB up to 600 GHz, and is between 10 and 17 dB for frequencies between 30 and 200 GHz.

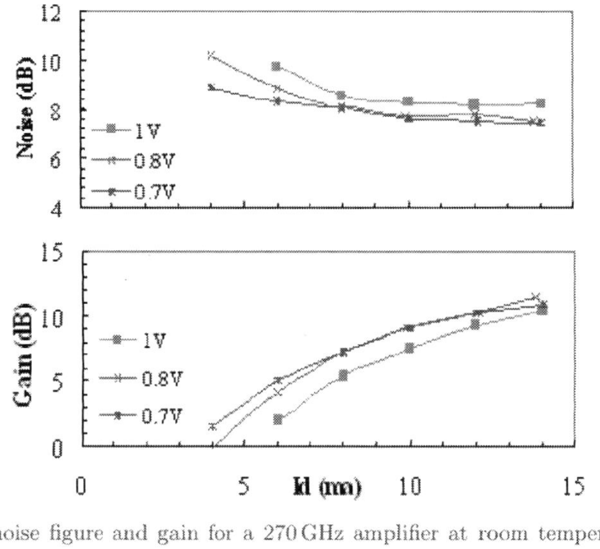

Figure 8. Measured noise figure and gain for a 270 GHz amplifier at room temperature. $NF = 7.5\,\mathrm{dB}$ at maximum gain of 11.5 dB.

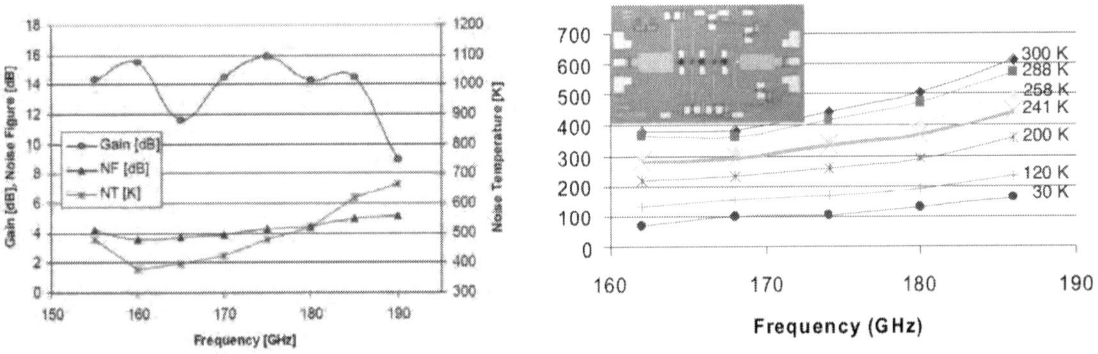

Figure 9. Measurements of 180 GHz amplifier (inset photo) at room temperature (*left*) and as a function of temperature down to 30 K (*right*). Noise is a factor of two lower than the best previous results at both room and cryogenic temperatures. The low cryogenic noise was obtained despite the fact that the DC bias performance was not optimum (a clear "kink effect" was seen in the cryogenic I-V curves; see § 2.2.3).

reduction that would be expected between 30 K and 20 K physical temperature. One might worry that cryogenic performance is being limited in the 35 nm gate devices by some factor not previously seen in other devices. An even more recent measurement shows that this is not the case.

Figure 10 shows the noise temperature for a W-band amplifier built by Eric Bryerton at NRAO cooled to about 15 K. From 300 K to 15 K the noise decreased by a factor of ten, showing that the 35 nm devices follow the trend of the best previous devices when they are cooled to cryogenic temperatures.

4.1.2. Noise Models and Predictions The key question is how close to the fundamental quantum limit can amplifiers be brought, and what is required to get there? A more immediate question is what is the full cryogenic potential of the 35 nm gate devices now being produced? Their performance significantly exceeds anything seen before, judged by room temperature performance at frequencies up to ∼350 GHz and the first cryogenic measurements. Even more specifically, we can ask what gains in performance can be realized at cryogenic temperatures over the range of frequencies important for CMB polarization?

A theoretical answer to the second question can be given by a noise model of the new devices. For nearly two decades, the standard model of low noise transistor noise has been that of Pospieszalski (1989), which uses two noise sources, one on the gate and one at the drain of the device. These noise sources are realized in practice by setting the physical temperatures of the R_g and R_{ds} to values that match our measured results. The correct temperature of the R_g is the physical temperature of the circuit (either 295 K or 20 K, depending on whether we model the room temperature or cryogenic performance of the amplifier). The r_{ds} models the drain noise of the transistor and is typically 15–20 times the physical temperature of the device, depending on the technology. Kangaslahti (private communication) has developed such a model, and with it obtains a noise figure of 3.3 dB for a single device at 270 GHz. If cascaded in a three stage amplifier with 11.5 dB of gain (assuming 1.5 dB losses in passives in each stage), this would translate to a noise figure of 6.7 dB for the amplifier in Figure 8, in good agreement with the measured value of 7.5 dB.

Using this noise model at 20 K gives the results shown in Figure 11, namely 7.9 K ($4q$) at 40 GHz, 11.5 K ($2.6q$) at 90 GHz, 19.5 K ($2.8q$) at 140 GHz, and 24 K ($2.9q$) at 165 GHz. All designs had more than 20 dB of gain, so these numbers would be very close to the final noise temperature of the actual receiver (i.e., later components in the receiver signal path would have little effect on the system noise). These design simulations are preliminary and not optimized.

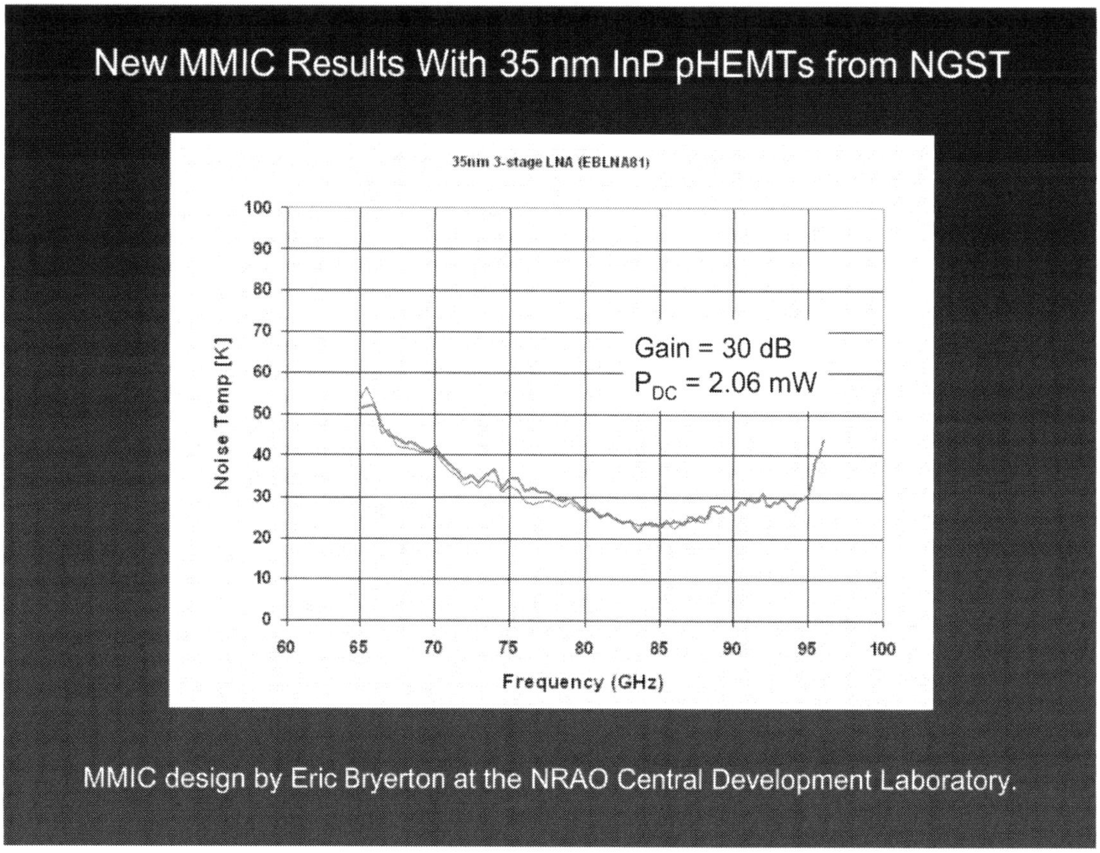

Figure 10. W-band MMIC amplifier cooled to $\sim 15\,\mathrm{K}$. The noise is the lowest ever measured for an amplifier at these frequencies, and an order of magnitude lower than at room temperature.

And although the predicted performance is still a factor of two better than measured for the amplifier in Figure 10 that was the very first W-band 35 nm amplifier measure cryogenically.

All of this suggests that **there is an excellent prospect of achieving noise levels of $\sim 2.5q$ over a range of frequencies important for CMB polarization, levels which only a year or two ago seemed far off in the future.** Figures 1 and 2 show that these levels would enable CMB polarization experiments and missions with the capabilities recommended by the TFCR.

4.2. Arrays and array performance

The experimental sensitivity needed for B-mode observations requires both low noise from individual detectors and many detectors. The detectors used for WMAP and Planck, whether amplifiers or bolometers, could not be scaled up to a level of hundreds of detectors per frequency. Over the last few years, designs for highly scalable coherent radiometers and polarimeters have been demonstrated and implemented in the QUIET experiment, a collaboration of 12 institutions (http://quiet.uchicago.edu). QUIET has deployed in Phase I a Q-band receiver of 19 elements, and is preparing a W-band receiver of 91 elements.

QUIET uses polarization modules built at JPL, shown schematically in Figure 12, which simultaneously measure both Q and U. A single Q-band module is shown on the right side of Figure 13; a 7-element W-band array with an OMT is shown in Figure 14.

Each module has 18 bias voltages that need to be set. An efficient means of optimizing these

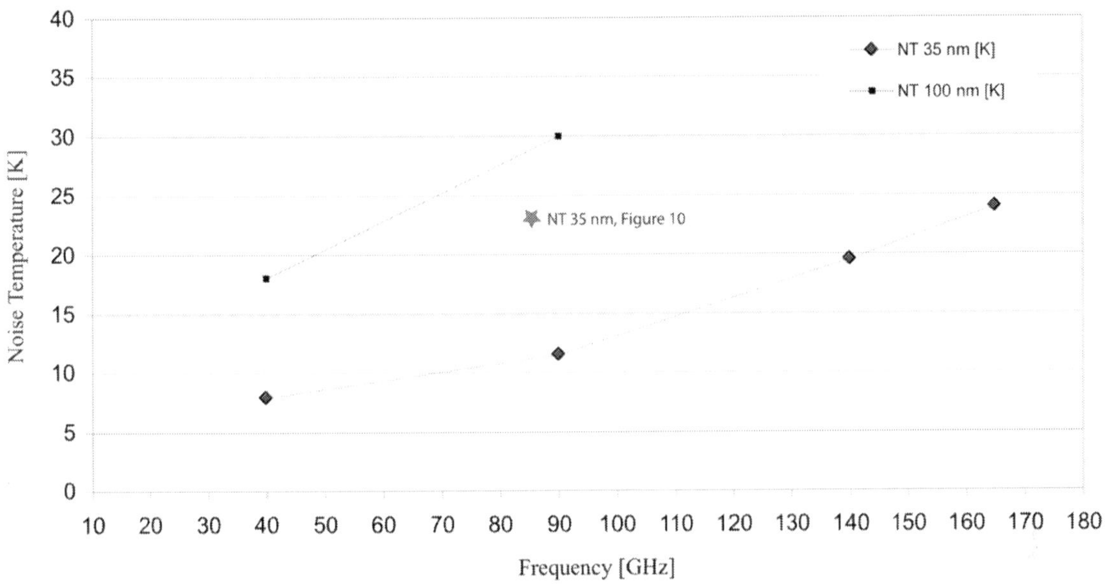

Figure 11. Noise temperatures for the 35 nm devices predicted at 40, 90, 140, and 170 GHz by the noise model extrapolating to lower frequencies and **cryogenic temperatures**. The improvement predicted over the best previous transistors (upper line in the plot) is dramatic.

values for the best polarization sensitivity was developed at The University of Chicago and used there for W-band and at Columbia for Q-band. The performance measured in the laboratory and scaled to the Chilean site (see Figure 1) gives array sensitivities for both receivers of order 60–70 $\mu K\, s^{1/2}$. In the case of Q-band, which is currently operating in Chile, this prediction has been confirmed at the site. These array sensitivities for Phase I of QUIET are as good as those for BICEP in its first incarnation, but are not as good as would be expected from the best current 100-nm devices as shown in Figure 11. We will return to this important point below. Phase II will have of order 1000 detectors and will take advantage of the improvements in amplifier performance discussed in this document.

Horn platelet arrays are a cost effective means of coupling the radiation into the modules. For QUIET, these have been developed at the University of Miami. The W-band array is shown on the left side of Figure 13. The arrangement of the receiver in its cryostat (designed and built at Columbia) is shown in Figure 15.

Three key points can be made about large arrays of coherent polarimieters suitable for B-mode instruments:

- The basic packaging technology for large arrays has been demonstrated. No technological breakthroughs are required. Savings in fabrication, assembly, integration, and testing of roughly two orders of magnitude over discrete component techniques used in the past are in hand. These techniques seem appropriate for arrays of the size that can be fit into the undistorted focal plane areas of CMB telescopes.
- Even more highly integrated packaging techniques, using multiple wafers stacked in 3-D, are being developed, and could be used for even larger arrays if there were a need. Set-up costs would be higher, but per unit costs could be much lower for large numbers.
- Although the basic arraying techniques have been well-demonstrated, there is one critical area in which current results are inadequate. In going from MMIC to module, a performance hit of a factor of roughly two is suffered. That is, the noise temperature of the module is

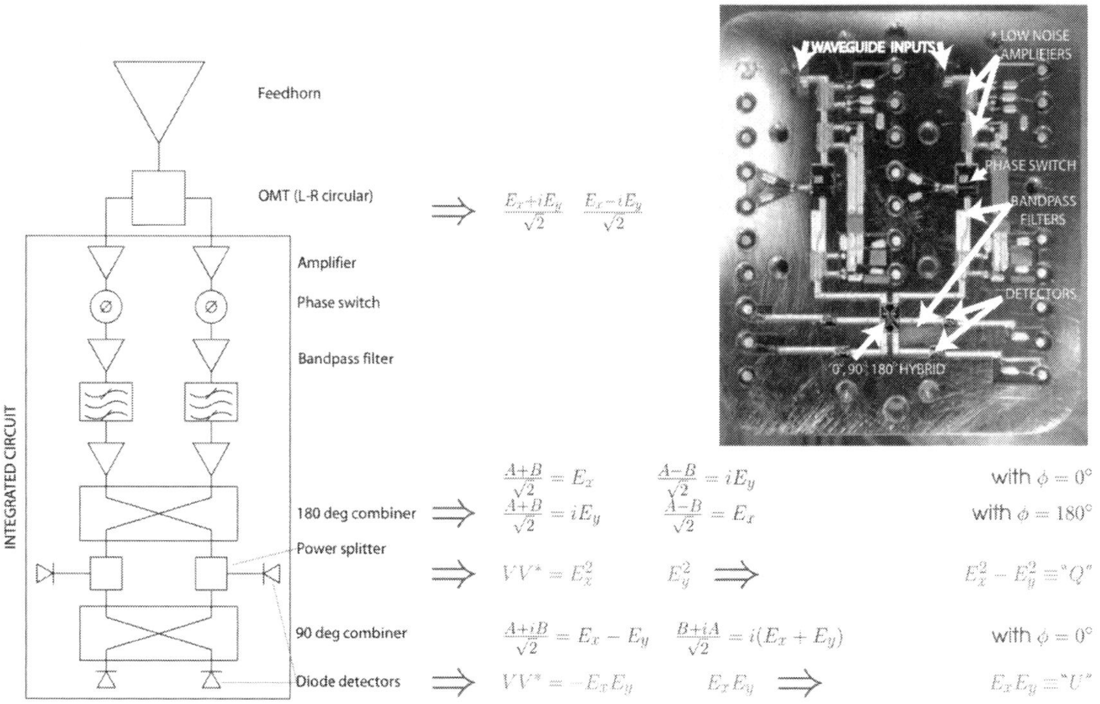

Figure 12. Schematic and photograph of a QUIET W-band polarimeter module, showing how Q and U are measured simultaneously through each feedhorn. The module, which contains all radiometer components following the OMT up to and including detectors, is about $1''.25 \times 1''$ in size, and fits into a standard circuit board socket, which supplies bias voltages and output signals. The modules are cooled to 20 K; bias supplies and signal processing electronics operate at room temperature. The modules can be assembled by machine, and can be built up into very large arrays.

roughly twice the noise temperature of the MMIC itself.

The last point is critical for the work that must be done. A factor of two in noise is simply too much to give up in going from individual amplifiers to array building blocks.

4.3. Development Plan

In the previous sections we have identified two critical areas of development needed to realize the full promise of coherent instruments for CMB polarization in space. These are:

- Improving the performance of individual transistors and MMICs to at least the level predicted to be achievable with 35 nm gate technology (Figure 11), and perhaps even further.
- Integrating high-performing MMICs into the building blocks of large arrays without loss of performance. Currently a factor of two or so in both noise and bandwidth is lost at this step.

We discuss these in §§ 4.3.1 and 4.3.2 below.

4.3.1. Device and MMIC level issues To achieve noise of $\leq 3 \times q$ from 30 to 150 or 200 GHz, a range important for CMB polarimetry, we can identify known issues and potential innovations that should be addressed.

Figure 16 shows a preliminary top level research plan. There are three main thrusts.

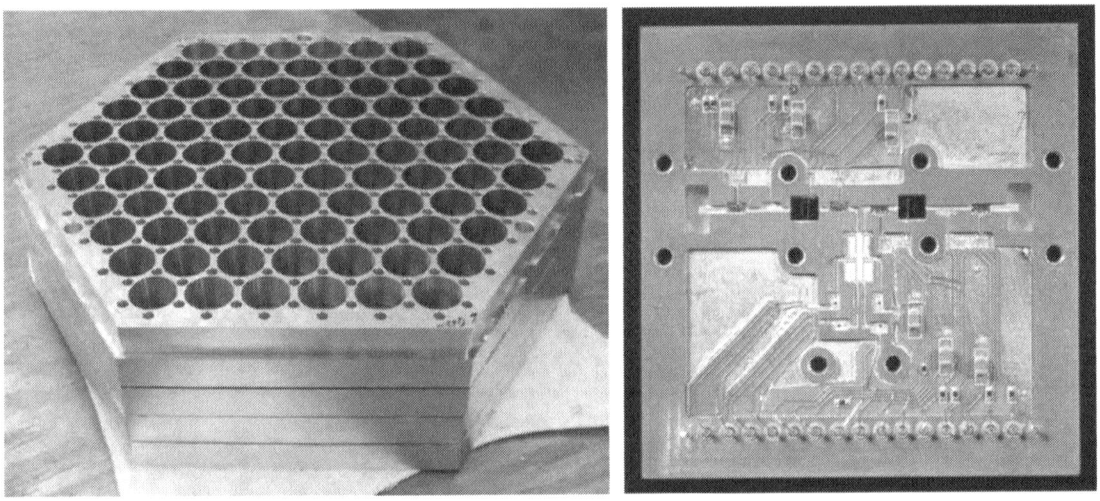

Figure 13. QUIET W-band 91-element feed array. This is a "platelet" array, in which an array of corrugated feed horns is built from many layers of aluminum with appropriately sized holes machined in them, all vacuum welded together. [Josh Gunderson, Miami] *Right*—QUIET Q-band module. Larger than the W-band module in Figure 13 because the array spacing for this wavelength is greater, the Q-band module combines some of the discrete components of the W-band module onto miniature circuit boards, making automated assembly even easier.

Figure 14. Test array of seven W-band modules mounted in standard sockets on a circuit board. One left-right circularly polarizing orthomode transducer (OMT) is attached to one module at the top. It is straightforward to build up large hex-packed arrays from these basic components. [Chicago]

- Measure and study existing devices and circuits for cryogenic performance and behavior to improve understanding of device physics, and to guide and focus the future device development.
- Develop new devices, with at least three iterations of experimental runs. At the end of each run, benchmarking and analysis inform the next run. The results from these runs will feed

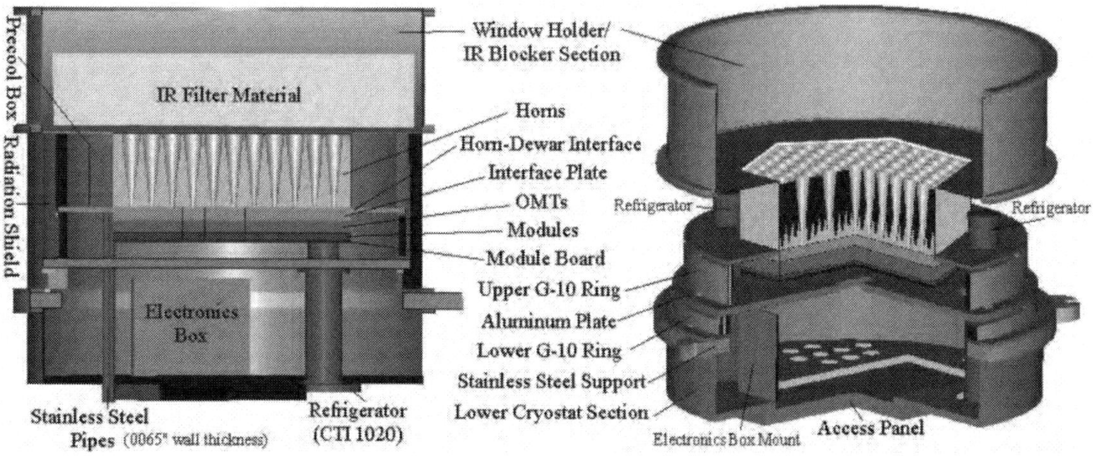

Figure 15. Sections of the QUIET cryostat, showing how a 91-feed platelet array, OMTs, and modules are arranged. The modules are placed in standard pin sockets on circuit boards. The size envelop for a given polarimeter module is limited by the spacing of the corrugated feeds, whose size is determined by the optical system. [Laura Newburgh, Columbia]

into the third main thrust

- Two amplifier design iterations and fabrication runs to demonstrate amplifiers with improved cryogenic performance. This type of comprehensive study to optimize HEMTs for cryogenic noise performance should enable reaching the ultimate limit for these devices, which may require new devices and materials not currently available.

Performance improvements will require addressing or solving the following issues, which are likely performance limiters.

Shot noise due to excessive gate leakage—In pursuit of the highest gain and transconductance InP HEMT devices, excessive leakage current at operating conditions limits the ultimate noise performance, especially at low frequencies where thermal noise is extremely reduced. Current research into alternate gate metals, barrier designs, gate recess chemistries and etch stops may improve both the turn-on voltage and reverse gate leakage for the InP HEMT devices while maintaining the other key dc and rf parameters. Of greatest interest is reduction of gate leakage at cryogenic operation which has not been studied or optimized carefully. NGST and JPL have observed wafer to wafer and lot to lot variation in the gate leakage. Although is less of an issue at higher frequencies, future improvements in device performance at higher frequencies will necessitate gate leakage below 1 μA to realize the target levels of noise performance.

High ohmic resistance at cryogenic temperatures—Recent investigations have shown that current InP HEMT ohmic contacts are not optimal and studies of the ohmic contact resistance to cryogenic temperatures may be important. This will have an impact not only on device gain, but also on the optimal dc bias voltage/current needed to achieve usable gain and its effect on drain temperature. Alternate ohmic contact and epitaxial schemes look promising for $> 2\times$ improvement, which may translate directly to cryogenic device noise improvements. Limits of epitaxial doping and design have also not been explored, especially for cryogenic operation.

Anomalous cryogenic HEMT behavior—Although not consistently observed, known potential issues such as IV kink at cryogenic temperatures (light sensitive), high output conductance, poor device pinchoff, low breakdown devices, high leakage and gain fluctuation need to be understood and avoided. Customized cryogenic tests can be developed to study these occurrences and how to avoid them. In some areas, promising room temperature solutions

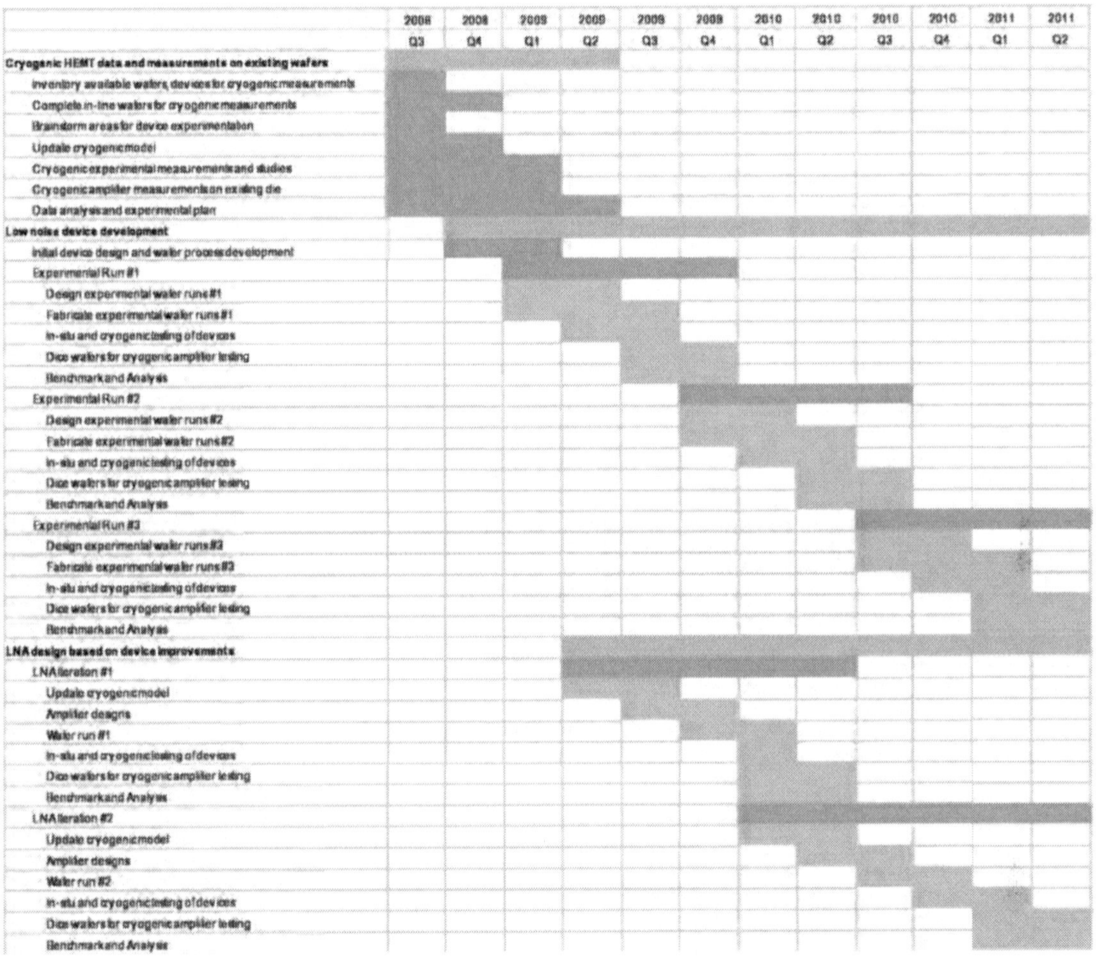

Figure 16. Top level program plan to achieve noise temperature goals.

have been developed and it will be beneficial to translate these improvements to cryogenic products.

InP HEMT yield limiters—Ohmic contact and sheet resistance, gate yield and defects, device breakdown, gate leakage, damaged airbridges, via hole yield, back metal adhesion, TFR damage, probe and metal scratch damage, dicing (splitting die), wafer breakage, and line errors are all factors.

Cryogenic device and noise models—Pospiesalski's cryogenic model is simple and predicts fairly well the noise parameters and overall performance of low noise amplifiers, especially for discrete device amplifiers. Updates to this and other alternately improved cryogenic HEMT noise models are important for future designs to improve ultimate noise performance

Over the past several years, NGST has developed advanced semiconductor materials, epitaxial designs, ohmic, gate and interconnect metals, device topologies, passivation thickness and interfaces, and backside wafer process improvements for its HEMT devices. The central focus for these advances at NGST continues to be to improve room temperature noise performance of LNAs with lowe dc power consumption and further system advantages in size and weight that can be derived through higher frequency implementations. These products are primarily aimed towards insertion into NGST's satellite communication payloads. To spur these innovations,

NGST has invested significant internal R&D funding for device and MMIC research (totaling more than \$10 M/year) and has consistently won contract R&D funding mainly from DoD services (also totaling more than \$10 M/year) over various semiconductor devices and products. NGST also continues to invest semiconductor equipment capital exceeding \$10 M/year especially to develop these novel materials and device processes.

What has not been studied recently is how these advances may spur improvements in the cryogenic operation of these devices. It can be projected that the room temperature improvements especially in noise performance may translate to improvements with cryogenic operation, but a careful engineering study and optimization have not been conducted to date. Listed below are key innovations both near and long term that can improve the state-of-art for cryogenic noise performance.

Nanometer-scale gate length reduction—Fabrication of short gate devices (70 nm and 35 nm) has achieved reasonable yield and reproducibility in combination with optimized epitaxial profile designs for each of these nodes. The use of these device technologies has focused mainly on a new generation of amplifiers from 140 GHz to 400 GHz, but extremely high device gain may also provide distinct low noise advantages at any frequency where the device and amplifier noise performance is potentially gain limited.

Atomic scale material growth and design—Advanced HEMT materials may offer higher frequency operation and low noise performance. Electron transport in InGaAs channels grown pseudomorphically on InP substrates has been improved with InAs channels (100% Indium) through the design of a composite channel epitaxy design. Extremely high room temperature mobilities of 16,000 $cm^2\,V^{-1}\,s^{-1}$, which represents a 30–50% increase compared to the baseline 60% InGaAs channels, have been observed. Further study of these devices is necessary with the potential of alloy scattering and impact ionization dominating with cryogenic operation. ABCS (Antimonide-based Compound Semiconductor) devices employing metamorphically grown InAs and potential InAsSb channels represent future material innovations that could push mobilities and electron velocities to even higher values. Limited cryogenic data have not achieved new state-of-art performance as current challenges in material quality, high leakage, and impact ionization still need to be addressed.

Device parasitics—Significant improvements have been made recently in ohmic contact and access resistance, in both epitaxial designs and new refractory ohmic metals. Contact resistance improvement by as much as 3× and sheet resistance improvement of 50% have been achieved on the most aggressive devices at room temperature, and should benefit cryogenic operation and low noise performance. Further development and exploration of sheet resistance limits through new epitaxial designs and smaller ohmic-to-gate spacings, including self-aligned gate device schemes, should be part of further work. Reduction of resistance in cryogenic operation is critical for lowering operating drain currents and voltages for high gain, which ultimately reduces drain temperature in the standard cryogenic FET model. Innovations for reduction of capacitance parasitics with both alternate low dielectric constant passivation films and thinner passivation films should be explored. The latter has been employed on recent high frequency devices and should be studied for potential benefits with cryogenic operation.

HEMT gate innovations—For HEMT devices, the gate process and metal-semiconductor junction remains the most critical in determining device performance. Barrier height and threshold voltage are crucial in optimizing device transconductance and gain, but breakdown and leakage current must be controlled to enable a useable device. Often these requirements conflict and the trades are even less understood at cryogenic operations where certain leakage currents are suppressed, while others are enhanced. The gate recess process formation is equally critical as it impacts breakdown, leakage, and access resistance. Many of the effects and variations with cryogenic operation have not been carefully studied and optimized. Several innovations are being explored. New refractory gate metals, especially Mo and TiW, are being explored that may provide lower leakage and superior cryogenic Schottky junctions. Controlled interdiffused junctions that could reduce $1/f$ noise for cryogenic operation have not yet been explored. Epitaxial growth is still critical, and although not studied to date, heterostructure and doping interface sharpness may be critical to determine limits on gate-to-channel separation, where we face tradeoffs between transconductance and gain vs. excess leakage and degraded Schottky junctions. Combined with new refractory metals, the trade space should be explored for optimal cryogenic performance. Tailoring the gate recess profile through etch stops and multiple recess steps may provide advantages where the designs are aimed towards cryogenic low noise operation. However, current research is aimed more towards higher power, higher device density circuits. Barrier layer epitaxial designs for both etch stop and bandgap engineering should be studied more carefully for cryogenic optimization. As an example, current ABCS HEMT device utilization is limited due to manufacturing issues of the gate barrier layer AlSb and GaSb.

New devices, materials, ideas—The advent of new transistors and materials as they come to some level of maturity and utility may be crucial to achieving the noise goal $\leq 3\times$ quantum limit or even beyond. Among the promising technologies being pursued are quantum wire devices, as well as carbon nanotube and graphene transistors. Nanolithography will also push the limits on device scaling and we anticipate research in smaller gate length nodes down to 20 nm or less to achieve amplifiers that could operate as high as 1 THz in the future. Optimizing these types of concepts for cryogenic noise operation may also reveal unexpected and further performance breakthroughs.

4.3.2. Module level issues The key is to realize the full performance potential of the transistors and MMICs in a unit cell package that enables massive arrays. This requires cryogenic measurement and characterization of MMICs and other components, isolation of the critical factors in performance, design and fabrication iterations, and exquisite control of fabrication.

An essential point that must be emphasized again is that while there is substantial commercial and military interest in coherent detectors for communications and imaging, there is little interest outside the astronomical world for *cryogenic* applications. The development for CMB polarimetry of array building blocks with no loss in MMIC performance is inherently cryogenic. The commercial world will not provide this development.

The work will involve both MMIC amplifier design and prototype multichip MMIC module development. A simple mask set with transistor test cells, simple MMIC test amplifiers, and space for new MMIC designs, must be developed. Multiple iterations of design, wafer processing, test and characterization, and module fabrication will be required. Based on recent history, there is no reason to expect that fundamental technological breakthroughs will be required, but careful attention to engineering details will be essential.

We emphasize that the the work requires three types of institutions: national labs, university groups, and industry. Industry provides materials and device design and processing capability. The national labs and universities provide the application-specific and cryogenic design and testing capability, as well as the critical experimental requirements and validation for ultimate performance. Experience shows that the participation of all three types of institutions, including postdocs and students, is essential.

5. HETERODYNE INTERFEROMETRY

5.1. Introduction

Heterodyne interferometry provides one of the most powerful techniques for making thermal-noise-limited observations of astrophysical and cosmological targets. It is particularly effective in eliminating many sources of systematic error that complicate observations with single antennas. Heterodyne interferometers have a long heritage in ground-based measurements of the CMB, with DASI making the first detection of polarization and CBI making some of the first measurements of the shape of the E-mode power spectrum. Both DASI and CBI also made detailed measurements of the CMB temperature power spectrum, detecting multiple peaks in the power spectrum prior to the launch of WMAP.

§ 4.4 discusses the advantages and disadvantages of heterodyne interferometry in more detail. For now, they are summed up well in a comment made by Tony Readhead, PI of the OVRO single-dish CMB experiment and the CBI, who said "In our experience over the last three decades with both single-aperture systems and interferometers, we found that it took many months to identify and eliminate a wide variety of sources of systematic error with single antennas, whereas such sources of contamination could be identified and eliminated in a matter of a few days with interferometers."

The key question to address immediately is this: *if heterodyne interferometers are so great, why have they not been seriously studied or proposed for B-mode experiments?* The answer is straightforward, and comes in three steps:

- For sensitivity, extremely low sidelobe levels, and frequency coverage to support component separation, a very large number of elements in a filled or nearly-filled array is required, resulting in of order a (very large number)^2 of baselines that must be correlated, requiring a very big correlator.
- Frequency coverage to realize the most accurate component separation can only be achieved from space.
- Historically, big correlators have been physically large and notorious power hogs (tens or hundreds of kilowatts!), far too demanding of mass and power to contemplate flying in space.

As a result, heterodyne interferometry, potentially the ultimate technique for control of systematic errors for the most demanding B-mode applications, was simply impossible. However, **THE SITUATION HAS CHANGED!!** Driven by the huge demands and needs of wireless communications, the capability of application specific integrated circuit (ASIC) technology has been increasing at a "Moore's Law" rate. In the next section we show that it has now reached a level where large interferometers in space are practical. The situation will only improve rapidly with time.

Interferometers with a large number of elements, broad bandwidth, and wide frequency coverage of the sort that would be needed in space to make definitive measurements of B-modes have not been studied seriously. Without enabling low-powered correlators, it made little sense to do so, and without such studies in hand, we cannot and do not claim that a space interferometer is the best way to measure B-modes. However, *we strongly believe that the time has come to study heterodyne interferometers seriously for the measurement of B-modes.* The rest of this section outlines some basic considerations of interferometers that lead us to that conclusion, starting with the status of low-powered correlators.

5.2. Low-Powered Correlators Enable Space Interferometers

The number of required correlations at a given frequency is:

$$N_{\rm cor} = \frac{N_{\rm feed}(N_{\rm feed}-1)}{2} \times 4N_{\rm sub} \approx 2N_{\rm feed}^2 N_{\rm sub},$$

where $N_{\rm feed}$ is the number of pixels, $N_{\rm sub}$ is the number of IF sub-bands, and the factor of 4 comes from the requirement to correlate LL, RR, LR, and RL, where R and L are the two polarization states. The number of sub-bands will be determined by the need to avoid chromatic aberration, and by the clock speed of the digitizer. The number of bits that must be correlated will be determined by the dynamic range required.

Recent technology developments with space-based cross-correlators for Fourier synthesis interferometers have been driven by Earth Science remote sensing missions, in particular the Lightweight Rainfall Radiometer (LRR) and the Geosynchronous Earth Orbit Synthetic Thinned Aperture Radiometer (GEOSTAR). In both cases, emphasis has been placed on reducing the power requirements, increasing the clock speed, and improving the radiation tolerance of the digitizers and the multipliers/accumulators through the use of ultra-low power CMOS ASICs based on a 0.5 V logic protocol with resistance by design to radiation-induced single event upsets. This protocol has extensive spaceflight heritage for other high speed digital signal processing applications in space.

The new multiplier/accumulator for the GEOSTAR project is being built using a 90 nm, 0.5 V, CMOS ASIC process. It is projected to be capable of complex cross-correlations of all possible pairs of 196 In-Phase and 196-Quadrature Phase 2-bit input signals, clocked at 1400 MHz, while drawing 1.68 W of DC power.

Scaling from these chips, the power for multiplier/accumulators for a CMB application given current mature technology would scale as

$$P = 98\left(\frac{N_{\rm el}}{196}\right)^2\left(\frac{\Delta\nu}{19.8\,{\rm GHz}}\right)\left(\frac{n_{\rm bits}}{2\,{\rm bits}}\right)\ {\rm W}.$$

Additionally we need to consider the digitizers. If the IF bands are 1.4 GHz, then the digitizer needs to run at 2.8 GHz for Nyquist sampling. Low-voltage digitizers that operate at this clock speed have not yet been demonstrated, but a 1-bit digitizer clocked at 392 MHz has been demonstrated that dissipates 4 mW. Assuming that the power dissipation scales with clock speed, the power dissipation would be

$$P = 40\left(\frac{N_{\rm el}}{2}\right)\left(\frac{\Delta\nu}{19.8\,{\rm GHz}}\right)\ {\rm W}.$$

Clearly, sampling and full correlation of many hundreds of elements over tens of gigahertz bandwidth can now be contemplated for less than 1 kW.

5.3. Interferometer Design

The way that interferometers achieve separation of E and B modes is shown schematically in Figure 17. A cross-correlation between the circularly polarized states, one from each antenna, is formed. The visibilities that correspond to the cross-correlations are related to the Stokes parameters as follows (Kovac et al., 2002).

$$V^{RR}(\mathbf{u}_i) = \alpha_i \int d\mathbf{x} A(\mathbf{x},\nu_i)[T(\mathbf{x}) + V(\mathbf{x})]e^{-2\pi i \mathbf{u}_i\cdot\mathbf{x}},$$

$$V^{LL}(\mathbf{u}_i) = \alpha_i \int d\mathbf{x} A(\mathbf{x},\nu_i)[T(\mathbf{x}) - V(\mathbf{x})]e^{-2\pi i \mathbf{u}_i\cdot\mathbf{x}},$$

$$V^{RR}(\mathbf{u}_i) = \alpha_i \int d\mathbf{x} A(\mathbf{x},\nu_i)[Q(\mathbf{x}) + iU(\mathbf{x})]e^{-2\pi i \mathbf{u}_i\cdot\mathbf{x}},$$

$$V^{RR}(\mathbf{u}_i) = \alpha_i \int d\mathbf{x} A(\mathbf{x},\nu_i)[Q(\mathbf{x}) - iU(\mathbf{x})]e^{-2\pi i \mathbf{u}_i\cdot\mathbf{x}},$$

from which the E and B components can be reconstructed:

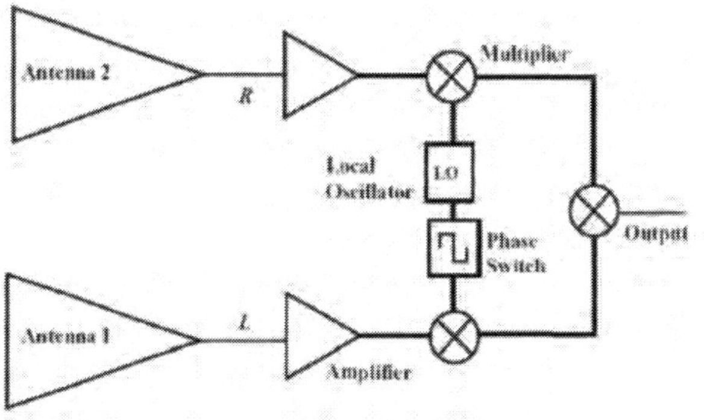

Figure 17. Schematic of the operation of an interferometer (from Staggs and Church). A single polarization state from one antenna (in this case circular) is correlated with the polarization state from a second antenna. Placing a OMT after the feedhorn to split the signal into left and right circular polarization states, with a separate amplifier chain for each, allows all four Stokes parameters to be measured.

$$\tilde{Q}(\mathbf{u}) = \cos(2\chi)\tilde{E}(\mathbf{u}) - \sin(2\chi)\tilde{B}(\mathbf{u}),$$

$$\tilde{U}(\mathbf{u}) = \sin(2\chi)\tilde{E}(\mathbf{u}) + \cos(2\chi)\tilde{B}(\mathbf{u}).$$

It is also possible to build an interferometer based on bolometric interferometry by combining the signals optically prior to detection on a bolometric array. Such an approach is discussed elsewhere in the workshop. However, coherent amplifier technology is more suited to interferometric measurements than bolometric technology because the former has the ability to replicate photons, once the quantum tax is paid, allowing the full polarization state of the incoming radiation from a single feed to be correlated with that from any number of other feeds without degrading the sensitivity. In contrast, a bolometric interferometer cannot simultaneously measure Q and U (and thus E and B), and cannot expand efficiently to a large number of interferometric elements. Instead the incoming polarization state must be optically modulated and split, a complicated technology development that is not required in a coherent amplifier interferometer and does not scale well to a large number of feeds/detectors.

The advantages of interferometry do not lessen the need for sensitivity. Large arrays of detectors operating near fundamental noise limits are required whether the detectors are treated independently or as part of an interferometer. Since a close-packed interferometer and a close-packed focal plane array with the same number of elements have the same sensitivity, a B-mode interferometer needs hundreds of array elements per frequency just like a focal plane array. Fortuitously, a close-packed interferometric array with hundreds of elements to satisfy the sensitivity requirement also has an extremely well-shaped synthesized beam with very low sidelobe levels. This is the kind of interferometer that we are discussing.

5.4. Systematics and Sensitivity: Advantages and Disadvantages of Interferometers

Because an interferometer directly measures the spatial correlation function of the incoming signal, the measurement technique is fundamentally different from that of focal plane arrays.

Features that make heterodyne interferometers particularly suited for B-mode polarization measurements include:

- insensitivity to large-scale offsets and drifts;
- accurate determination of the synthesized beam shape, which can be calculated with high precision based on the geometry of the interferometric elements;

Journal of Physics: Conference Series **155** (2009) 012002 doi:10.1088/1742-6596/155/1/012002

- simultaneous measurement with low $1/f$ noise of temperature (I) and polarization (Q and U);
- sensitivity of each baseline to a particular Fourier mode on the sky, so measurements are made direcly in Fourier space
- simultaneous measurement of the real and imaginary parts of the signal, which can be combined in a straightforward way to decouple the E and B mode signals accurately.

Advantages:

- Coherent-amplifier interferometric methods allow certain systematics to be greatly reduced or eliminated alltogether without the need for complicated multi-parameter modeling. Table 3 shows a list of systematics and a comparison of their effect on an interferometer with a system based on a focal plane array of bolometers. Note that many of these advantages are lost in a system based on bolometric interferometry.
- The cross-correlation of a single polarization state from two different antennas measures a single Fourier mode on the sky. Uncorrelated signals, such as amplifier $1/f$ noise, are strongly suppressed.
- There is inherent differencing involved in the cross-correlation process. The fringe pattern that is formed on the sky filters the images in ℓ-space in a well-understood way.
- A single Fourier mode is measured with just two detectors, which simplifies the mode-by-mode calibration in ℓ-space. Note that this advantage is lost in a bolometric interferometer, where the fringes are projected onto a bolometer array and many detectors are required to measure a single mode.
- With hundreds of close-packed corrugated horn elements, the synthesized beam will be both extremely clean and calculable with high precision. Moreover, the radiation pattern of individual corrugated feeds, which adds up to the envelope of the "primary beam", is better than that of any telescope. Beamshape uncertainties with telescopes are a major systematic affecting current experiments. WMAP, for example, continues to make beam corrections at the 1% level or so that affect science results directly. *The impossibility of fully characterizing the feed+telescope patterns with high dynamic range in the post-launch, cold environment of space may provide an ultimate limitation to telescope+focal plane array experiments. An interferometer seems likely to avoid this problem (detailed analysis is needed to quantify this), and as a result seems to be the ultimate in low systematics for CMB polarization.*
- The noise modeling is simplified by the interferometric strategy because the noise from a single baseline is localized to a specific point in Fourier space.
- Because the angular resolution requirements are modest, corrugated feed horns can be used as the antennas, with the result that sidelobes and cross-polarization will be the lowest that is possible for any of the proposed B-mode experiments at the present time.
- A heterodyne interferometer can be configured in a natural way as a broadband spectrometer by splitting the IF band into $N_{\rm sub}$ spectral bands. Depending on the requirements that this places on the downlink speed, this spectral information can be retained and used as a check on systematics from foregrounds.

Disadvantages:

- An interferometer filters out all angular scales larger than the primary beam. This sets a lower ℓ limit to the interferometric measurement; however, the total power signal from the feeds could be used directly to measure the lowest multipoles. Work would be required to figure out the best way to do this, and this complication makes it appropriate to consider this a disadvantage of an interferometer. The silver lining is that the low ℓs would be measured without a telescope, which will reduce systematics.

Considerations & Practicalities:

- Crosstalk or mixing between polarization states in the same antenna can lead to leakage of T into Q and U. This can be mitigated by phase switching the two states at different frequencies. Unlike the complex phase modulation required for bolometric inteferometer, a simple 180 degree phase switch—a mature, well understood technology—is used. Crosstalk that occurs between the antennas and the phase switch is mitigated by enclosing the low-noise amplifiers in separate metal housings. Crosstalk between antennas can be mitigated by scanning the instrument across the sky to modulate the sky signal with respect to the more slowly varying cross-talk.
- The number of pixels that can be accommodated in a space-based interferometer will be limited primarily by power dissipation and the capabilities of the correlator.
- At 150 GHz and above, the noise temperature of the heterodyne system will be significantly higher than a bolometric system, but the foreground signals, which are the primary target at higher frequencies, are also rising. As discussed in § 3.2, preliminary indications from modeling of component separation are that optimum results are obtained when noise per channel is proportional to total signal, including foregrounds.

5.5. Technical Design and Technology Development Program

In order to achieve the noise levels required to measure B-modes, both the sensitivity and the number of individual detectors/interferometer elements must be increased. The sensitivity of a filled-array interferometer and a close-packed focal plane array with the same angular resolution and sensitivity per detector are the same. Therefore, from a detector standpoint the same basic requirements apply to interferometers and focal plane arrays, namely, better detectors, and more detectors. The number of interferometer elements must be increased from the dozen or so of past CMB interferometers (e.g., VSA, DASI, CBI) to hundreds. This simultaneously produces a synthesized beam with extremely low and symmetric sidelobes. Figure 18 shows the sort of building blocks that could be used in a large-N interferometer.

The speed and power dissipation of sampling and multiplying chips will continue to advance at a rapid rate, driven by the needs of wireless communication. At present, development of 90 nm-technology digitizers is lagging behind that of multipliers. Based on the modeling projections made for the 90 nm multiplier/accumulator ASICs, near-term development can be expected to result in digitizers with maximum clock rates in the neighborhood of 1400 MHz. A 90 nm ultra-low power 0.5 CMOS ASIC digitizer can be expected to reduce the power dissipated by two orders of magnitude.

Table 3 shows the TRL level of essential components of a CMBPol interferometery.

5.6. Technological Readiness

A great deal of design and engineering work would be required to build a space CMBPol interferometer. As with any experiment to measure B-modes, control of systematics and the effectiveness of the overall scheme must be demonstrated in suborbital observations. Most of the key technologies are applicable beyond interferometry. The most important technology developments required are in ultra-low-noise cryogenic amplifiers and modules discussed in § 4, and in the correlator digitizers and multipliers. The correlator technologies are driven by commercial demand, and science can piggyback on their development.

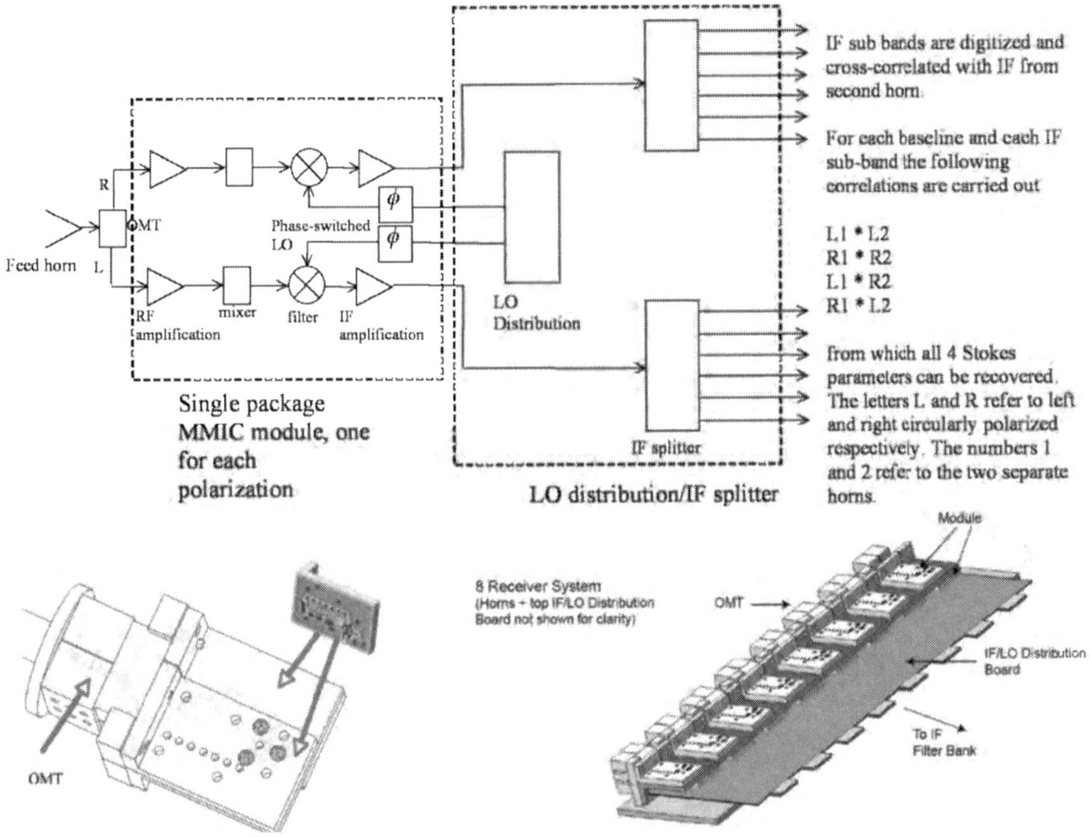

Figure 18. Schematic of a single pixel suitable for use in an interferometer to measure CMB polarization. Two MMIC modules (lower left) are used, one per polarization. The modules are mounted on the IF/LO distribution board (lower right), which further amplifies the IF signals. There are two boards, one each for left and right circular polarization, but only one is shown for clarity.

Acknowledgments
We thank John Kovac and Andrew Lange for data used in Figures 1 and 2, Lorene Samoska and Pekka Kangaslahti for Figures 6–9 and 11, Matt Morgan for Figure 10, Josh Gundersen for the left panel of Figure 13, the University of Chicago QUIET team for Figure 14, and Laura Newburgh for Figure 15, and Amber Miller and Bruce Winstein for helpful comments.

Table 3. Key technologies for a CMBPol interferometer.

Element	Heritage	Development Needed for CMBPol	TRL	Comments
Feed horn array	WMAP, Planck	Test crosstalk on very small baselines	6^+	Individual horns very well understood. QUIET will demonstrate latelet arrays.
	Platelet arrays for QUIET	Weight reduction		
Orthomode transducers	WMAP, Planck, QUIET	Weight reduction, assembly of many units.	6^+	
20 K Cooler	Planck		8	
Heterodyne MMIC amplifier modules	QUIET, GeoSTAR	Prototype pixel fabrication	4	Prototype single polarization heterodyne module is under development. Paper design exists, fabrication expected by end 2008.
	Platelet arrays for QUIET	Weight reduction		
IF electronics		Low power amplifiers		Use HEMTs?
Correlator	GeoSTAR	low power ASIC multipliers and digitizers	4–6^+	

Appendix: Sensitivity Comparison of Bolometric and HEMT-based systems

We give here details of the method used in § 3.1 to compare the sensitivities of HEMT and bolometric systems for both space and ground.

Appendix A.1. Sensitivity of a bolometric system

The sensitivity of a bolometer to a single mode of polarization can be calculated from first principles based on the techniques in Zmuidzinas (2003). The Noise Equivalent Power (NEP) is given by:

$$\mathrm{NEP} = \sqrt{2\left(\frac{P_{\mathrm{det}}^2}{\Delta\nu} + P_{\mathrm{det}}h\nu\right)} \qquad \mathrm{W\,Hz^{-1/2}}. \tag{A.1}$$

P_{det} is the power incident on the detector and is given by

$$P_{\mathrm{det}} = \Delta\nu h\nu \left(\eta n_{\mathrm{CMB}} + \sum_i \epsilon_i n_{\mathrm{back},i}\right), \tag{A.2}$$

where η is the optical efficiency of the detection system, ϵ is the emissivity of each contribution to the background, and $\Delta\nu$ is the bandwidth of the system. The quantities n_{CMB} and n_{back} are the photon occupation numbers of the signal (CMB) and background, respectively, given by

$$n(T) = \frac{1}{\exp(h\nu/kT) - 1.} \tag{A.3}$$

For the Figures 1 and 2, the following assumptions have been made:

- For a perfect bolometric system from the ground, it is assumed that the signal is the CMB and the background is from atmospheric and telescope emission. A ground-based telescope is assumed to have a temperature of 280 K and an emissivity of $\sim 2.9\%$ (giving a telescope temperature of 8 K). The atmospheric emissivity is calculated for a high dry site.
- For a space-based system, the telescope is assumed to be cold so that its contribution to the background power can be ignored.

To convert NEP to sensitivity to Q (or U) involves the following steps:

- First the bolometer NEP is converted to an NET

$$\frac{\text{NET}}{T_{\text{CMB}}} = \frac{\exp(x) - 1}{x \exp(x)} \times \frac{1}{\sqrt{2}} \times \frac{\text{NEP}}{P_{\text{CMB}}} \tag{A.4}$$

  where $x = h\nu/kT_{\text{CMB}}$ and the factor of $1/\sqrt{2}$ converts $\text{W Hz}^{-1/2}$ to $\text{W s}^{1/2}$.
- It is assumed that the detected signals from two bolometers, each sensitive to an orthogonal direction of polarization, are differenced to get Q. Then:

$$\text{NEQ} = \frac{\text{NET}}{\sqrt{2}} \tag{A.5}$$

Appendix A.2. Sensitivity of a HEMT-based system

A similar procedure is followed for a HEMT system. The sensitivity of a single quantum-limited amplifier can be calculated from first principles (Zmuidzinas, 2003), and is shown to have an NEP of

$$\text{NEP} = \sqrt{\frac{2}{\Delta\nu}}(P_{\text{det}} + h\nu\Delta\nu). \tag{A.6}$$

We don't normally think of amplifier sensitivities in terms of NEP, but there is no reason not to do so. To see this in the more familiar terms of system temperature, we divide by $k\Delta\nu$ and convert from $\text{W Hz}^{-1/2}$ to $\text{W s}^{1/2}$to get

$$\Delta T_{\text{RJ}} = \frac{1}{\sqrt{\Delta\nu}}\left(T_{\text{back,RJ}} + \frac{h\nu}{k}\right), \tag{A.7}$$

where $P_{\text{det}} = k\Delta\nu\Delta T_{\text{back,RJ}}$ defines the Rayleigh-Jeans temperature of the radiation that is incident on the amplifier. The second term in Equation A.7 is then identified as the quantum noise introduced by the amplifier and defines a noise temperature of $T_{\text{amp}} = h\nu/k$ at the quantum limit. The sensitivity of a system with an amplifier that has noise higher than the quantum limit can be accomplished by modifying Equation A.7 to be

$$\Delta T_{\text{RJ}} = \frac{1}{\sqrt{\Delta\nu}}\left(T_{\text{back,RJ}} + q\frac{h\nu}{k}\right), \tag{A.8}$$

where q is the ratio of the amplifier noise temperature to the quantum noise limit (e.g., a "five times quantum" amplifier would have $q = 5$.) This expression can then be converted back to NEP, giving

$$\text{NEP} = \frac{2}{\sqrt{\Delta\nu}}(P_{\text{det}} + qh\nu\Delta\nu), \tag{A.9}$$

from which the NET for CMB measurements can be calculated using Equation A.4.

The NEQ of a HEMT-based system, whether it is a correlation radiometer, or a interferometer, is given by:

$$\text{NEQ} = \frac{\text{NET}}{\sqrt{2}} \tag{A.10}$$

The difference is that these systems can measure Q and U simultaneously without noise penalty, so it is conventional when making comparisons between bolometric and HEMT based systems to include an extra factor of $\sqrt{2}$ so that:

$$\text{NEQ}_{\text{bolo}} = \frac{\text{NET}}{\sqrt{2}} \tag{A.11}$$

$$\text{NEQ}_{\text{HEMT}} = \frac{\text{NET}}{2} \tag{A.12}$$

This convention has been adopted in the figures.

REFERENCES

Blum, E. 1959, *Ann. Astrophys.* **22**, 140.

Brandt, W. N., Lawrence, C. R., Readhead, A. C. S., Pakianathan, J., and Fiola, T. 1994, *Ap. J.*, **424**, 1.

Dawson, D., Samoska, L. Fung, A. K., Lee, K., Lai, R., Grundbacher, R., Liu, P. H., Raja, R., 2005, *Beyond G-band: a 235 GHz InP MMIC amplifier, IEEE Microwave and Components Letters*, **15, Issue 12**, 874.

Deal, W. R., Din. S., Radisic., V., Padilla., J., Mei, X. B., Yoshida, W., Liu, P. H., Uyeda, J., Barsky, M., Gaier, T., Fung, A., Samoska, L., Lai, R. 2006, *Demonstration of a Sub-Millimeter Wave Integrated Circuit (S-MMIC) using InP HEMT with a 35-nm Gate, IEEE Compound Semiconductor Integrated Circuits Conference*, 33.

Deal, W. R. 2007, *Demonstration of a 270 GHz MMIC Amplifier using 35 nm InP HEMT Technology, IEEE Microwave and Components Letters*, **17, Issue 5**, 391.

Faris, J. 1967, *J. Res. Natl. Bur. Stand. (Eng. and Instrum.*, **71C**, 153.

Gaier, T., Samoska, L., Fung, A., Deal, W. R., Radisic, V., Mei, X. B., Yoshida, W., Liu, P. H., Uyeda, J., Barsky, M., Lai, R. 2007, *Measurement of a 270 GHz Low Noise Amplifier with 7.5 dB Noise Figure, IEEE Microwave and Wireless Components Letters*, July, 2007.

Holder, G. P., Haiman, Z., Kaplinghat, M. & Knox, L. 2003, *The Reionization History at High Redshifts. II. Estimating the Optical Depth to Thomson Scattering from Cosmic Microwave Background Polarization, ApJ*, **595**, 13.

Hu, W. & White, M. 1997, *A Polarization Primer, New Astronomy*, **2**, 323.

Jones, R. C. 1941, *New Calculus for the Treatment of Optical Systems, J. Opt. Soc. Am.*, **31**, 488.

Jones, R. C. 1941, *New Calculus for the Treatment of Optical Systems, J. Opt. Soc. Am.*, **31**, 500.

Jones, R. C. 1942, *New Calculus for the Treatment of Optical Systems, J. Opt. Soc. Am.*, **32**, 486.

Kangaslahti, P., Pukala, D., Gaier, T., Deal, W. Mei, X., and Lai, R., 2008, *Low Noise Amplifier for 180 GHz Frequency Band, IEEE MTT-S International Microwave Symposium Digest*, in press.

Page, L. et al. 2007, *Three-Year Wilkinson Microwave Anisotropy Probe (WMAP) Observations: Polarization Analysis, ApJ*, in press.

Pospieszalski, M. W. 1989, *Modeling of noise parameters of MESFETs and MODFETs and their frequency and temperature dependence, IEEE Transactions on Microwave Theory and Techniques*, **37**, 1340.

Predmore, C., Erickson, N. Huguenin, G., and Goldsmith, P. 1985, *IEEE Transactions on Microwave Theory and Techniques*, **33**, 44.

Pukala, D., Samoska, L., Gaier, T., Fung, A., Mei, X. B., Yoshida, W., Lee, J., Uyeda, J., Liu, P. H., Deal, W. R., Radisic, V., 2008, *Submillimeter-Wave InP MMIC Amplifiers From 300–345 GHz, IEEE Microwave and Wireless Components Letters*, **18, Issue 1**, 61.

Samoska, L. 2006, *Towards terahertz MMIC amplifiers: present status and trends, IEEE MIT-S International Microwave Symposium Digest*, 336.

Samoska, L., Deal, W. R., Chattoapadhyay, G., Pukala, D., Fung, A., Gaier, T., Soria, M., Radisic, V., Mei, X. B., Lai, R. 2008, *A Submillimeter-wave HEMT Amplifier Module with Integrated Waveguide Transitions Operating Above 300 GHz, IEEE Transactions on Microwave Theory and Techniques*, in press.

Seljak, U. & Zaldarriaga, M. 1997, *Signature of Gravity Waves in Polarization of the Microwave Background, Phys. Rev. Lett.*, **78**, 2054.

Tinbergen, J. 1996, *Astronomical Polarimetry*, Cambridge: Cambridge University Press, Sections 4.2–4.4.

TFCR: Weiss, R. et al. 2005, *Final Report of the Task Force on Cosmic Microwave Background Research*, NASA, NSF, DOE, www.nsf.gov/mps/ast/tfcr_final_report.pdf.

Weinreb, S., Lai, R., Erickson, N., Gaier, T., and Wielgus, J. 1999, *W-Band InP wideband MMIC LNA with 30 KJ noise temperature*, *IEEE MIT-S Int. Microwave Symp. Dig.*, **1**, 101.

Zmuidzinas, J. 2003, Applied Optics, 42, 4989.

Adding interferometry for CMBPol

Peter T. Timbie[1] **and Greg S. Tucker**[2]

[1] Department of Physics, University of Wisconsin, Madison, WI 53706
[2] Department of Physics, Brown University, Providence RI 02912

E-mail: pttimbie@wisc.edu, gst@brown.edu

Abstract. Interferometry offers an alternative to imaging of the CMB. Some systematic errors may be easier to control than in the imaging case. Adding interferometry is capable of correlating signals from a large number of antennas, more than is currently possible with traditional multiplying interferometers. Many of the technologies required for a space-based adding interferometer for CMB studies are the same as for imaging systems. We evaluate those critical components which are different from imagers.

1. Introduction

Interferometers have been used for many years for studying the CMB temperature and polarization power spectra and the Sunyaev-Zel'dovich effect. In fact, the first detection of CMB E-mode polarization was made by an interferometer: DASI [1]. There are many reasons to consider using interferometers for measurements of the B-mode signal. The key reason is to control systematic effects [2].

Recently, several groups have studied the possibility of building future interferometers specifically to search for the small polarization signals in the CMB. Compared to existing interferometers, these new instruments would have to do the following to: 1) collect more modes of radiation from the sky by including more (single-mode) antennas, 2) operate with broader bandwidth, and 3) operate over a broader range of frequencies, at least up to 90 GHz, to be able to detect and reject astrophysical foreground sources. The most significant challenge to increasing the number of antennas is correlating the large number of baselines. There are two approaches — *multiplying* interferometry and *adding* interferometry.

Another white paper [3] for this workshop addresses means of applying traditional multiplying interferometry (sometimes called heterodyne interferometry because the RF signal is typically mixed to a lower frequency before the correlator) to the B-mode search. These methods use coherent receivers (SIS or HEMT) and are currently limited by the correlator, both in bandwidth and in the number of baselines. We focus here on adding interferometers, which have the ability to use either coherent receivers or incoherent detectors (bolometers) and for which correlators with large bandwidths and large numbers of inputs appear feasible.

We focus on systems that provide modest angular resolution ($\sim 1°$) and large fields of view ($\sim 10°$), appropriate for measurements of the recombination and reionization peaks in the B-mode power spectrum. In this case a compact interferometer array can be formed from a cluster of circular horn antennas (similar to DASI).

2. History and Advantages of Interferometry

Interferometers have proved to be powerful tools for CMB observations (see [4] for a comprehensive list). The Sunyaev-Zel'dovich effect has been imaged by the Ryle [5], OVRO and BIMA interferometers [6] and the SZA [7] at centimeter wavelengths. The CMB temperature anisotropy has been imaged by the CAT [8], VSA [9], DASI [10] and CBI [11] interferometers, also at centimeter wavelengths. DASI was the first instrument to detect the CMB polarization [1; 12], and CBI has detected CMB polarization at smaller angular scales [13; 14]. These measurements were all made by "traditional" interferometers that use coherent receivers and correlate signals from each pair of antennas in the array by multiplying the amplified electric fields together. The correlated signals form the visibility. There are $N(N-1)/2$ such pairs (baselines).

The main reason for building interferometers instead of traditional imaging systems for studying the CMB is to control systematic effects, which in some cases are more manageable than in imaging systems. There are additional factors, especially aperture size, that favor interferometric approaches over imaging for space-based systems.

2.1. Angular Resolution

For a monolithic dish of diameter, D, equal to the length of a two-element interferometer baseline, B, the interferometer has angular resolution (fringe spacing) roughly twice as good as that of the monolithic dish. The reason for this difference in angular resolution is that the filled dish is dominated by spacings that are much smaller than the aperture diameter. The full width to the first zero for a uniformly illuminated circular aperture of diameter D is $2.4\lambda/D$. The full width to the first zero for a two-element interferometer, when the baseline B is much larger than the individual aperture diameter, is λ/B. It is helpful to consider the difference between the systems in l-space as well. For an interferometer the window function peaks at $l = 2\pi B/\lambda$. For an imaging system with a Gaussian beam the window function is $W_l = e^{-l^2\sigma^2}$. The beamwidth $\sigma = 0.42$ FWHM and FWHM $= (1.02+0.0135T_e)\lambda/D$ where T_e is the edge taper of the antenna in dB [15]. For an edge taper of -40 dB (typical for CMB instruments), FWHM $= 1.51\lambda/D$, $\sigma = 0.66\lambda/D$ and the window function falls to 10% of its peak value at $l = 2.29D/\lambda$, which is less than half of the peak l-value for an interferometer baseline of the same size.

This angular resolution factor is important because the size of the aperture is a cost-driver for the CMBPol mission. Angular resolution is important for CMB polarization measurements in two ways. First, imperfections in the shape and pointing of beams couple the CMB temperature anisotropy into false polarization signals. These problems can be reduced significantly if the CMB is smooth on the scale of the beam size, which happens for beams smaller than $\sim 10'$ [16]. Second, removing contamination of the tensor B-mode signal by B-modes from weak lensing requires maps of the lensing at higher angular resolution than the scale at which the tensor B-modes peak [17].

2.2. No Rapid Chopping and Scanning

Imaging systems with either coherent or incoherent detectors typically use some form of "chopping," either by nutating a secondary mirror or by steering the entire primary at a rate faster than the $1/f$ noise in the atmosphere and detectors. Similar approaches are used with arrays of detectors. When using an imaging system to form a two-dimensional (2D) map with minimal striping or other artifacts, the scan method must move the beam (or beams) on the sky at a rapid rate. Interferometers provide direct 2D imaging and do not require such scanning strategies. In the interferometer, only correlated signals are detected, so it has reduced sensitivity to changes in the total power signal absorbed by the detectors [18].

2.3. Clean Optics

The simplicity of an interferometric optical system eliminates numerous systematic problems that plague imaging optical systems. Instead of a single reflector antenna, the interferometers discussed here use arrays of corrugated horn antennas. These antennas have extremely low sidelobes and have easily calculable, symmetric beam patterns. Furthermore, there are no reflections from optical surfaces to induce spurious instrumental polarization, an unavoidable problem for any system with imaging optics [19; 20]. In principle, one could construct an altenate instrument without reflective optics — an array of horn antennas, each coupled directly to a polarimeter, could view the sky directly [21]. Each horn aperture would be sized to provide the required angular resolution. However, such a system uses the aperture plane inefficiently. A single horn antenna in such an imaging system will have angular resolution $\sim 2\lambda/D$, where D is the horn diameter. An N - element interferometric horn array that achieves the same angular resolution will have a maximum baseline length of $B = D$ (and require the same aperture size), but will collect N modes of radiation from the sky and hence be more sensitive.

Another advantage over an imaging system is the absence of aberrations from off-axis pixels: all feed elements are equivalent for the interferometer. In contrast to an imaging system, the field-of-view (FOV) of an interferometer is determined by the primary beamwidth of the array elements, not by beam distortion and cross-polarization at the edge of the focal plane. One can choose to increase the sensitivity of the instrument by collecting more modes (optical throughput) of radiation from the sky. In the interferometer this can be done by adding additional antennas; the only limitation is the size of the aperture plane rather than optical aberrations in the focal plane. The largest usable FOV for an off-axis Gregorian reflector is approximately $7°$ [22]. See Table 1 for a comparison of imaging and interferometric optical systems.

2.4. Direct Measurement of Stokes Parameters

Interferometry solves many of the problems related to mismatched beams and pointing errors raised by Hu *et al.* [16]. This advantage arises because interferometers measure the Stokes parameters directly, without differencing the signal from separate detectors.

An interferometer measures the Stokes parameters by correlating the components of the electric field captured by each antenna with the components from all of the other antennas. If the output of each antenna is split into E_x and E_y by an orthomode transducer (OMT), on the baseline formed by two antennas, 1 and 2, the interferometer's correlators measure $\langle E_{1x}E_{2x}\rangle$, $\langle E_{1y}E_{2y}\rangle$, $\langle E_{1x}E_{2y}\rangle$, and $\langle E_{1y}E_{2x}\rangle$. The first two are used to determine I and the latter two measure U. Rotating the instrument allows a measurement of Q. Stokes V can be recovered in a similar manner. Alternatively, the antenna outputs can be separated into left- and right-circular polarization components by a combination of an OMT and a polarizer. Correlating these signals also allows recovery of all four Stokes parameters. DASI uses a switchable polarizer to accomplish this [23].

Hu *et al.* [16] have reviewed systematic effects relevant to CMB polarization measurements, mainly in the context of imaging instruments. Bunn [24] performed similar calculations for interferometers. Table 2 outlines a variety of systematic errors and how they can be managed in imaging and interferometric instruments. The relative importance of these effects in interferometric systems is different from imaging systems; some sources of systematic error in imaging systems are dramatically reduced in interferometers. As examples we consider the effects of pointing errors and mismatched antenna patterns.

In a traditional imaging system, the Stokes parameters Q and U are determined by subtracting the intensities of two different polarizations. For example, Q might be measured by splitting the incoming radiation into x and y polarizations, determining the intensities I_x and I_y of the two polarizations, and subtracting. In such an experiment, any mismatch in the beam

Journal of Physics: Conference Series **155** (2009) 012003 doi:10.1088/1742-6596/155/1/012003

Table 1. Comparison of three optical designs for CMBPol. To achieve the same angular resolution each instrument requires different aperture diameters, D (column 2). The aperture diameter required by each system to achieve an angular resolution of 1° at wavelength $\lambda = 3$mm is given in column 3. Each instrument can provide the field-of-view (FOV) listed in column 4 and hence provide different amounts of throughput. Throughput is proportional to the number of modes (column 5) and hence determines the instrument sensitivity. We make the following assumptions. For the Gregorian, the edge taper of the illumination on the primary mirror is assumed to be −40dB. The diameter of the usable FOV is given for an optimized system [19; 20]. The number of modes is approximately [FOV/(angular resolution)]^2, assuming all the modes reaching the focal plane are coupled to detectors. For the horn array ('feed farm'), the horn diameter = D and determines both the angular resolution and the FOV. For the interferometric horn array, $D = B$, the diameter of a close-packed array of horns, each of diameter d, and the number of modes is given by the number of horns $\sim (D/d)^2$. In principle, the field of view of the interferometer, and hence the number of modes collected, can be made arbitrarily large by the use of small antenna apertures. However, for interferometers which do not subdivide their bands the usable bandwidth is $\sim d/D$. (See section on *'Bandwidth Smearing'* below. Bandwidth smearing is not a limitation for coherent adding interferometers because the bands are easily subdivided. Here we assume a bolometric adding interferometer with $D/d = 8$.) The interferometer makes the most efficient use of aperture size – for a fixed aperture diameter , the interferometer has higher angular resolution and more throughput than the other systems.

Instrument	*Angular resolution* (FWHM)	*Aperture D* (m)	*FOV* (rad.)	*# Modes*
Gregorian telescope	$1.51\lambda/D$	0.26	~ 0.12	49
Horn array	$2\lambda/D$	0.34	$2\lambda/D$	1
Interfer. horn array	$\lambda/2D$	0.09	$2\lambda/d$	64

patterns used to determine I_x and I_y (including differential pointing errors as well as different beam shapes) will cause leakage from total power (T) into polarization (Q, U).

In an interferometer, these signals are multiplied together to obtain a visibility. In such a system, mismatched beams do not lead to leakage from temperature into polarization. Suppose that the signal entering each horn of an interferometer is split into horizontal and vertical polarizations. Working in the flat-sky approximation, let $E_{ix}(\hat{r})$ and $E_{iy}(\hat{r})$ stand for the x and y components of the electric field of the radiation entering the ith horn from position $\hat{r}$ on the sky. The signals coming out of each horn are averages of the incoming electric fields weighted by some antenna patterns $G_{i(x,y)}(\hat{r})$. To measure the Stokes parameter U, for example, we would multiply the x signal from horn i with the y signal from horn j to obtain the visibility

$$V_{ij}^U = \int d\hat{r}_1\, d\hat{r}_2\, G_{ix}(\hat{r}_1) G_{jy}(\hat{r}_2) \langle E_{ix}(\hat{r}_1) E_{jy}^*(\hat{r}_2)\rangle.$$

The angle brackets denote a time verage. The electric fields due to radiation coming from two different points on the sky are uncorrelated, and the product of x and y components of the electric field gives the Stokes U parameter:

$$\langle E_{ix}(\hat{r}_1) E_{jy}^*(\hat{r}_2)\rangle = U(\hat{r}_1) e^{2\pi i \vec{u}\cdot\hat{r}_1} \delta(\hat{r}_1 - \hat{r}_2),$$

so the visibility is

$$V_{ij}^U = \int d\hat{r} G_{ix}(\hat{r}) G_{jy}(\hat{r}) U(\hat{r}) e^{2\pi i \vec{u}\cdot\hat{r}}.$$

Journal of Physics: Conference Series **155** (2009) 012003 doi:10.1088/1742-6596/155/1/012003

Table 2. A Comparison of Systematic Effects

Systematic Effect	*Imaging System Solution*	*Interferometer Solution*
Cross-polar beam response	Instrument rotation & correction in analysis	Instrument rotation & non-reflective optics
Beam ellipticity	Instrument rotation & small beamwidth	No T to E and B leakage from beams; inst. rot'n
Polarized sidelobes	Correction in analysis	Correction in analysis
Instrumental polarization	Rotation of instrument & correction in analysis	Clean, non-reflective optics
Polarization angle	Construction & characterization	No T to E and B leakage from beams; construction & characterization
Relative pointing	Rotation of instrument & dual polarization pixels	No T to E and B leakage from beams; inst. rot'n
Relative calibration	Measure calibration using temperature anisotropies	Detector comparison not req'd for mapping or measuring Q and U
Relative calibration drift	Control scan-synchronous drift to 10^{-9} level	All signals on all detectors
Optics temperature drifts	Cool optics to $\sim$3 K & stabilize to $< \mu$K	No reflective optics
$1/f$ noise in detectors	Scanning strategy & phase modulation/ lock-in	Instant. measurement of power spectrum without scanning
Astrophysical foregrounds	Multiple frequency bands	Multiple frequency bands

Note that the visibility V_{ij}^U does not contain any contribution from the total intensity (Stokes I), *even if* the two antenna patterns are different. This means that differential pointing errors and different beam shapes for different antennas do not cause leakage from T into E and B. Antenna pattern differences do cause distortion of the observed polarization field, so errors in modeling beam shapes and pointing may cause mixing between E and B.

Coupling between intensity and polarization will arise if the beams have cross-polar contributions. In that case, the visibility V_{ij}^U, which is supposed to be sensitive to just polarization, will contain contributions proportional to $\langle E_x E_x^* \rangle$ and $\langle E_y E_y^* \rangle$, to which Stokes I does contribute.

The same considerations apply if the incoming radiation is split into circular rather than linear polarization states. The visibility V_{ij}^{RL}, obtained by interfering the right-circularly-polarized signal entering horn i with the left-circularly-polarized signal entering horn j, contains only contributions from Q and U if the beams are co-polar, even if the two horns have different beams. Again, cross-polarity induces leakage from intensity into polarization. In short, in an interferometer, beam mismatches are less of a worry than for imaging systems.

2.5. Separation of E and B Modes

A significant challenge in CMB polarization measurements is separation of the very weak B modes from the much stronger E modes. Unless a full-sky map is made with infinite angular resolution, the two modes "leak" into each other [25; 26]. It has been shown [27; 28], however, that an interferometer can separate the E and B modes more cleanly than can an imaging experiment although detailed calculations of this advantage in realistic simulations remain to be done.

2.6. Foreground Removal
Foreground removal can occur in visibility space. Removing foregrounds directly from visibility data has been studied in another context [29]. CMBPol will undoubtedly have to measure foregrounds itself, without relying on other instruments. A set of interferometer modules scaled in proportion to wavelength and operating from ~ 30 to 300 GHz would provide a set of visibility measurements with identical $u - v$ coverage so that foregrounds could be studied and removed in visibility space.

3. Multiplying Interferometry
In a simple 2-element radio interferometer, signals from two telescopes aimed at the same point in the sky are correlated so that the sky temperature is sampled with an interference pattern with a single spatial frequency. The output of the multiplying interferometer is the visibility (defined in the last section). With more antennas these same correlations are performed along each baseline. To recover the full phase information, complex correlators are used to measure simultaneously both the in-phase and quadrature-phase components of the visibility. All interferometers used for CMB studies so far are multiplying interferometers and use use coherent receivers. They can use either analog or digital correlators.

Analog correlators. DASI and CBI use analog correlators. They first amplify the RF signals from each of the N antennas, downconvert in frequency with a mixer, and then split the signals $N - 1$ ways. The correlator then combines these signals in a pairwise fashion to measure visibilities for $N(N-1)/2$ baselines. For DASI and CBI $N = 13$, and the number of baselines is 78. But for $N = 100$, however, there would be 4950 baselines and this type of correlator is not yet feasible (although correlators could be made to correlate only a fraction of the possible baselines).

Digital correlators. Coherent interferometers typically downconvert the RF signal to an IF signal, digitize it, and then correlate pairwise with a digital correlator. Such correlators are under development for combining IF signals with $N \sim 200$ antennas with low power requirements [3]. The sensitivity of such interferometers is limited by their bandwidth; the bandwidth of these correlators is currently 1.4 GHz.

4. Adding Interferometry: Overview
An alternate approach is to use adding interferometry. Adding interferometry has the advantage that beam combiners that can be scaled to large bandwidths and large N are feasible and either coherent receivers or incoherent detectors (bolometers) can be used. Bolometers have the advantage of operating over the entire range of millimeter wavelengths of interest for CMB studies. In addition, they have comparable sensitivity to coherent receivers below ~ 90 GHz and better sensitivity at higher frequencies. The high-frequency sensitivity advantage improves in low background environments (balloons and space). When used without amplifiers, the main challenge to adding interferometry is combining the signals from the multiple antennas without sacrificing signal-to-noise. The beam combination is necessarily performed by analog correlators.

In adding interferometers the electric field wavefronts from two antennas are added and then squared in a detector [30]. (See Figure 1 for an example of a 2-element adding interferometer.) The result is a constant term proportional to the intensity plus an interference term. The constant term is an offset that is removed by phase-modulating one of the signals. Phase-sensitive detection at the modulation frequency recovers both the in-phase and quadrature-phase interference terms and reduces susceptibility to low-frequency drifts ($1/f$ noise) in the detector and readout electronics. The adding interferometer recovers the same visibility as a multiplying interferometer.

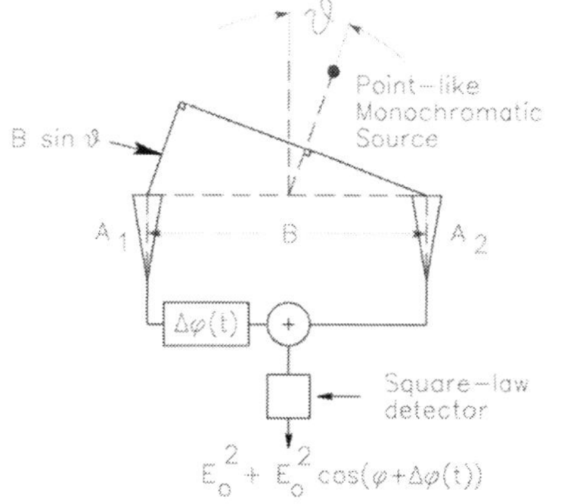

Figure 1. Adding interferometer with $N = 2$ antennas. At antenna A_2 the electric field is E_0, and at A_1 it is $E_0 e^{i\phi}$, where $\phi = kB \sin\alpha$ and $k = 2\pi/\lambda$. B is the length of the baseline, and α is the angle of the source with respect to the symmetry axis of the baseline, as shown. (For simplicity consider only one wavelength, λ, and ignore time dependent factors.) In a multiplying interferometer the in-phase output of the correlator is proportional to $E_0^2 \cos\phi$. For the adding interferometer, the output is proportional to $E_0^2 + E_0^2 \cos(\phi + \Delta\phi(t))$. Modulation of $\Delta\phi(t)$ allows the recovery of the interference term, $E_0^2 \cos\phi$, which is proportional to the visibility of the baseline.

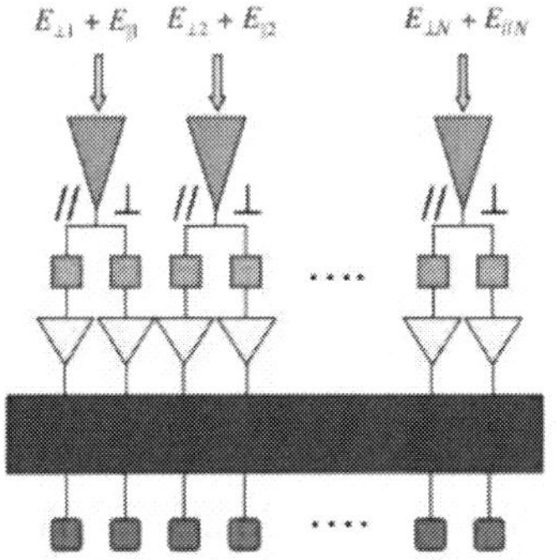

Figure 2. Block diagram of an adding interferometer with $N > 2$. Each phase shifter is modulated in a sequence that allows recovery of the interference terms (visibilities) by phase-sensitive detection at the detectors. The signals are mixed in the beam combiner and detected. The beam combiner can be implemented either using guided waves (e.g. in a Butler combiner) or quasioptically (Fizeau combiner). The top triangles represent corrugated conical horn antennas. Orthomode transducers (OMTs) allow all the Stokes parameters to be determined simultaneously. For the case of an interferometer using coherent receivers, amplifiers and/or mixers could be placed before the beam combiner. (Figure modified from Charlassier [31].)

In an interferometer with an array of $N > 2$ antennas, the signals are combined in such a way that interference fringes are measured for all possible baselines ($N(N-1)/2$ antenna pairs). This combination can occur in two different ways: "pairwise" combination (analogous to a Michelson stellar interferometer) or "all-on-one" combination [32].

Pairwise combination involves splitting the power from each of the N antennas in the array $N-1$ ways, adding the signals in a pairwise fashion, and then squaring the signals and separating out the interference term as described above. In optical systems this approach is analogous to Michelson stellar interferometry. This approach has the disadvantages of producing extremely low signal levels at each detector and requiring $N(N-1)/2$ extremely sensitive detectors.

In all-on-one combination the signals from each of the antennas are split and then combined in such a way that linear combinations of all the antenna signals are formed at each of the outputs of the combiner (Figure 2). This scheme avoids the problem of large numbers of detectors and low signal levels. To allow all the Stokes parameters to be determined simultaneously, orthomode transducers (OMTs) are inserted after the antennas. An example of a beam combiner that uses guided waves is the Butler combiner. The Butler combiner delivers the signals from $2N$ antenna outputs to $2N$ detectors. Each detector squares these amplitudes, creating interference signals from all baselines simultaneously on each detector. Effectively, the signals from all baselines are multiplexed onto each of the $2N$ detectors. Only $2N$ detectors are required, rather than the $2N(2N-1)/2$ detectors required for pairwise combination. Butler combiners are commonly used for phased array antennas with coherent systems using either waveguide or coaxial techniques.

Journal of Physics: Conference Series **155** (2009) 012003 doi:10.1088/1742-6596/155/1/012003

The optical analog is Fizeau combination, which is typically used for incoherent systems at optical wavelengths and has lower loss than Butler combiners. In a coherent system, with amplifiers, the Butler or other guided-wave approaches are attractive options for forming a large-N interferometer.

In the Fizeau/Butler approach, the signals from redundant baselines can be added together to improve the signal - to - noise ratio at each bolometer compared to the pairwise combination case [31]. The signals reaching each bolometer are multiplexed in such a way that a portion of the visibility of each baseline appears at each bolometer. When the signals are combined the resultant sensitivity is comparable to that of a filled-dish with an array of bolometers coupled to the same number of modes (N) on the sky [33].

These systems actually interfere antenna signals in two modes of operation. In one mode, signals from different antennas are interfered to measure the visibility for each baseline. Each visibility selects a narrow range of l values and has no response to very low multipoles. In the second mode, signals from each antenna are combined with other signals from the same antenna (autocorrelation) to form a correlation polarimeter. This latter mode has lower angular resolution than the first, but is essential for measuring large spatial features (low-l).

5. Adding Interferometry: Details

Here we discuss each component of an adding interferometer and explain how it recovers the Stokes visibilities. Much of this section comes directly from Charlassier *et al.* [31] and refers to Fig. 2.

5.1. *Horns*

We assume that the instrument observes the sky through N input horn antennas placed in a coplanar array Each points towards the same direction on the sky.

5.2. *Equivalent baselines*

If two baselines b and b' are such that $\vec{u}_b = \vec{u}_{b'}$, then the phase shifts associated with the two baselines are equal. All baselines b such that $\vec{u}_b = \vec{u}_\beta$ form a class of equivalent baselines associated with mode $\vec{u}_\beta$ in visibility space. For all baselines b belonging to the same class β, the phase difference between the two horns i and j is the same. The number of different classes of equivalent baselines depends on the array, and the number of different baselines in an equivalence class also depends on the particular class.

5.3. *Polarization splitters*

There is an OMT at the output of each horn which separates the radiation into two orthogonal components noted $\parallel$ and $\perp$. Each horn therefore has two outputs measuring the electric field integrated through the beam in the two orthogonal directions. The OMT can split out either two linear polarizations or two circular polarizations.

5.4. *Phase-modulators*

Phase-modulators placed on each of the outputs allow the phase of the electric field to be shifted by a given angle that can be chosen and controlled externally. The signals are modulated in order to measure the various visibilities on each detector.

5.5. *Amplifiers*

In the case of an adding interferometer that uses coherent receivers, low noise amplifiers (e.g. HEMT amplifiers) would be placed before the beam combiner.

Journal of Physics: Conference Series **155** (2009) 012003 doi:10.1088/1742-6596/155/1/012003

5.6. Beam combiner
In the beam combiner, $2N$ input channels are combined to create $N_{\rm out}$ output channels that are linear combinations of the inputs. To conserve the input power in an ideal lossless device, the number of output channels $N_{\rm out}$ has to be at least equal to the number of input channels $2N$. The beam combiner could use either guided waves (waveguides or planar transmission lines), or a Fizeau combiner, which uses quasi-optical techniques. One type of guided-wave beamformer (for a review see [34]) is the Butler combiner, in which signals from N input ports are combined with fixed phase relationships to create signals at $\geq N$ output ports. Another guided-wave combiner is the Rotman lens (Figure 3). Guided-wave structures can form beams in one-dimension or in two dimensions [35]. Building guided-wave combiners with low-loss and well-controlled phases is difficult at millimeter-wavelengths — they probably would only be suitable for an adding interferometer that uses coherent amplifiers to overcome the beam combiner losses. The Fizeau combiner has lower loss and is more easily scaled to large numbers of inputs. An example of such a system is described in Tucker *et al.* [36] and in Section 7.

5.7. Total power detector
The signal from each of the outputs of the combiner is detected (with bolometers if amplifiers are not used) through its total power averaged on time scales given by the time constant of the detector. The time series from each detector contains voltages proportional to the visibilities from each of the baselines formed by the array. They also contain signals from the autocorrelation of the two polarizations measured by single horns. Each visibility and autocorrelation signal is modulated (by the phase shifters) in such a way that they can be separated from each other by demodulation of the detector signals. The demodulated visibility and autocorrelation signals from each detector are co-added. The sensitivity of an adding interferometer that uses bolometers as detectors has been calculated in detail [33]. The bolometric interferometer has sensitivity comparable to that of an imaging system that uses bolometers and couples to the same number of modes on the sky (i.e. that has the number of detectors equal to the number of antennas in the interferometer array).

6. Adding Interferometry: Systematic Effects and Challenges
Some of the advantages of interferometry for controlling systematic effects were discussed in Section 2. Here we focus on some systematic effects and challenges that are specific to adding interferometers.

6.1. Phase modulation
Perhaps the most significant technical challenge for the adding interferometer is the phase modulator. The difference in loss in the different phase states must be small and stable or else the phase modulation will couple a portion of the total power signal on the bolometers into Q, U, and V when the bolometer signals are demodulated. This challenge represents the most significant difference between adding interferometers and traditional multiplying interferometers.

In addition, the phase modulation must allow the signals from redundant (equivalent) baselines to be read out simultaneously when the bolometer signals are demodulated. Charlassier *et al.* [31] and Hyland *et al.* [37] use phase shift sequences which are a type of time-division multiplexing. For large arrays the number of steps in the switching sequence can become very large. The sequence length is minimized when the phase modulator can switch between multiple discrete phase angles between 0 and 360 degrees. For example, for an $8\times$ square array with phase modulators capable of switching between 15 equally spaced phase angles requires a sequence of 675 phase states. In order to cycle through the full sequence faster than the 1/f knee in the detector response requires rapid phase modulation ($\sim$1 kHz) and hence detectors with short time constants. Alternatively, with phase modulators capable of continuous phase shifts,

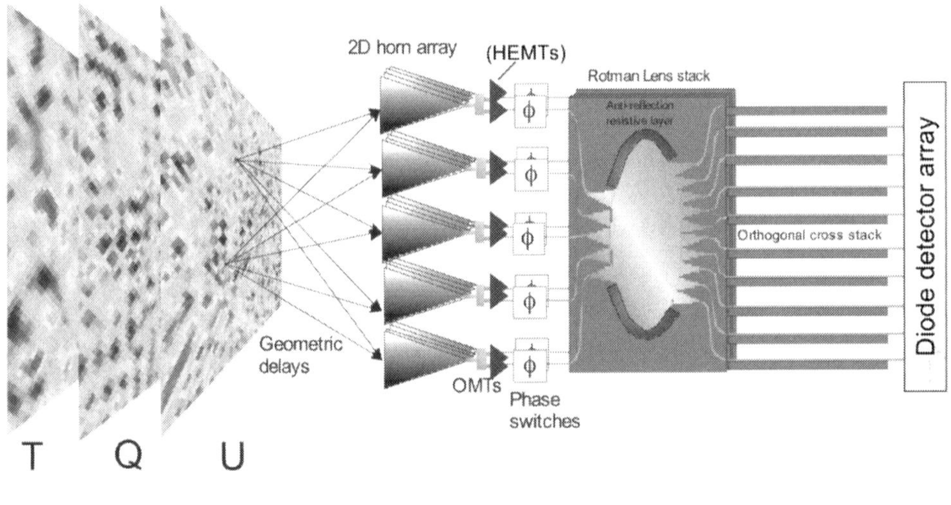

Figure 3. Idealised Rotman interferometer scheme showing input TQU maps and the paths through to the detectors. The signal from each sky element reaches each horn with different geometric delays. OMTs split the signals into different polarization components which are then phase-switched independently before being input to the first Rotman stack. For a system using coherent amplifiers, HEMT amplifiers could be placed just after the OMTs. Rotman lenses combine the input signals from each column of horns. An orthogonal stack of Rotman lenses combines the outputs of the first stack. Demodulation of the detected signals recovers the visibilities from all baselines. The Rotman lens is a planar guided-wave structure formed by two parallel conducting plates. Input signals launched into the combiner couple to output ports with fixed phase relationships. Figure courtesy of L. Piccirillo and R. A. Watson.

frequency-division multiplexing is possible. In this case each phase is modulated sinusoidally at a different frequency. The frequencies can be chosen so that signals from equivalent baselines are modulated at the same frequency on all bolometers [38]. In this scheme all visibility signals can be modulated at frequencies much higher than the detector $1/f$ knee or satellite scan frequency.

6.2. *Bandwidth smearing*

The sensitivity of a receiver to broadband signals increases as the square root of the bandwidth. For interferometers, the bandwidth restricts the angular range, θ, over which fringes are detected [39], [40]. If we assume the path lengths for a source at the center of the FOV are equal, then the path length difference for a source at an angle θ from the center along the baseline axis is θB, where B is the baseline distance (see 1.) If this path length difference is small compared to the coherence length of the light, then the fringe contrast is not affected. For a point source, the coherence length is $\lambda^2/\Delta\lambda$, and the FOV is determined by $\theta_{\rm FOV} \leq (\lambda/\Delta\lambda)(\lambda/B)$. This equation indicates that for angles of the order of the product of the spectral resolution times the angular resolution, the fringe smearing is important. This relation imposes restrictions on the ratio between the maximum baseline achievable by the interferometer and the spectral bandwidth of the receiver. However, for a diffuse source (i.e.of the CMB) the constraint on the bandwidth is relaxed. We have performed a simulation of the fringe smearing for the longest baseline of an interferometer with an 8×8 packed array of 8° FWHM horns. With a bandwidth of 20%, fringe smearing decreases the sensitivity by 40%.

6.3. Bandwidth and ℓ-space resolution

Another consequence of interferometers with large bandwidths is that the ℓ-space resolution for a measurement of a bandpower is $\Delta\ell/\ell = \Delta\lambda/\lambda$. This problem is overcome with coherent interferometers (both multiplying and adding) by splitting the RF or IF band into sub-bands before the correlation or detection occurs. In principle, band sub-division could be used with bolometric interferometers but at the expense of increasing the number of bolometers and associated readout electronics.

6.4. Simulations

Simulations will be essential for optimizing interferometer designs and observing schemes and for fully understanding the impact of systematic effects on estimates of the power spectrum. We are aware of two programs to develop such simulations — one at APC/University of Paris VII [31] and another at Manchester [38]. The APC simulation is capable of recovering the Stokes parameter visibilities for realistic adding interferometer designs that incorporate a Fizeau beam combiner. The following design parameters can be adjusted: number and location of the input antennas (horns) in the aperture plane, number and location of the detectors in the "fringe plane," the phase shifter sequences used for recovering the visibilities, and the focal length of the Fizeau combiner. Bandwidth effects such as those mentioned above are now being included. Studies that could be carried out include effects of asymmetric beam patterns on the sky, low-frequency stability, spectral band shape, cross-talk between antennas, calibration, instrumental polarization, etc. The Manchester simulation performs similar calculations but for the case of guided-wave beam combiners.

6.5. Fringe rotation

Interferometers with independently mounted, tracking antennas (like VLA or SZA) enjoy a modulation of the signal caused by the rotation of the earth. This modulation is different for sources in the sky than for sources on the ground and provides a powerful tool for interferometers to reject ground-spill. On the other hand, co-mounted interferometers (DASI and CBI) do not have this advantage. Large arrays with hundreds of antennas will almost certainly have to be co-mounted.

6.6. Cross-coupling

Some coupling between the antennas in a close-packed array will occur. This effect will lead to correlated signals that will be modulated by the phase modulators, and hence to an offset in the demodulated signals. DASI used cylindrical baffles around its horn antennas to reduce the cross-coupling. Further study is needed to determine the implications of this systematic effect.

7. EPIC Mission Concept Study

The EPIC mission concept study for the Einstein Inflation Probe focused on the possibility of using a bolometric adding interferometer. Figure 4 shows a possible configuration for a

Journal of Physics: Conference Series 155 (2009) 012003 doi:10.1088/1742-6596/155/1/012003

bolometric interferometer module for EPIC. The array views the sky through a close-packed cluster of corrugated horn antennas. The two polarizations (either linear or circular) are split by an ortho-mode transducer and individually phase-modulated (Fig. 5). The beams are then combined with a Fizeau combiner in the form of a cold, compact, on-axis Cassegrain telescope. Interference fringes formed by the various antenna baselines appear on the bolometer array in the focal plane of the telescope. The superimposed fringes are separated from each other using a phase modulation sequence that uniquely encodes each visibility (Figs. 6, 7). A prototype, the Millimeter-wave Bolometric Interferometer (MBI) has been constructed and is undergoing testing [36].

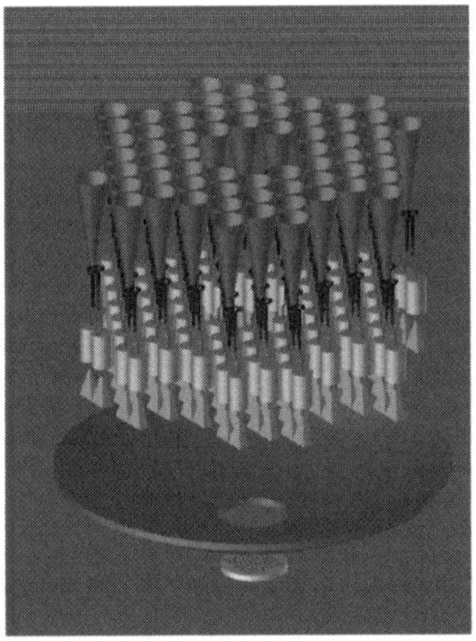

Figure 4. A three-dimensional view of 64 corrugated horn antennas arranged in a close-packed array illuminating a Fizeau combiner. The detector array sits behind the primary mirror of the beam combiner. Note that the distances between the antennas, primary mirror and detector array are not to scale. EPIC could be made of a cluster of these fundamental modules, with multiple copies operating at frequencies from 30 GHz to 300 GHz

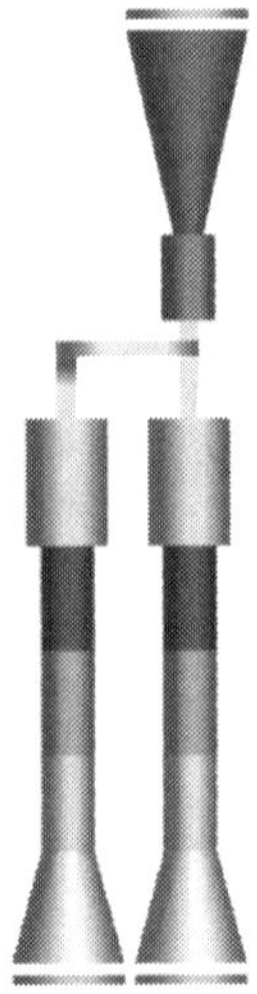

Figure 5. Input unit (IU) of the EPIC interferometer. The two polarizations are separated using an orthomode transducer and are rotated in waveguide (WG) so that the two polarization vectors are aligned. A $\pm 90°$ phase modulation is introduced in one of the arms and the two signals are directed at the Fizeau combiner. The interference of the two signals from an IU results in a correlation receiver, instantaneously sensitive to the Stokes U parameter. The interference of signals from different IUs results in an interferometer.

The EPIC mission concept includes multiple close-packed arrays of horn antennas that are co-aligned and pointed directly at the sky, with no intervening lenses or reflectors. Each array is configured as an adding interferometer using the beam combination scheme of Figure 4. The interferometer measures the visibilities from all baselines in the array. In addition, the phase modulators can be operated in such a way that the signals from each antenna interfere with themselves. In this mode the system acts as an array of correlation polarimeters, sensitive to Q and U averaged over a single antenna beam. The correlation polarimeter mode is used to measure the lowest spherical harmonics, while the interferometric mode recovers the higher-order multipoles. Both of these modes can operate simultaneously.

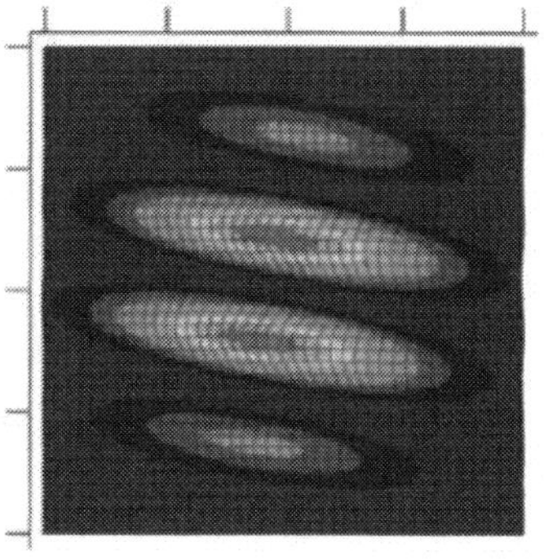

Figure 6. Simulation of fringe patterns formed in the focal plane of the Fizeau beam combiner from a single baseline.

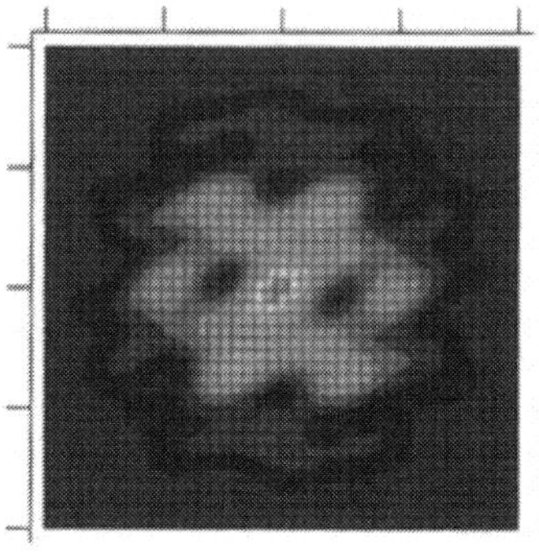

Figure 7. Superposition of fringes from 6 baselines (as expected in MBI). Fringes are separated by phase modulation sequence.

In an interferometer, each individual pointing covers a large sky area and samples many different baselines simultaneously, potentially reducing systematic errors in map-making. The relatively simple configuration of the EPIC instrument may allow for an additional degree of freedom in the scan using rotation of the instrument. If the low-l modes are recovered by using the instrument in a correlation receiver configuration, then scan-strategy issues similar to those of an imaging system may arise.

Figure 8 displays the sensitivity for one possible configuration under study for EPIC. There is a total of 16 arrays, each including 64 close-packed corrugated horn antennas, for a total of 1024 horns. Each horn has a beam width of 15°. Each array operates in wide bands (~20%) centered at frequencies between ~30–300 GHz. There are 8 arrays sensitive to 90 GHz, the primary science channel. The other 8 arrays are for measuring and removing foregrounds; they are not included in the sensitivity estimate. The instrument and observing patterns have not been optimized. The exact band placements and number of bands will be chosen to optimize the removal of foreground contamination. The detectors are cold (~100 mK), background-limited superconducting transition-edge sensors (TES) read out by SQUID multiplexers. The instrument efficiency is taken to be 50%. Emission from the cryostat window dominates the optical loading on the detectors. EPIC surveys the full sky with a combination of instrument rotation and precession.

8. Technology Readiness Assessment

Although no adding interferometers have been used for CMB measurements the technologies required for building such an instrument are not very different from those required for imaging systems. We list here the critical components and some of the required specifications. See Table 3. Most of these components are discussed in more detail in other white papers for this workshop.

Horn arrays: Close-packed horn antenna arrays with ~ 100 elements are required for each wavelength, from ~ 30 GHz to 300 GHz. Lightweight platelet arrays of corrugated horns are an attractive option [43]. Recent developments in smooth-walled horns (similar to Potter horns) may offer comparably symmetric beams with low sidelobes and low cross-polar response with lower mass and easier fabrication [44]. Currently these horns are limited to ~ 15 % bandwidth.

Phase modulators: As mentioned above, these components are critical to the success of adding interferometry. Differential loss between the different phase states must be small and stable to reject the total power signal on the bolometers. Rapid switching and settling is necessary to

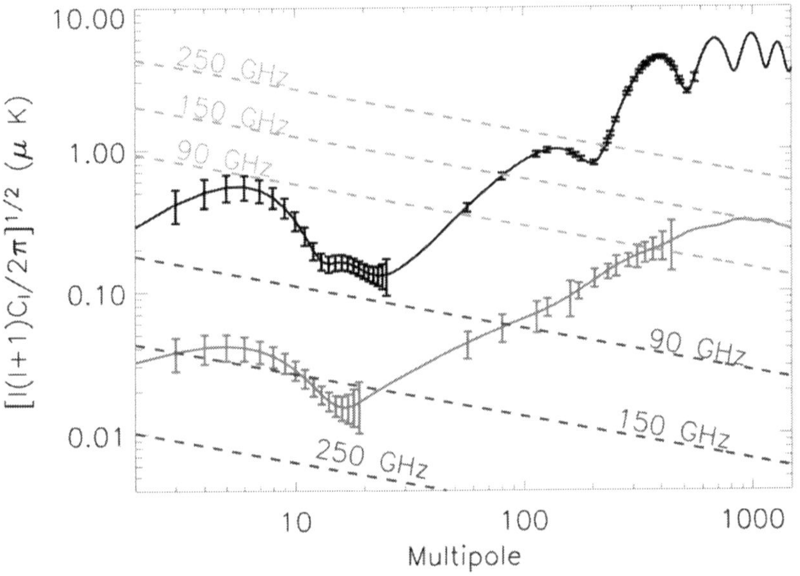

Figure 8. Expected sensitivity of EPIC, a mission concept for the Einstein Inflation Probe, to E (black) and B (red) polarization. The power spectra are based on the best-fit model from *WMAP* [41]. The tensor-to-scalar ratio is taken to be 0.01. Errors (1σ) assume one year of integration sampling the full sky uniformly. The estimates are representative of the capabilities of possible designs for the CMBPol. The configuration assumed here includes 1024 feed horns, with 512 sensitive to 90 GHz, the primary science channel; the other 512 feed horns are for measuring and removing foregrounds and are not included in this estimate. The dotted lines show the expected levels of polarized dust emission and the dashed lines show the expected levels of polarized synchrotron emission at 90 GHz, 150 GHz and 250 GHz based on the WMAP results [42]. EPIC operates both as an imaging instrument and an interferometer; low-l points come from operating the interferometer as single-beam correlation radiometers while high-l points come from operating the instrument as an interferometer.

accommodate long switching sequences. Low power dissipation is also required. Ferrite rotation modulators [45] [36] are one possibility. Other promising options include MEMs devices [46] and varactor diodes [47]. MEMs and varactor diode phase modulators most naturally are implemented in a planar transmission line structure (e.g. microstrip or coplanar waveguide) which in some interferometer schemes would require undesirable transitions into and out of waveguide. Alternatively, these devices could be implemented in finline, for which wide-band and low-loss transitions to waveguide are well developed.

Beam combiners: At millimeter wavelengths only quasi-optical beam combiners offer low enough loss to be used with bolometric detectors. Guided wave combiners are suitable for adding interferometers with amplifiers.

Amplifiers: For a coherent adding interferometer low noise HEMT amplifiers are required. The WMAP and PLANCK programs have advanced these to a high TRL. The power and cooling requirements for HEMT receivers are discussed in Lawrence *et al.* [3].

Detectors : For bolometric adding interferometers, bolometer arrays operating at the background limit are required. Because the number of detectors for each interferometer $\sim 4\times N_h$, the power loading on each bolometer is $\sim 1/4$ the loading from a single mode looking at the CMB. Hence, detector noise must be even lower than for bolometers used for imaging systems.

Journal of Physics: Conference Series **155** (2009) 012003 doi:10.1088/1742-6596/155/1/012003

Table 3. Technology Readiness Levels for Adding Interferometers

Component	*TRL*	*Heritage*
Horn Antennas		
— corrugated horn antennas	9	WMAP & COBE
— platelet arrays	5	QUIET
— smooth-wall horn arrays	5	
OMT (< 110 GHz)	9	WMAP
OMT (150 GHz)	4/5	CℓOVER
Phase Modulator		
— ferrite phase modulator (90 GHz)	6	BICEP & MBI
— MEMs/SIS phase modulators	2/3	
Beam Combiner		
— quasi-optical beam combiner	5	MBI
— guided-wave beam combiner	5	
Focal Plane Arrays		
— NTD Ge bolometers	8	Planck & Herschel
— TES bolometers	6	SCUBA, GBT, EBEX, GISMO
LHe cryostat	9	Spitzer, ISO, Herschel, COBE
Sub-K cooler: single-shot ADR	9	ASTRO-E2

Arrays of ~ 400 detectors are required. In order to capture all of the radiation arriving at the fringe plane, these must be absorber-coupled detectors (rather than antenna-coupled). Excellent examples would be the BUG arrays developed for GISMO [48] or spider web bolometers similar to those used for PLANCK [49].

Cryogenics: For bolometric adding interferometers the detectors must be cooled to ~ 100 mK in order to be limited by photon noise. Suitable coolers include ADRs and dilution refrigerators [50]. For a coherent adding interferometer using HEMT amplifiers the cooling requirements are of course considerably easier to handle; passive cooling in space may be sufficient.

9. Conclusion

Adding interferometry is a viable approach to B-mode searches and offers an attractive alternative to imaging techniques. The most critical technology in need of development is phase modulation. The other necessary technologies are similar to those required for CMB imaging systems.

Acknowledgments

We thank the members of the MBI and BRAIN collaboration, who are responsible for most of the ideas presented here. This work has been partially supported by NASA Grants NNX07AG82G and NNG04GI15G and by the Rhode Island Space Grant and Wisconsin Space Grant.

References

[1] Kovac J M, Leitch E M, Pryke C, Carlstrom J E, Halverson N W and Holzapfel W L 2002 *Nature* **420** 772–787 (*Preprint* astro-ph/0209478)

[2] Tucker G S and Timbie P T 2008 *J. Phys.: Conf. Series - these proceedings*

[3] Lawrence C R, Church S, Gaier T, Lai R, Ruf C and Wollack E 2008 *J. Phys.: Conf. Series - these proceedings*

[4] Timbie P T, Tucker G S, Ade P A R, Ali S, Bierman E, Bunn E F, Calderon C, Gault A C, Hyland P O, Keating B G, Kim J, Korotkov A, Malu S S, Mauskopf P, Murphy J A, O'Sullivan C, Piccirillo L and Wandelt B D 2006 *New Astronomy Review* **50** 999–1008

[5] Birkinshaw M 1999 *Physics Reports* **310** 97–195 (*Preprint* astro-ph/9808050)

[6] Carlstrom J E, Holder G P and Reese E D 2002 *ARA &A* **40** 643–680 (*Preprint* astro-ph/0208192)

[7] Loh M, Carlstrom J E, Cartwright J K, Greer C, Hawkins D, Hennessy R, Joy M, Lamb J, Leitch E, Miller A, Mroczkowski T, Muchovej S, Pryke C, Reddall B, Richardson G, Runyan M, Sharp M and Woody D 2005 *American Astronomical Society Meeting Abstracts* **207** 41.01–+

[8] Baker J C, Grainge K, Hobson M P, Jones M E, Kneissl R, Lasenby A N, O'Sullivan C M M, Pooley G, Rocha G, Saunders R, Scott P F and Waldram E M 1999 *MNRAS* **308** 1173–1178 (*Preprint* astro-ph/9904415)

[9] Dickinson C, Battye R A, Carreira P, Cleary K, Davies R D, Davis R J, Genova-Santos R, Grainge K, Gutiérrez C M, Hafez Y A, Hobson M P, Jones M E, Kneissl R, Lancaster K, Lasenby A, Leahy J P, Maisinger K, Ödman C, Pooley G, Rajguru N, Rebolo R, Rubiño-Martin J A, Saunders R D E, Savage R S, Scaife A, Scott P F, Slosar A, Sosa Molina P, Taylor A C, Titterington D, Waldram E, Watson R A and Wilkinson A 2004 *MNRAS* **353** 732–746 (*Preprint* astro-ph/0402498)

[10] Halverson N W, Leitch E M, Pryke C, Kovac J, Carlstrom J E, Holzapfel W L, Dragovan M, Cartwright J K, Mason B S, Padin S, Pearson T J, Readhead A C S and Shepherd M C 2002 *ApJ* **568** 38–45 (*Preprint* astro-ph/0104489)

[11] Readhead A C S, Mason B S, Contaldi C R, Pearson T J, Bond J R, Myers S T, Padin S, Sievers J L, Cartwright J K, Shepherd M C, Pogosyan D, Prunet S, Altamirano P, Bustos R, Bronfman L, Casassus S, Holzapfel W L, May J, Pen U L, Torres S and Udomprasert P S 2004 *ApJ* **609** 498–512 (*Preprint* astro-ph/0402359)

[12] Leitch E M, Kovac J M, Halverson N W, Carlstrom J E, Pryke C and Smith M W E 2005 *ApJ* **624** 10–20 (*Preprint* astro-ph/0409357)

[13] Readhead A C S, Myers S T, Pearson T J, Sievers J L, Mason B S, Contaldi C R, Bond J R, Bustos R, Altamirano P, Achermann C, Bronfman L, Carlstrom J E, Cartwright J K, Casassus S, Dickinson C, Holzapfel W L, Kovac J M, Leitch E M, May J, Padin S, Pogosyan D, Pospieszalski M, Pryke C, Reeves R, Shepherd M C and Torres S 2004 *Science* **306** 836–844 (*Preprint* astro-ph/0409569)

[14] Cartwright J K, Pearson T J, Readhead A C S, Shepherd M C, Sievers J L and Taylor G B 2005 *ApJ* **623** 11–16 (*Preprint* astro-ph/0502174)

[15] Goldsmith P F 1998 *Quasioptical Systems* (IEEE Press)

[16] Hu W, Hedman M M and Zaldarriaga M 2003 *Phys. Rev. D* **67** 043004–+ (*Preprint* astro-ph/0210096)

[17] Knox L and Song Y S 2002 *Phys. Rev. Lett.* **89** 011303–+ (*Preprint* astro-ph/0202286)

[18] White M, Carlstrom J E, Dragovan M and Holzapfel W L 1999 *ApJ* **514** 12–24 (*Preprint* astro-ph/9712195)

[19] Carretti E, Tascone R, Cortiglioni S, Monari J and Orsini M 2001 *New Astronomy* **6** 173–187 (*Preprint* astro-ph/0103318)

[20] Carretti E, Cortiglioni S, Sbarra C and Tascone R 2004 *A &A* **420** 437–445 (*Preprint* astro-ph/0403493)

[21] Page L A 2008 *J. Phys.: Conf. Series - these proceedings*

[22] Hanany S and Marrone D P 2002 *Appl. Opt.* **41** 4666–4670 (*Preprint* astro-ph/0206211)

[23] Leitch E M, Kovac J M, Pryke C, Carlstrom J E, Halverson N W, Holzapfel W L, Dragovan M, Reddall B and Sandberg E S 2002 *Nature* **420** 763–771 (*Preprint* astro-ph/0209476)

[24] Bunn E F 2007 *Phys. Rev. D* **75** 083517–+ (*Preprint* arXiv:astro-ph/0607312)

[25] Lewis A, Challinor A and Turok N 2002 *Phys. Rev. D* **65** 023505–+ (*Preprint* astro-ph/0106536)

[26] Bunn E F 2003 *New Astronomy Review* **47** 987–994 (*Preprint* astro-ph/0306003)

[27] Park C G, Ng K W, Park C, Liu G C and Umetsu K 2003 *ApJ* **589** 67–81 (*Preprint* astro-ph/0209491)

[28] Park C G and Ng K W 2004 *ApJ* **609** 15–21 (*Preprint* astro-ph/0304167)

[29] Conway J E, Cornwell T J and Wilkinson P N 1990 *MNRAS* **246** 490–+

[30] Rohlfs K and Wilson T L 2004 *Tools of Radio Astronomy* (Springer)

[31] Charlassier R, Hamilton J C, Bréelle É, Ghribi A, Giraud-Héraud Y, Kaplan J, Piat M and Prêle D 2008 *ArXiv e-prints* **806** (*Preprint* 0806.0380)

[32] Zmuidzinas J 2003 *Optical Society of America Journal A* **20** 218–233

[33] Hamilton J C, Charlassier R, Cressiot C, Kaplan J, Piat M and Rosset C 2008 *ArXiv e-prints* **807** (*Preprint* 0807.0438)

[34] Hall P S and Veterlein S J 1990 *Inst. Elect. Eng. Proc.* **137** 293–303

[35] Remez J, Segal A and Shansi R 2005 *IEEE Antennas Wireless Propag. Letters* **4** 293–296

[36] Tucker G S, Korotkov A L, Gault A C, Hyland P O, Malu S, Timbie P T, Bunn E F, Keating B G, Bierman E, OSullivan C, Ade P A R and Piccirillo L 2008 *Millimeter and Submillimeter Detectors and Instrumentation for Astronomy IV. Edited by Zmuidzinas, Jonas; Holland, Wayne S.; Withington, Stafford; Duncan, William D. to appear in Proceedings of the SPIE (2008).* Presented at the Society of Photo-Optical Instrumentation Engineers (SPIE) Conference

[37] Hyland P, Follin B and Bunn E F 2008 *ArXiv e-prints* (*Preprint* 0808.2403)

[38] Watson R A 2008 Personal Communicaton

[39] Thompson A R, Moran J M and Swenson Jr G W 2001 *Interferometry and Synthesis in Radio Astronomy, 2nd Edition* (Interferometry and synthesis in radio astronomy by A. Richard Thompson, James M. Moran, and George W. Swenson, Jr. 2nd ed. New York : Wiley, c2001.xxiii, 692 p. : ill. ; 25 cm. "A Wiley-Interscience publication." Includes bibliographical references and indexes. ISBN : 0471254924)

[40] Böker T and Allen R J 1999 *ApJS* **125** 123–142 (*Preprint* arXiv:astro-ph/9903490)

[41] Page L, Hinshaw G, Komatsu E, Nolta M R, Spergel D N, Bennett C L, Barnes C, Bean R, Doré O, Dunkley J, Halpern M, Hill R S, Jarosik N, Kogut A, Limon M, Meyer S S, Odegard N, Peiris H V, Tucker G S, Verde L, Weiland J L, Wollack E and Wright E L 2007 *ApJS* **170** 335–376 (*Preprint* arXiv:astro-ph/0603450)

[42] Spergel D N, Bean R, Doré O, Nolta M R, Bennett C L, Dunkley J, Hinshaw G, Jarosik N, Komatsu E, Page L, Peiris H V, Verde L, Halpern M, Hill R S, Kogut A, Limon M, Meyer S S, Odegard N, Tucker G S, Weiland J L, Wollack E and Wright E L 2007 *ApJS* **170** 377–408 (*Preprint* arXiv:astro-ph/0603449)

[43] Gundersen J and Wollack E J 2008 *J. Phys.: Conf. Series - these proceedings*

[44] Kittara P, Jiralucksanawong A, Yassin G, Wangsuya S and Leech J 2007 *International Journal of Infrared and Millimeter Waves* **28** 1103–1114

[45] Keating B G 2008 *J. Phys.: Conf. Series - these proceedings*

[46] Kogut A 2008 *J. Phys.: Conf. Series - these proceedings*

[47] Kim H, Ho S J, Yen C C, Sun K O and van der Weide D W 2005 *IEEE Microwave and Wireless Components Letters* **15** 147–+

[48] Allen C A, Benford D J, Miller T M, Moseley S H, Staguhn J G and Wollack E J 2008 *Journal of Low Temperature Physics* **151** 266–270

[49] Yun M, Beeman J W, Bhatia R, Bock J J, Holmes W, Hustead L, Koch T, Mulder J L, Lange A E, Turner A D and Wild L 2003 *Society of Photo-Optical Instrumentation Engineers (SPIE) Conference Series* (*Society of Photo-Optical Instrumentation Engineers (SPIE) Conference Series* vol 4855) ed Phillips T G and Zmuidzinas J pp 136–147

[50] Shirron P 2008 *J. Phys.: Conf. Series - these proceedings*

Journal of Physics: Conference Series **155** (2009) 012004 doi:10.1088/1742-6596/155/1/012004

Multiplexed Readout of CMB Polarimeters

Matt Dobbs[1], Mark Halpern[2], Kent D. Irwin[3], Adrian T. Lee[4], J.A.B. Mates[5], and Benjamin A. Mazin[6]

[1]Department of Physics, McGill University, Montréal, QC, Canada

Matt.Dobbs@mcgill.ca

[2]Department of Physics, Department of Physics and Astronomy, University of British Columbia, 6224 Agricultural Road, Vancouver, B.C Canada V6T1Z1

halpern@physics.ubc.ca

[3]National Institute of Standards and Technology, 325 Broadway, Boulder, CO

irwin@nist.gov

[4]Department of Physics, U. California, Berkeley, CA, USA

Adrian.Lee@berkeley.edu

[3]National Institute of Standards and Technology, 325 Broadway, Boulder, CO, and Department of Physics, University of Colorado at Boulder, Boulder, CO 80309

John.Mates@colorado.edu

[6]Department of Physics, University of California, Santa Barbara, CA, USA

bmazin@physics.ucsb.edu

Abstract. This paper describes contributions to the workshop, "Technology Development for a CMB Probe of Inflation," held at NIST in Boulder CO, Aug. 25-28, 2008 concerning technologies to read out direct detectors (including bolometers and microwave kinetic inductance detectors) in a CMBPol satellite mission. The large number of polarimeters required for a satellite mission will likely make it necessary to multiplex the output signals into a small number of readout channels at the cold state. We describe both the cryogenic components and the present-generation warm readout electronics, consider the benefits and challenges of each option, and analyze their technology readiness level and needed additional investments.

Journal of Physics: Conference Series **155** (2009) 012004 doi:10.1088/1742-6596/155/1/012004

1. Introduction

The first technology recommendation of the report of the Task Force for Cosmic Microwave Background Research was "technology development leading to receivers that contain a thousand or more polarization sensitive detectors, and adequate support for the facilities that produce these detectors." The recommendation also identified that "It is important to keep open a variety of approaches until a clear technological winner has emerged. Nevertheless, highest priority needs to be given to the development of bolometer-based polarization sensitive receivers." For arrays of a thousand or more bolometer-based polarization-sensitive receivers, constraints on complexity and heat load make it difficult to route separate leads from each bolometer to the warm readout electronics. It is necessary to multiplex the signal from many bolometers at the cold stage into a smaller number of output channels. One of the challenges in fielding a thousand or more bolometer-based polarization sensitive receivers is the development of appropriate readout technology.

In this paper, we present an analysis of the state of the art of three leading multiplexed readout technologies for polarization-sensitive bolometers: time-division multiplexing (TDM), MHz frequency-division multiplexing (FDM), and GHz frequency-division multiplexing with superconducting microresonators. All three techniques can be used to read out superconducting transition-edge sensor (TES) bolometers [1], and the third can be also used with microwave kinetic inductance detectors (MKIDs) [2]. In each case we present a technology overview, an analysis of benefits and challenges, and a discussion of technology readiness level (TRL) and needed investment.

2. Time-Division SQUID Multiplexers

Principal section authors: Kent D. Irwin and Mark Halpern

The superconducting transition-edge sensor is a leading bolometer technology for a CMBPol satellite mission. The low noise, low power dissipation, and low impedance of Superconducting Quantum Interference Devices (SQUIDs) make them the preamplifier of choice for TES bolometers.

In time-division multiplexing (TDM), many SQUID-coupled TES bolometers are read out through a single set of wires by turning the SQUIDs on sequentially (Fig. 1a). With proper design, the multiplexed SQUID amplifiers do not appreciably contribute to the system noise. SQUID TDM is a mature technology that is being deployed in many astronomical instruments with multiplexing factors up to 40:1. It is being used in arrays with sizes greater than that required for CMBPol.

Time-division multiplexed (TDM) SQUID amplifiers systems have been developed for TES bolometers by NIST, NASA/Goddard Space Flight Center, and the University of British Columbia (UBC). The first multiplexed TES bolometer instrument tested on a telescope was the TDM FIBRE instrument in 2001[3]. There are now a number of TDM instruments for Cosmic Microwave Background (CMB) measurements deployed in the field and in various stages of development. The Atacama Cosmology Telescope [4] (ACT), in particular, has acquired a season of data in 2007 with 900 functioning pixels in one 32×32 array, and is observing in 2008 with three 32×32 arrays. Other instruments for CMB measurements include the SPIDER balloon-borne CMB polarimeter [5], BICEP-2 [6], the Keck Array [6], and Clover [7]. The X-ray calorimeters that are being developed by NASA for the International X-Ray Observatory (IXO) use similar TDM SQUID multiplexers. A full complement of SQUID multiplexers has also been tested for the SCUBA-2 submillimeter camera [8], which has 10,240 pixels.

The warm readout electronics for TDM has gone through multiple generations of development at NASA/GSFC, NIST, and the University of British Columbia. The present generation of control electronics is the Multi-Channel Electronics [9][10] (MCE) developed at the UBC (Fig. 1b). Each MCE module controls up to 1,280 TES pixels (the 10,240 pixels in SCUBA-2 are controlled by 8

MCE modules). The MCE sets the detector biases, controls the SQUID multiplexer stages, and reads out the signal from one array of up to 41 × 32 pixels through 32 output channels. The MCE provides automatic optimization of operating points for the bolometers and SQUID amplifiers, and implements a digital Proportional Integral Differential (PID) servo loop to apply feedback to the switched first-stage SQUIDs to keep them in a linear regime at optimal gain. The MCE, originally developed for SCUBA-2, is also in use in ACT, SPIDER, BICEP-2, the Keck Array, and Clover.

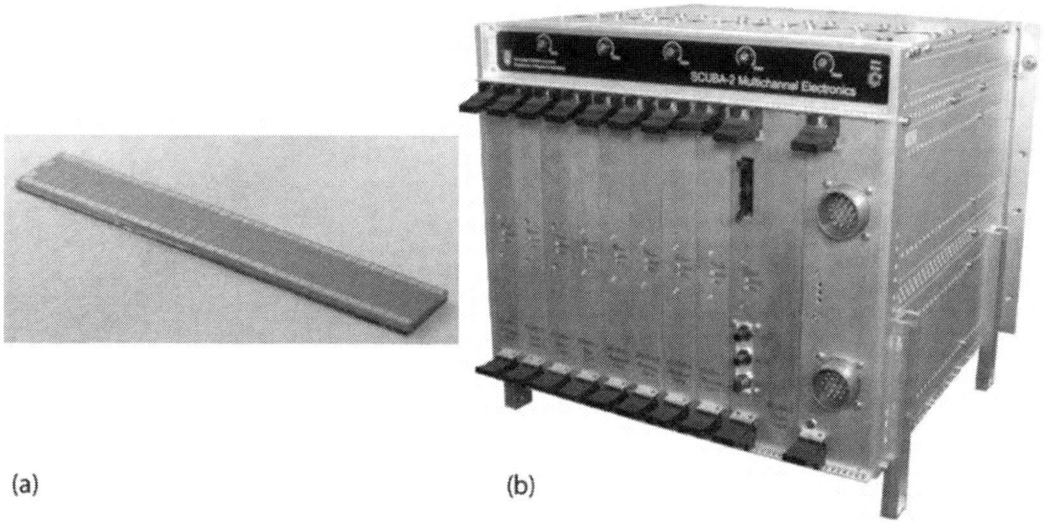

Fig. 1. (a) A 32-channel SQUID multiplexer chip. The chip dimension is 3 mm × 20 mm. The 32 SQUIDs are turned on sequentially and read out through one output channel. (b) A Multi-Channel Electronics (MCE) module fabricated by the University of British Columbia. This module can instrument up to 32 output channels, and read out up to 1,280 TES bolometer channels.

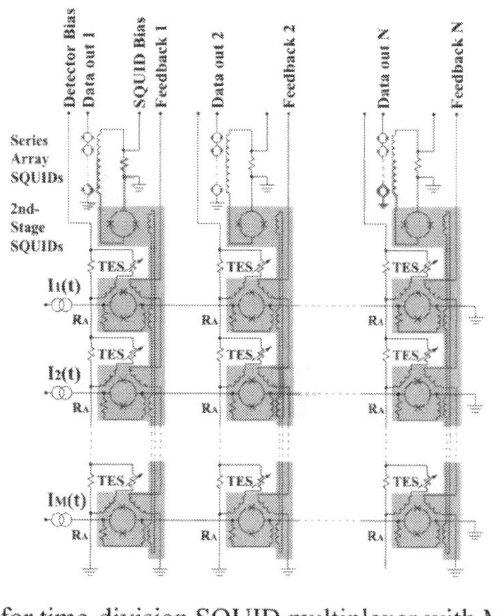

Fig. 2. Circuit diagram for time-division SQUID multiplexer with M × N pixels. The first-stage SQUIDs in each column are coupled through a common transformer to the second stage. A single common feedback line is used to linearize all of the first-stage SQUIDs in each column.

2.1. SQUID TDM: Technology Overview

When a superconducting transition-edge sensor is biased at a constant voltage, the current through the device drops when optical power is absorbed. In TDM, each TES is instrumented by a separate first-stage SQUID that measures this current drop. The bandwidth of the TES is limited by a one-pole low-pass L/R filter formed by the inductance of the SQUID's input coil (and possibly an extra inductor) and the resistance of the TES. A two-dimensional (M rows × N columns) array of pixels is read out by sequentially turning on the first-stage SQUIDs in every column, one row at a time (Fig. 2). The row of SQUIDs is turned on by a set of M row-select currents ($I_1 \ldots I_M$) from 0-1V, 14-bit video DACs in the address-card (AC) in the MCE. The AC firmware controls up to 41 DACs with user-specifiable order, rate, and voltage settings. The typical 'on' time in each row is about 1 μs. Thus, in an array with a 41:1 multiplexing factor, every pixel in the array is revisited with a frequency of about 20 kHz (the 'frame rate'), which is well above the signal frequency in the bolometers.

A ~1 ohm address resistor, R_A, shunts each first-stage SQUID. The current through the address resistor is inductively coupled to a second-stage SQUID shared by all the first-stage SQUIDs in a column. The coupling to the second stage occurs through a transformer coil that is common to all of the first-stage SQUIDs. A feedback flux is provided to the switched first-stage SQUIDs to linearize them. Since only one SQUID in each column is on at a time, one feedback coil can be common to all first-stage SQUIDs in the column (Fig. 2). The feedback current is provided by a PID feedback servo loop implemented in firmware in the MCE. The feedback signal in the SQUIDs, which compensates for changes in current through the TES bolometers, constitutes a measurement of the optical input power to each TES.

The output from each column of SQUIDs is amplified by a series array of SQUIDs (Fig. 2) that can be located at either the base temperature or at 4 K, and then by a preamplifier followed by a 14-bit, 50 MHz video ADC in a readout card (RC) of the MCE. The PID loop is also implemented in the RC. Each RC couples to eight columns and handles independent feedback loops for up to 8 × 41 TES pixels; each MCE can accommodate 4 RCs. The PID output is based on the reading from the previous visit to a given row. It is applied to the first-stage SQUID feedback lines using a 14-bit video DAC with 0-1 V range. Since the frame rate of 20 kHz far exceeds the detector bandwidth, a 4-pole Butterworth IIR low pass filter with a cutoff consistent with the detector thermal time constant is implemented in RC firmware after the signals have been demultiplexed. To aid in diagnosing new arrays, the MCE can output unfiltered data at the frame rate of approximately 20 kHz or for short bursts at the ADC rate of 50 MHz.

A combination of software and firmware commands has been developed for the MCE that autonomously characterizes a 1280 pixel array and chooses optimized biases and feedback currents for the full array (about 2100 free parameters) in a few minutes.

A clock card (CC) in each 1,280-pixel MCE module communicates with the computer through a dual fiber-optic link. It also dispatches incoming control commands through the backplane to the appropriate cards and generates the master clock for the modules. The data-acquisition computer sends commands to the MCE through its fiber-optic link and receives data packages. An external controller with a 25 MHz clock synchronizes the data acquisition of multiple MCEs. It also provides numeric tags that are written to each data frame for data synchronization during analysis.

The power dissipation of an MCE module used to control a 1,280-pixel array is presently about 175 W. Approximately 40% of this power budget is used to run the video analogue-to-digital converter (ADC) in each column. New boards are in production in which these have been replaced with 50 Mhz serial video ADCs that consume 15 times less power.

Journal of Physics: Conference Series **155** (2009) 012004 doi:10.1088/1742-6596/155/1/012004

2.2. SQUID TDM: Benefits and Challenges

A. Benefits

Warm electronics: maturity, power dissipation, and path towards flight qualification

The warm electronics for SQUID TDM are relatively simple and mature. The MCE uses commercial components (DACs, ADCs, and FPGAs) that have been readily available for some time. While a large investment was required at UBC to develop the MCE, this was due to the complexity of the firmware, not because of challenges in the performance of individual components. The power dissipation of the MCE (about 175 W for 1,280 pixels) is also quite low and work is under way to reduce it further.

The MCE operates at 20 kHz frame rates that are extremely fast compared to the bolometer thermal response times. In a fully optimized system for CMBPol, it would be possible to operate the warm electronics an order of magnitude slower. This would reduce the power dissipation, and also make it possible to use legacy electronic components that have already been flight qualified. As part of the IXO satellite program, there is already an effort to explore a system that can be flight qualified. However, unlike CMBPol, IXO requires high frame rates, so it does not have the potential of using legacy components.

The MCE have been tested for cosmic ray induced upsets by exposing them to a neutron flux corresponding to a 30-day stratospheric balloon flight. The electronics showed no loss of function during the test.

Wiring length

SQUID TDM has a significant engineering budget for wiring length. This is because the delay time of the propagating feedback signal can be longer than the row switching time, since the PID algorithm is implemented based on information from the previous frame. Thus, the wiring between the cold stage and the warm electronics can be several meters long.

Demonstrated Multiplexing Factor

TES instruments multiplexed with SQUID TDM have been demonstrated to operate with a 40:1 multiplexing factor, with no appreciable degradation to the bolometer performance from aliased detector noise, amplifier noise, or switching transients. The theoretical limit on the number of channels that can be multiplexed in each column is set by bandwidth per pixel, available bandwidth, and by aliasing of SQUID noise. Present implementations are very far from these theoretical limits. However, unlike superconducting microresonators, practical constraints on geometry are likely to limit the multiplexing factor to somewhere near 100:1, so TDM technology will not in the long term be as scalable as GHz microresonators.

Low-frequency noise

SQUID TDM systems tend to be robust against low-frequency noise from SQUIDs and amplifiers. Because SQUID noise (but not bolometer noise) is degraded by aliasing during sampling, the TES bolometers are significantly overcoupled to the SQUIDs to prevent loss in bolometer performance. Thus, even though the amplifier / SQUID low-frequency knee is typically above 10 Hz, when properly optimized, the bolometer noise is typically above the amplifier / SQUID noise down to very low frequencies (tens of millihertz).

Compact filter elements

The filter elements in analog cryogenic multiplexers can be the physically largest part of the cryogenic multiplexer circuits. However, SQUID TDM has very compact filter elements. In SQUID TDM, the bandwidth is limited by a one-pole filter formed by the SQUID inductor and the resistance of the TES. Since a TES bolometer used for TDM can be biased at a very low resistance (e.g. 2 mΩ), a bandwidth of 5 kHz can be achieved by a lithographically fabricated 60 nH inductor, which can fit in an area of less than 0.1 mm^2.

In MHz FDM, the bandwidth is limited by an LC resonant filter, which requires physically larger filter components. If the bandwidth is limited to 5 kHz, a simultaneous optimization of the L and C elements drives the TES bias resistance to about 0.5 Ω. Then, a 16 μH inductor and 1.6 nF capacitor are required for a 1 MHz resonance with 5 kHz bandwidth. The inductor is typically lithographically fabricated with an area of about 2 mm^2. The capacitor is often a component soldered onto a circuit board.

The high operational frequency of GHz superconducting microresonators makes it possible for them to use compact resonant filter elements. The frequency band is defined by quarter-wave coplanar waveguide stubs that are typically 5-10 mm long, and meandered into a fairly compact configuration.

<u>In-focal-plane multiplexing</u>
Because the cryogenic filter elements for SQUID TDM are very compact, and the SQUIDs have very low power dissipation, it is possible to integrate them into the focal plane (as is done in SCUBA-2). This is even more straightforward in superconducting microresonators, since they operate at GHz frequencies with relatively small filter elements. In contrast, it would be difficult to integrate the large filter elements into the focal plane for SQUID FDM: leads are usually routed out of the focal plane from every pixel to the filter elements.

B. Challenges
<u>SQUID fabrication</u>
In a TDM SQUID multiplexer circuit, a SQUID must be fabricated and tested for each TES pixel. (In contrast, MHz FDM SQUID multiplexers use only one SQUID series array per multiplexed set of pixels). However, many TDM SQUID channels are integrated onto each chip, and the fabrication challenge for TDM SQUID multiplexers for CMBPol is manageable. There is now a mature process at NIST that can fabricate and test sufficient TDM SQUID multiplexer chips for CMBPol.

<u>Power dissipation on the sub-K stage</u>
In TDM SQUID multiplexers, the SQUID amplifiers dissipate power at the base temperature. In contrast, the SQUIDs in MHz FDM SQUID multiplexers are located at a higher temperature stage. GHz Superconducting microresonator multiplexers (both MKIDs and microwave SQUID multiplexers) dissipate negligible power at the base temperature.

In the present generation of TDM SQUID multiplexers, a power budget of about 10 nW is allocated for each multiplexed column of up to 40 SQUIDs. In the case that this power dissipation is too high, SQUIDs with much lower power dissipation have already been demonstrated in TDM SQUID multiplexer chips, which would drop the power budget to about 1 nW per multiplexed column, or about 30 nW per kilopixel.

<u>Achievable Multiplexing Factor</u>
While a large multiplexing factor is used with TDM SQUID multiplexers (40:1), and while the fundamental limit on TDM SQUID multiplexers can be higher than 1000:1, practical constraints are likely to limit the multiplexing factor to about 100:1. A 12:1 multiplexing factor has been demonstrated with MHz FDM, but this is expected to increase significantly in the future. Because of the multi-GHz available bandwidth and resonator quality factor, GHz superconducting microresonator multiplexers can potentially be scaled to much higher multiplexing factors than either MHz multiplexing technology.

2.3. SQUID TDM: Technology Readiness and Needed Investment

TDM SQUID multiplexers and the MCE electronics are now at Technology Readiness Level (TRL) in the high 4s, having been demonstrated in a relevant environment at the kilopixel scale (e.g. ACT / SCUBA-2). It remains to demonstrate TDM SQUID multiplexers in a relevant environment simulating a satellite to achieve TRL 5. TDM SQUID multiplexers have attractive advantages, and their performance in most ways is already at the level that would be required for a CMBPol satellite mission. However, investment is still needed in several areas:

Power dissipation

Power dissipation at the cold stage (including engineering margin) is presently 10 nW per multiplexed column. This power budget can readily be reduced by a factor of 10, if required. Single channels with lower power designs have been demonstrated on TDM chips. If a design study indicates the lower power is required, a full demonstration should be prioritized.

Yield

Multiplexer pixel yields of greater than 90% can be routinely achieved. The Atacama Cosmology Telescope was populated with 96 TDM 32-channel chips, most of which had a 100% yield. Furthermore, over 10,000 TDM pixels have been screened for SCUBA-2. A flight instrument would be required to start with 100% yield on all of the multiplexer chips. Some investment in improving the yield of the process would improve the rate at which perfect chips could be produced.

SQUID flight qualification

SQUIDs have already been flight qualified for the Gravity-Probe B experiment. Radiation hardness tests should also be conducted on TDM SQUID chips.

Systematic error requirements and magnetic shielding

The systematic error specifications for a CMBPol mission are extremely stringent. A full systematic error budget needs to be developed. The characteristics of the multiplexers are an important part of this study. One key issue that must be considered is the sensitivity of the SQUID multiplexers to scan-synchronous magnetic fields from the instrument. The present generation of TDM SQUIDs are gradiometric, and thus insensitive to first order to uniform fields and field gradients. However, magnetic shielding is still required. Systematic error concerns are likely to place the strongest constraint on the characteristics of the required magnetic shielding, including its weight.

Warm electronics development

The present warm electronics are based on FPGAs. Although the particular FPGAs used in the MCE have been demonstrated to be sufficiently robust for ground-based and balloon operations, they are not space qualified. Non-FPGA designs can be made for three of the MCE cards, but radiation-robust solutions based either on older, previously space qualified FPGAs or newly qualified parts must be found for the RC and CC functions.

3. MHz Frequency Domain SQUID Multiplexers

Principal section authors: Matt Dobbs and Adrian Lee

A key technology for deploying large format Transition Edge Sensor (TES) bolometer arrays on satellite platforms is SQUID-based multiplexed readout systems. The Frequency domain multiplexed readout (fMUX) was developed for mm-wavelength observations using large arrays of TES bolometers by LBNL, U.C. Berkeley, and McGill. The system, with its original "analog" backend electronics [11][12][13], targeted ground based telescopes and is deployed on the APEX-SZ instrument [14] and the South Pole Telescope (SPT) [15]. These instruments have achieved unprecedented on-sky noise performance. The analog backend draws too much power for balloon or satellite applications. In the course of the analog system's development, fast ADCs and FPGAs with substantially increased gate-count and reduced power consumption became available. This allowed for the development of a new digital backend [16] for the fMUX system that dissipates an order of magnitude less power, making it amenable for stratospheric balloon payloads. This Digital fMUX (DfMUX) system is being deployed for the EBEX balloon-borne CMB polarimeter [17] and the ground-based POLARBEAR [18] instrument. The DfMUX also provides a path to further substantial power reductions by grouping many more bolometers together in a multiplexer module. With the digital fMUX system, the power consumption is roughly proportional to the number of multiplexer modules and not to the number of bolometer pixels.

The frequency-domain multiplexer reads out many TES bolometers on a single set of wires without appreciably contributing to the system noise. The detectors are low impedance ($\approx\frac{1}{2}\ \Omega$) devices cooled to sub-Kelvin temperature. The sky signals are modulated in the bandwidth 0.05-100 Hz by the motion of the telescope or optics. The low-frequency noise specification places strict requirements on all aspects of the system and distinguishes the firmware and digital algorithms employed in this system from other modulation/demodulation applications such as software-defined radio. The electronics system also tunes the detectors to the optimum bias point by adjusting their voltage bias and tunes the SQUID pre-amplifiers using bias currents to obtain the best noise performance and dynamic range.

Advantages of the DfMUX system include: (1) bolometer signals are modulated above microphonic and low frequency noise, (2) there is no fundamental limit on the number of detectors that can be multiplexed in a module, other than available bandwidth, (3) no heat dissipation on the sub-Kelvin stage, (4) the system is highly modular, and (5) individual bolometer biases and the detector readout bandwidth can be software configured.

The fMUX system is complementary to the time domain multiplexed system [19], [20] developed at NIST and UBC. Recently the NIST group began work on a system that frequency-multiplexes SQUIDs (rather than bolometers) at microwave frequencies [21][22]. The DfMUX electronics described here is a lower frequency version of what is needed for the backend of this future RF-SQUID multiplexing system.

3.1. MHz FDM: Technology Overview

A. System Description

The frequency domain multiplexer (DfMUX) is shown schematically in Fig. 3. For the EBEX system, the bolometer sensors R_{bolo} are biased with sinusoidal voltages in the frequency range from 300 kHz to 1 MHz. Each bolometer is biased at a different frequency. Intensity variations from the sky-signal change the bolometer resistance and amplitude modulate the bolometer current such that the sky-signal from each bolometer is transferred to a sideband adjacent to its carrier. Thus, the signals from different bolometers within a module are uniquely positioned in frequency space, so they can be summed and connected through a single wire to a SQUID preamplifier operating at 4K. Each bolometer is connected through a series resonant LC circuit* that defines the bias frequency. This allows the bias frequencies for all bolometers in a module to be applied through a single wire, as the tuned circuit selects the appropriate frequency for each bolometer. Only two wires are needed to connect the bolometers of a readout module on the sub-Kelvin stage to the 4K stage on which the SQUIDs are mounted. The tuned circuits also limit the bandwidth of the bolometer Johnson noise, which would otherwise contribute to the noise in all other channels of the module. Refer to [13] for a detailed description of the readout system's cold circuits.

* The first generation superconducting inductors for this system were fabricated at TRW which has closed its superconducting fab facility. The present generation is being fabricated at NIST. The capacitors are off-the-shelf ceramic chip devices.

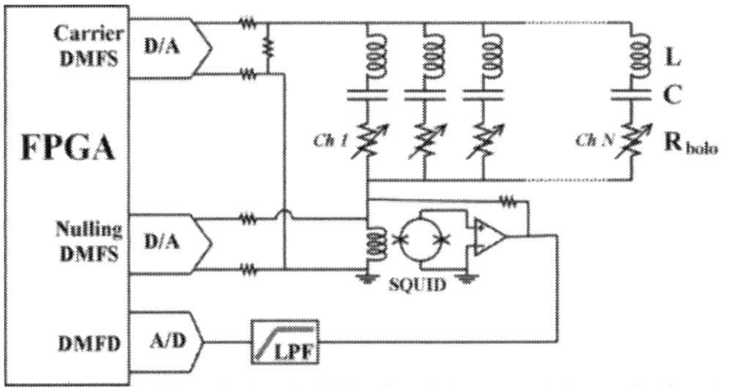

Fig. 3: The digital frequency domain multiplexer system is shown schematically.

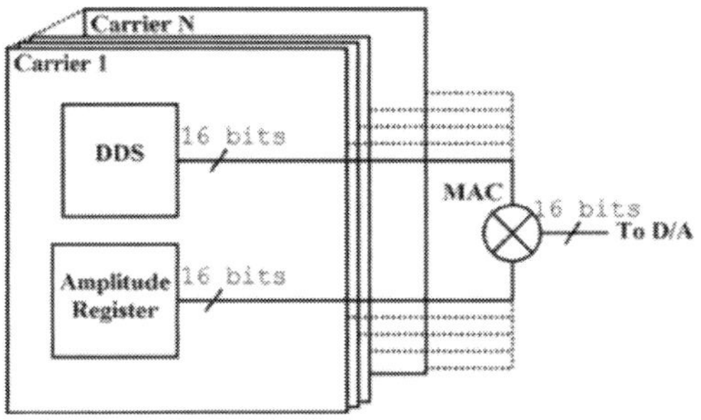

Fig. 4: The Digital Multi-Frequency Synthesizer algorithm is shown schematically.

The comb of bias carriers is synthesized with a *Digital Multi-Frequency Synthesizer* (DMFS). It produces a comb of sine waves using an algorithm implemented in firmware. It converts the signal to analog using a 16-bit D/A operating at 25 MHz. The DMFS firmware is shown schematically in Fig. 4. Each sinusoidal carrier is synthesized using a Direct Digital Synthesizer algorithm [23]. A 'comb' of carriers is created by multiply-and-accumulating the DDS outputs with an amplitude control register for each sine wave. The comb is sent to the cryostat as a differential signal on a shielded twisted pair cable.

The sky-modulated signals from the bolometers are pre-amplified with a SQUID as it has the necessary noise temperature and has low input impedance. To maintain constant voltage bias across the TES, the input resistance of the SQUID must be small compared with the TES resistance. This is achieved by operating the SQUID with shunt feedback from the output of the room-temperature amplifier that follows. The feedback amplifier has a high gain × bandwidth product, so connections between the SQUID and the room temperature electronics must be short. Negative feedback also linearizes the SQUID response, reducing intermodulation between the bias carriers.

Since the carrier amplitudes are orders of magnitude larger than the sky signals, we cancel the carriers at the SQUID input with a second comb (synthesized with a second DMFS), referred to as the nulling signal. The nulling comb is an inverted version of the original carrier comb, and serves to remove the large carrier signals. It does not affect the carrier sidebands, which contain the information of the sky-signals. The use of the nulling signal dramatically reduces the dynamic range requirements of the system. Nulling factors of 10^3 are routinely achieved.

Journal of Physics: Conference Series **155** (2009) 012004 doi:10.1088/1742-6596/155/1/012004

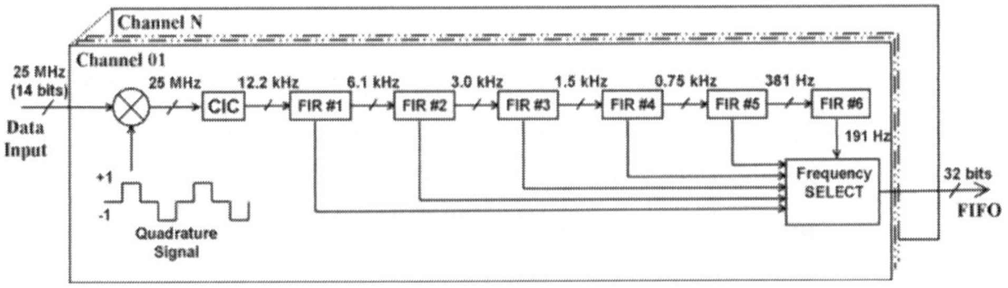

Fig. 5: The Digital Multi-Frequency Demodulator algorithm is shown schematically.

The SQUID amplifiers are 100-element series-array devices [24] manufactured by NIST in Colorado. Each SQUID device is operated in shunt-feedback with a low-noise bipolar transistor op-amp located on a custom room temperature SQUID controller circuit board. These boards also include digital control electronics and DACs to provide the bias currents and tuning functionality.

After amplification, the comb of sky-signal-modulated carriers output by the SQUID controller is transmitted to the Digital Multi-Frequency Demodulator (DMFD) on a twisted pair cable. The comb is digitized with a 14-bit A/D converter operating at 25 MHz. The over-sampling improves the resolution beyond 14 bits. Since the sky-signals occupy only a small fraction of the waveform's total bandwidth, it is not feasible to store the entire waveform on disk. The signals are processed in real time.

Inside the FPGA, the DMFD input data is sent down a set of parallel algorithm pipelines, each of which consists of a quadrature mixer, Cascade-Integrator-Comb (CIC) [25] low-pass filter, and a chain of band-defining FIR filters. There is one pipeline for each detector channel in the comb. A basic schematic of the configuration is shown in Fig. 5.

The low-pass filtering of the waveform is challenging. Low-frequency (typically 0.05-100~Hz) signals need to be maintained while filtering and sub-sampling the waveforms by a factor of roughly 10^5 to reduce the 25 MHz sampling rate to the ~200 Hz data rate that will be recorded on disk. The user can easily change the output data rate and hence the bandwidth of the system by bypassing some of the FIR filters with software commands. This is a powerful tool for debugging during the integration phase of the instrument's commissioning, where typically a larger bandwidth is desired to measure detector properties such as electrical and thermal time constants.

Four DfMUX modules are contained on a single 6U VME board, shown in Fig. 6. Each motherboard can handle 4 multiplexer modules. The digital circuits, including a powerful Xilinx Virtex4 LX160 FPGA, reside on this FPGA motherboard, while the low-noise analog converters, amplifiers, and filters ride 'piggy-back' on two mezzanine boards. The total number of detectors that can be demodulated by a single FPGA is determined by its gate count, but this gate count has been increasing faster than Moore's law over the last decade. A SQUID controller board, which is capable of handling 8 SQUID modules, forms an intermediary between the cold SQUIDs and the analog signals coming from the DfMUX backend.

The electronics attaches timestamps to the bolometer data from either an IRIG-B-encoded GPS signal or Manchester-encoded system clock. High-level commands ("set up squids", "set up bolometers") are encoded as TCP/IP packets and sent to the boards over standard Ethernet. Bolometer data is streamed off the board as UDP Ethernet packets. Each board has an embedded processor running μCLinux allowing it to run autonomously. It has its own watchdog circuit and temperature monitors. A monitoring system is being developed that will recognize configuration bit-flips due to single event upsets and correct for them.

Fig. 6: The digital fMUX backend electronics. An FPGA motherboard is shown (large lower blue board) with two of its analog converter mezzanine boards attached (smaller upper red boards) and the copper convective heat pipe and conductive heat sink.

The SQUID controller and cold components of the system were developed at LBNL/UC Berkeley and are described in [14], [15]. The room temperature digital backend electronics developed at McGill are described in [16]. The heat dissipation scheme being employed for EBEX was developed at U. Minnesota. Performance validation of the digital system in the laboratory can be found in [16] and [26].

B. Channel spacing

The number of detectors that can be grouped together in a multiplexer comb is defined by the SQUID pre-amplifier system bandwidth divided by the carrier spacing.

Our present SQUID system has a bandwidth of just above 1 MHz, limited by phase shifts along the wires that connect the 4K SQUID to the room temperature amplifier that forms part of the flux-locked loop. There are several suggestions for SQUID systems that could greatly improve this bandwidth while maintaining the necessary system loop gain and linearity. One example is the Linearized SQUID Array (LISA) that is outlined in [13] and has recently been further developed. Bandwidths of 5-10 MHz have been achieved with LISA.

The detector time constants define the minimum LCR filter width and ultimately the channel spacing. While detectors employed with this system typically have optical time constants >1 ms, the TES sensor itself can respond to much faster thermal signals and the filter needs to be wide enough to allow stability across the full TES bandwidth. A careful reduction of the TES time constant to just wider than the optical time constant would allow closer channel spacing. The system presently uses 16μH inductors with ½ Ω bolometers, resulting in an L/R bandwidth of 5 kHz.

Once the L/R bandwidth is defined, the frequency spacing of adjacent carriers in the bias comb is specified by the requirement that Johnson noise from neighboring channels be attenuated to a negligible level. For EBEX, the carriers occupy the bandwidth from 0.3-1 MHz with a spacing of ~50 KHz, resulting in 12 detectors per multiplexer module.

Journal of Physics: Conference Series **155** (2009) 012004 doi:10.1088/1742-6596/155/1/012004

It should be emphasized that there is nothing fundamental limiting the module channel count—by optimizing the detector time constants and implementing new SQUID technology, a factor of many should be achievable without the development of any new technology. A commensurate reduction in power consumption and sub-Kelvin heat load would be achieved. If the SQUID to room temperature feedback loop can be eliminated, the 4K-300K wire length could be greatly increased, greatly reducing the heat load there as well.

C. System configuration for EBEX
For the EBEX science flight, 12 detectors will be multiplexed per module. The 1536 bolometer system has 32 DfMUX backend boards, 16 SQUID controller boards, and 128 SQUIDs. The SQUID controllers are mounted in a Faraday cage directly on the receiver cryostat. The DfMUX boards are housed in two 6U VME racks.

D. Power and thermal considerations
The power consumption is about 4W per multiplexed module. For EBEX (12 detectors per multiplexer module) the power dissipation is ~500W for 1536 channels. By increasing the number of multiplexed detectors per module as described in the channel spacing section, this power consumption could be substantially decreased.[†]

Heat dissipated on the boards is removed through a two-stage system designed by the Hanany group at U. Minnesota. Heat is conducted from the board and ICs through thermal grease to a pair of crossed rectangular heat pipes made of copper and a convective fluid (Tradename: "Nano Spreader", developed for laptop computers by Celsia Technologies). The heat pipes, visible in Fig. 6, are terminated in a copper plug that brings the heat through the backplane to the gondola frame. The system has been thermally modeled and the results verified with a thermal-vacuum chamber at NASA's Palestine balloon facility. The heat is radiated away from the instrument using panels mounted on the bottom of the gondola.

3.2. MHz FDM: Benefits and Challenges

A. Benefits
Sky signals modulated above microphonics and low frequency electronic noise
With frequency multiplexing, all detectors are continuously read out without interruption or switching transients. Sky-signals are modulated at ~MHz frequencies, greatly reducing susceptibility to microphonic pickup and amplifier/SQUID low-frequency noise.

No power dissipation on sub-Kelvin stage
In the fMUX system, there is one 4K SQUID per multiplexed module (rather than one SQUID per detector) and there is no power dissipation from the readout system on the sub-Kelvin stage.

No fundamental limit in the multiplexing factor
While existing fMUX systems use multiplexing factors of just 8 or 12 channels, there is nothing fundamental limiting this number. This number reflects the relatively small investment that has been made in this technology to date. A path to substantially larger channel counts using existing technology exists.

[†] As the detector count goes up, FPGAs with larger gate-counts are needed to handle the demodulation. Fortunately the density of this technology is increasing very rapidly. We believe that much more efficient demodulation algorithms are possible as well.

Modularity

For the DfMux system, the bias and demodulation functions for a multiplexer module of detectors are provided by the same electronics board (e.g. there are no rows of bias and columns of demodulation). This provides modularity, such that if an electronics board or cryogenic wire fails, it brings down only those combs. Each board is autonomous in its execution of setup scripts, and needs only high level commands from a control computer.

Configurable bandwidth

The configurable firmware demodulator allows the bandwidth of bolometer data recorded to disk to be changed by factors of 2 with a simple software command. For EBEX, we plan to record sky data with a bandwidth of 180 Hz (381 Hz sampling), but can command the system to record data with a bandwidth up to 6 kHz anytime without changing the system tuning. This is useful for mapping out the bolometer response functions or debugging the system.

Other advantages of the fMUX system include:

- bias voltage can be configured for each bolometer separately,
- the bolometer to SQUID wiring can be interrupted with small resistances ($R \ll R_{bolo}$) allowing the use of connectors with copper contacts,
- low power dissipation on the 4K stage (one SQUID per multiplexer module).
- FPGA Vendors such as Xilinx are actively pursing satellite applications and large-gate-count devices, similar to the ones used for this system, are already space qualified.
- DSP power in FPGAs is increasing quickly—faster than Moore's law by some estimates—this will result in even lower power consumption and make processing of drastically higher channel counts possible.

B. Challenges

Stray inductance

To maintain good voltage bias across the TES, the loop that includes the LC filter, SQUID input, and bolometer should be dominated by the bolometer impedance. Stray inductance between the sub-Kelvin and 4K stages spoils this voltage bias. For the analog system, low-inductance lead-coated copper strip-lines are used for the majority of the distance from the SQUIDs to detectors, with a heat-gap created by several inches of Nb twisted pairs. The inductance is dominated by the twisted pair. While this inductance is low enough for sub-MHz operation, it would not work at substantially higher frequency. The short length of twisted pair adds extra heat loading to the sub-Kelvin stage. The Hanany group at U. Minnesota are developing Nb superconducting strip-lines for EBEX by rolling thin Nb wire flat and suspending the traces between Kapton films. This is expected to provide a simple and cost effective solution that is both low inductance and low thermal conductivity.

4K–room temperature wire lengths

The SQUID flux locked loop, as it is presently implemented, includes a room temperature amplifier in the feedback loop. To maintain stability at high loop gain, the wire length between these devices must be kept short (<20cm), loading the 4 K stage. We note that a system with cold feedback, like the LISA pre-amplifier described above, would remove this constraint.

Readout white noise

white noise sources that do not modulate the carrier (such as SQUID and readout electronics noise) are enhance by a factor $\sqrt{2}$ post-demodulation. This is true of any AC-biased bolometer system. For a system with carefully optimized bolometer parameters, this is usually not an issue, as this noise is, even after the enhancement, small compared to other noise sources.

Low frequency D/A noise

The stability of the DMFS waveforms is extremely important to maintain good low frequency performance for temperature anisotropy measurements. The DMFS carrier, DMFS nuller, and DMFD are all clocked with the same crystal oscillator such that clock jitter cancels out to first order. The largest contributor is the low frequency noise from the transistors in the D/A converter output ladder. This low frequency transistor noise is modulated up to the carrier tone frequency by the D/A switching and appears as sidebands on the tones. While the older analog fMUX system suffered from low frequency noise with a low frequency knee typically at 1 Hz, this has been greatly improved for the digital system, where the knee sits at ~0.1 Hz. The location of the knee is determined by the amplitude of the bolometer bias, and so will be lower yet for low thermal conductivity bolometers such as those that will be used on satellite platforms.

Inter-modulation distortion

Any SQUID system is inherently non-linear. Since the SQUIDs in the fMUX system must handle many large carriers, inter-modulation distortion products are produced. Accurate nulling greatly reduces this effect. If the carriers were located at arbitrary frequencies, this distortion would create a forest of inter-modulation distortion products from these tones. Fortunately, there is some flexibility in the specification of the carrier frequency for each LCR resonance. By specifying that every carrier frequency is a multiple of (i.e.) 117 Hz, these distortion products are forced to live post-demodulation at either DC or a 117 Hz. It is easy to notch out this frequency off line.

3.3. MHz FDM: Technology Readiness and Needed Investment

Overall, MHz FDM is presently at a Technology Readiness Level (TRL) in the high 4s, having been demonstrated at close to the kilopixel scale in the South Pole Telescope. The DfMUX system is presently at TRL 4. Components have been prototyped and strung together end-to-end in the laboratory with TES bolometers to demonstrate performance, including noise [16,26]. Two areas need more attention and investment: space qualification for FPGAs, and pre-amplifier bandwidth.

Space qualification for FPGAs

The system makes heavy use of DSP implemented with Xilinx Virtex-4 FPGAs. The vendor has identified satellite and aerospace applications as an important market, and has an active program of space qualification. Recently Xilinx announced a new space grade version of the Virtex-4 line called Virtex-4QV [27]. This line includes a model that has 25% more processing power than the device presently used in the digital backend.

The McGill team has modest funding from the Canadian Space Agency and NSERC to work with an industrial partner (COM DEV) to explore technology that corrects configuration bits in FPGAs that have been flipped by single event upsets (SEUs). The strategy is to allow and expect SEUs, but recognize and correct them quickly. COM DEV has experience in this regime from systems it flew on the MAESTRO payload.

Pre-amplifier Bandwidth

Presently, the number of multiplexed detectors and hence the power consumption and heat load on the sub-Kelvin stage is limited by the bandwidth of the SQUID pre-amplifier system. Very little research, time, or funding has been invested to improve this bandwidth. A relatively small investment here will likely provide substantial returns. The goal is to produce a high-bandwidth (>~10 MHz) SQUID pre-amplifier system that does not use warm components inside the feedback loop.

3.4. Summary and future prospects

Recent developments in the processing power of FPGAs have allowed for the development of digital backend electronics for a SQUID-based frequency domain multiplexer system that operates with large

Journal of Physics: Conference Series **155** (2009) 012004 doi:10.1088/1742-6596/155/1/012004

arrays of sub-Kelvin Transition Edge Sensor bolometers. This new technology has sufficiently low power consumption to allow for the readout of large focal plane arrays on stratospheric balloon platforms. The system will be deployed on the EBEX instrument in 2008/9.

A number of important advantages, discussed above, make this system attractive—while its primary disadvantages are addressable with existing technology.

Substantial reductions in power consumption could be achieved by improving the bandwidth of the SQUID pre-amplifier flux-locked loop. An increase in bandwidth of a few MHz would result in power reduction by a factor of many. Space qualification of the large gate-count Virtex-4 FPGAs employed in the system has already been undertaken the vendor, Xilinx.

To bring the system up to the maturity level necessary for satellite platforms, an investment similar to that made for the UBC TDM electronics is probably necessary. Fortunately, there are no fundamental limits or boundaries that stand in the way of this development—the number of detectors grouped together in a multiplexed module is limited only by the bandwidth of the analog electronics so large advances are in principle possible.

4. Frequency-Division Multiplexing for Superconducting Microresonators

Principal section authors: Benjamin A. Mazin, J.A.B. Mates, and Kent D. Irwin

A large community has converged on the idea of using high quality factor superconducting microresonators to achieve dense frequency domain multiplexing at microwave frequencies. This technique uses wide bandwidth high electron mobility transistors (HEMTs) to provide several GHz of bandwidth with noise temperatures below 5 Kelvin, potentially allowing thousands of detectors to be read out through a single transmission line. This technique, developed originally for microwave kinetic inductance detectors [28], is also applicable to transition edge sensors coupled to resonators through rf SQUIDs [22], and Normal-Insulator-Superconductor detectors [29].

4.1. GHz FDM: Technology Overview

In a MKID [2], shown to the right in Figure 7, a high Q (10^4-10^6) superconducting resonator with a resonant frequency between 1-20 GHz is fabricated lithographically. Incident photons (panel b) change the surface impedance of the superconductor, causing a change in the amplitude (panel c) and phase (panel d) of a microwave probe signal send past the resonator at the resonant frequency. Since the transmission far away from the resonance is unity, we can multiplex many MKIDs off a single transmission line by setting each MKID's resonant frequencies to be slightly different with lithography. This is analogous to the different tones produced by different length pipes in a pipe organ. Using this approach, groups have recently demonstrated resonator-to-resonator frequency scatter below 1 MHz (Figure 2).

Journal of Physics: Conference Series **155** (2009) 012004 doi:10.1088/1742-6596/155/1/012004

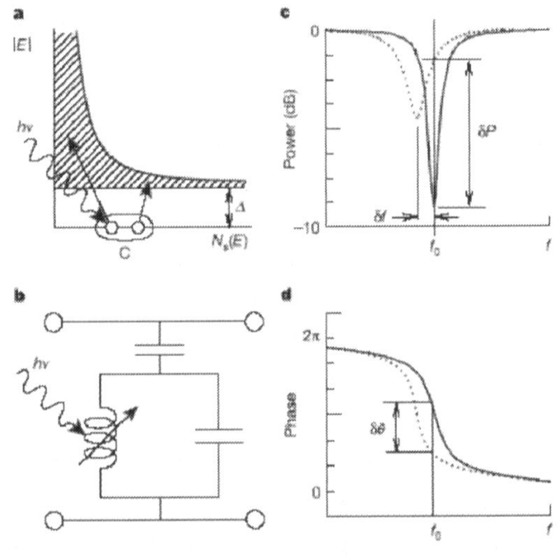

Figure 7. The mechanics of a MKID.

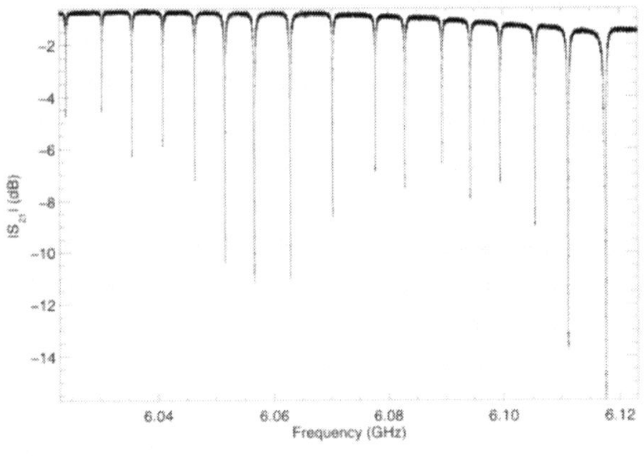

Figure 8. Transmission past a microwave MKID array.

Superconducting transition-edge sensors can also be read out in superconducting microresonators by coupling them through an array of microwave SQUIDs. In a microwave SQUID array, each high-Q resonator is loaded with a dissipationless rf-SQUID. The flux-variable inductance tunes the resonance frequency without affecting Q. Any detector that a SQUID can read out can be multiplexed this way, in particular the TES. To linearize the response one can apply flux feedback if the number of detectors is small. If the number of detectors is large one can utilize the SQUID flux periodicity to modulate the signal, for example by applying a common flux ramp to all SQUIDs in the array. This is a form of phase modulation and it both linearizes the SQUID response and up-converts the detector signal to higher frequencies in the bandwidth of the resonance. It requires more bandwidth per detector and additional room-temperature computation, but makes it possible to modulate above the 1/f noise of the resonators and the HEMT. The lithography for defining closely spaced resonances is essentially the same as for MKIDs, and the SQUIDs can be fabricated on the same wafer as the TESs.

Journal of Physics: Conference Series **155** (2009) 012004 doi:10.1088/1742-6596/155/1/012004

In order to read out an MKID or microwave SQUID array, a comb of frequencies is generated with a sine wave at the resonant frequency of each individual resonator. This comb is then sent through the device, where each sensor imprints a record of its illumination on its corresponding sine wave. The comb is then amplified with a cryogenic HEMT and brought outside the cryostat. The comb is then digitized, and the phase and amplitude modulation of each individual sine wave is recovered in room temperature electronics. Aside from a HEMT amplifier, there are no cryoelectronics. Compared to existing low-frequency TES SQUID multiplexers, much of the complexity is moved from the base temperature to room temperature, where the full power of modern microwave electronics is available. In the case of the microwave SQUID array, the signal will be entirely in the phase direction, and higher readout data rates will be required to sample the external magnetic modulation signal required to linearize the SQUID response.

The technique described above, where a comb of frequencies is created, modified, then digitized and analyzed, is very common in modern wireless communications, where it is usually referred to as software-defined radio (SDR). An implementation suitable for reading out detectors is shown in Fig. 9. In this implementation, dual 400 MHz, 16-bit digital to analog converters play back a pre-computed waveform to generate the comb. Since two D/As are used, we can use an IQ modulator, which allows us to produce signals within a 400 MHz wide band centered on our LO frequency (usually 2-6 GHz). After the comb passes through the detector, it is mixed back down to baseband with another IQ modulator, low pass filtered, then digitized with dual 400 MHz, 14-bit analog to digital converters. After digitization, the signals are passed to a fast field programmable gate array (FPGA). There are many algorithms that can be run in the FPGA to demodulate the signals. The simplest is a direct digital downconverter (DDC) that simply digitally multiplies the complex input signal by sine wave at the desired frequency. This shifts the frequency of interest to 0 Hz. A digital filter followed by decimation gives the desired output data stream.

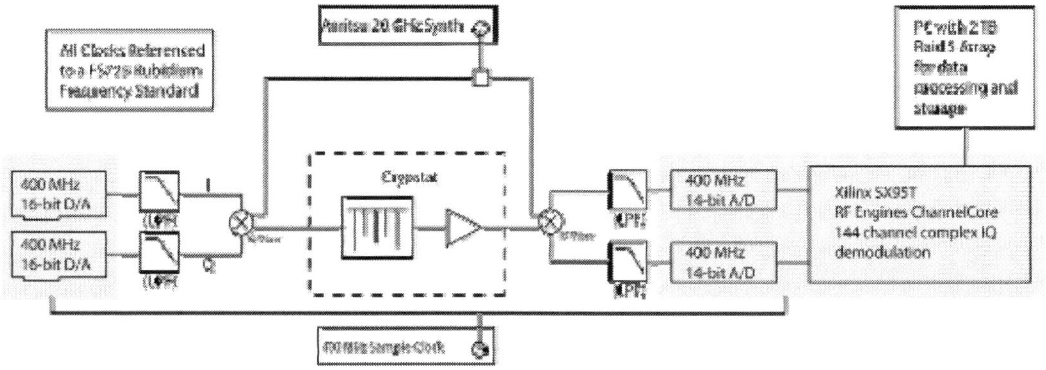

Figure 9. A block diagram of a SDR readout.

4.2. GHz FDM: benefits and challenges

A SDR readout allows a drastic simplification of the focal plane. This simplification cascades through the system, lowering costs at every stage. For instance, a MKID based focal plane will only need several wires going between 4K and the array, while a TDM TES multplexer will require hundreds. This drastically simplifies the construction of the focal plane, and also significantly lowers the cooling requirements.

The advantage of moving the readout complexity from 4 Kelvin to room temperature cannot be overstated. The entire hardware set developed to allow SCUBA-2 is extremely powerful, but large and very expensive due to the requirement of having one SQUID per pixel. Also, since these multiplexers are completely custom they do not gain much from external developments. In the two years since experimentation on these readouts first started, there has been a factor of 4 improvement in the speed of 14-bit A/D converters. The ability of SDR readout power to scale with Moore's law ensures that they will grow significantly more powerful with time.

The drawback of these systems for space applications is that they require flying fast, precise A/Ds, D/As, and FPGAs. Flight ready versions of these components tend to lag their ground-based counterparts, so the bandwidth of space-based SDR readouts will be lower than ground-based systems. The complex signal processing may also consume a significant amount of power at room temperature, in the worst case as high at 0.1 Watts/pixel.

4.3. GHz FDM: Technology Readiness and Needed Investment

The first SDR readout for LTDs was demonstrated in the lab in 2006 [28]. In April 2007, Caltech and the University of Colorado brought a SDR readout to the Caltech Submillimeter Observatory (CSO). This MKID demonstration camera proved the technology viable for ground-based operations [30]. This work has brought the readout to a TRL of the low 4s, and has enabled Caltech and Colorado to successfully propose to the NSF for a much larger camera for the CSO. The MKID camera on the CSO will read out approximately 2400 pixels in 8 GHz of bandwidth, making it by far the largest microwave FDM readout. This camera will push the readout to a TRL in the high 4s, having demonstrated a kilopixel-scale implementation, and lacking a demonstration in a relevant environment simulating a satellite to achieve TRL 5.

5. Conclusions

All three multiplexing techniques described in this paper are at a TRL of 4 or higher, and are progressing rapidly. Each of them are being deployed in suborbital cameras of the kilopixel scale, and both TDM and MHz FDM are being deployed in balloon-borne CMB polarimeter experiments (SPIDER using TDM, and EBEX using MHz FDM). If appropriate additional investment is made, we do not anticipate that the readout techniques and electronics will be a source of high technological risk for a CMBPol space mission. SQUID TDM and FDM are rapidly approaching a level of TRL 5. While each has its benefits and challenges, it is likely that either would work well for a CMBPol satellite. GHz FDM based on superconducting microresonators are also evolving rapidly, and they promise focal-plane simplification and scalability to even larger array sizes.

References

[1] K. D. Irwin, Appl. Phys. Lett. 66, 1998 (1995).
[2] Day, P. K., LeDuc, H. G., Mazin, B. A., Vayonakis, A., Zmuidzinas, J. A broadband superconducting detector suitable for use in large arrays. Nature, vol. 425, pp. 817–821, 2003.
[3] D. J. Benford et al., in Proceedings of the 9th International Workshop on Low Temperature Detectors (AIP, 2001), Vol. 605, pp. 589.
[4] J. Fowler et al., App. Optics 45, 3746 (2006).

[5] T.E. Montroy et al., Proceedings of the SPIE, 6267, 62670R (2006).
[6] J.C. Kovac et al., Bulletin of the American Astronomical Society, 38, 913 (2006).
[7] M.D. Audley et al., Proceedings of the SPIE, 6275, 627524 (2006).
[8] W. S. Holland et al. Proceedings of the SPIE, 6275, 62751E (2006).
[9] E. Battistelli et al., Journal of Low Temperature Physics, Volume 151, Issue 3-4, pp. 908-914.
[10] E Battistelli, Proceedings of the SPIE 7020, 7020xP (2008).
[11] H. Spieler, "Frequency Domain Multiplexing for Large Scale Bolometer Arrays", Monterey Far-IR, Sub-mm and mm Detector Technology Workshop proceedings, 2002, pp. 243-249.
[12] T.M. Lanting et al., "Frequency domain multiplexing for bolometer arrays", Nuclear Instruments and Methods in Physics Research A vol. 520, 2004, pp. 548-550.
[13] T.M. Lanting, "MUX Readout of Superconducting Bolometers", 2007 UC Berkeley Ph.D Thesis.
[14] M. Dobbs, N. Halverson et. al., "APEX-SZ first-light and instrument status", New Astronomy Reviews vol. 50, 2006, pp. 960-968.
[15] John E. Ruhl et al., "The South Pole Telescope", Proc. SPIE Int. Soc. Opt. Eng. Vol. 5543, 2004. [astro-ph/0411122]
[16] M. Dobbs, E. Bissonnette, and H. Spieler, "Digital Frequency Domain Multiplexer for mm-Wavelength Telescopes", IEEE Transactions on Nuclear Science, TNS-00230-2007.R2, 2008 [arXiv:0708.2762v1].
[17] P. Oxley et al., "The EBEX Experiment", Proc. SPIE Int. Soc. Opt. Eng. Vol. 5543, 2004, pp. 320-331 [astro-ph/0501111v1]
[18] POLARBEAR Experiment [Online]. Available: http://bolo.berkeley.edu/polarbear/
[19] J.A. Chervenak, K.D. Irwin, E.N. Grossman, J.M. Martinis, C.D. Reintsema, and M.E. Huber, "Superconducting Multiplexer for Arrays of Transition Edge Sensors", Applied Physics Letters, vol. 74, 1999, pp. 4043-4045.
[20] W. Holland et al., "SCUBA-2: a 10,000 pixel submillimeter camera for the James Clerk Maxwell Telescope", Proceedings of the SPIE, vol. 6275, pp. 62751E (2006).
[21] K.D. Irwin, KW Lehnert, "Microwave SQUID multiplexer", Applied Physics Letters 85, 2107 (2004).
[22] Mates, J. B., Hilton, G. C., Irwin, K. D., Vale, L. R., Lehnert, K. W. Demonstration of a multiplexer of dissipationless superconducting quantum interference devices. App. Phys. Lett., 92, 023514.
[23] Xilinx Logicore DDS v5.0 Product Specification, 2005 [Online]. Available: http://www.xilinx.com/ipcenter/catalog/logicore/docs/dds.pdf
[24] M.E. Huber et al., "DC SQUID Series Array Amplifiers with 120 MHz Bandwidth (Corrected)", IEEE Transactions on Applied Superconductivity, vol. 11 (2), 2001, pp. 4048-4053.
[25] E. B. Hogenauer. "An economical class of digital filters for decimation and interpolation", IEEE Transactions on Acoustics, Speech and Signal Processing, ASSP-29(2) vol. 155, 1981.
[26] J. Hubmayr et al., (EBEX), "Design and characterization of TES bolometers and SQUID readout electronics for a balloon-borne application", submitted to the proceedings of SPIE, 2008.
[27] http://www.xilinx.com/publications/prod_mktg/virtex4qv_flyer.pdf
[28] Mazin, B. A., Day, P. K., Irwin, K. D., and Reintsema, C. D. Digital readouts for large microwave low-temperature detector arrays. Proceedings of LTD-11, vol 599, pp. 799–801, 2006.
[29] Schmidt, D. R., Duncan, W. D., Irwin, K.D., Lehnert, K.W., Miller, N. A., Ullom, J. N. Normal metal-insulator-superconductor junction technology for bolometers. Proceedings of

LTD-11, vol 599, pp. 516–518, 2006.
[30] Glenn, J., Day, P. K., Ferry, M., Gao, J., Golwala, S. R., Kumar, S., LeDuc, H. G., Mazin, B. A., Schlaerth, J., Vaillancourt, J., Zmuidzinas, J. A microwave kinetic inductance camera for sub/millimeter astrophysics. Proceedings of SPIE, in press.

Journal of Physics: Conference Series **155** (2009) 012005 doi:10.1088/1742-6596/155/1/012005

Optical Coupling

James J Bock,[1] J Gundersen,[2] Adrian T Lee,[3] P L Richards[4] and E Wollack[5]

[1]Jet Propulsion Laboratory, 4800 Oak Grove Drive, Pasadena, CA 91109

James.Bock@jpl.nasa.gov

[2]Department of Physics, University of Miami, Coral Gables, FL 33146

gunder@physics.miami.edu

[3]Department of Physics, University of California, Berkeley CA 94720
Physics Division, Lawrence Berkeley National Laboratory, Berkeley CA 94720

Adrian.Lee@berkeley.edu

[4]Department of Physics, University of California, Berkeley CA 94720

Richards@cosmology.berkeley.edu

[5]Observational Cosmology Laboratory, NASA/GSFC, Greenbelt, MD 20771

Edward.j.wollack@nasa.gov

Abstract. This paper describes contributions to the CMBpol Technology Study Workshop concerning optical coupling structures. These are structures in or near the focal plane which convert the free space wave to a superconducting microstrip on a SI wafer, or to the waveguide input to a HEMT receiver. In addition to an introduction and conclusions by the editor, this paper includes independent contributions by Bock on "Planar Antenna-Coupled Bolometers for CMB Polarimetry", by Gunderson and Wollack on "Millimeter-Wave Platlet Feeds", and by Lee on "Multi-band Dual-Polarization Lens-coupled Planar Antennas for Bolometric CMB polarimetry."

1. Introduction

The Workshop has chosen the term Optical Coupling to mean structures in, or near to the focal plane which convert the free space wave from the optics to a superconducting microstrip on a Si wafer, or to the waveguide input to a HEMT receiver module. Other terms, such as feedhorn or antenna with orthomode transducer (OMT) are in widespread use for such structures, but they tend to be technology specific in a way that is not appropriate for this Workshop.

The mapping speed for a detector array in a given band depends on the throughput, AΩ of the system and the detector sensitivity. It also depends on the Optical Coupling. A directly illuminated close packed detector array with pixel separation (dimension) fλ/2 gives the fastest mapping speed. However, this scheme is little used for CMB measurements at millimeter wavelengths. It has several disadvantages including a large numbers of detectors and essentially no beam definition. Much of the throughput of each detector views the interior of the cryostat. This imposes stringent conditions on blackness, low temperature and temperature stability. It also lacks the RF shielding provided by some optical coupling schemes.

The optical coupling schemes described below all provide significant beam definition by expanding the diffraction aperture to several wavelengths. They can apodize the diffraction by tapering the illumination. They typically reduce both the number of detectors required and the mapping speed. For example, a close packed array with pixel separation (dimension) of 2fλ requires an order of magnitude fewer detectors, but sacrifices a factor ~3 in mapping speed compared with the ideal directly illuminated array with the same throughput.

Most, but not all of the CMBpol mission concepts plan to use a cold aperture, or Lyot stop to supplement the beam definition provided by the optical coupling. This choice and the question of how aggressive the stop should be, depend on a number of system design choices and tradeoffs. In order to proceed with system design, it is necessary to understand the performance that can be achieved from the candidate technologies for optical coupling. Fortunately, much detailed information will be available since each of the optical coupling schemes discussed here has been used, or will soon be used in a major CMB polarization measurement.

In this paper, we present three quite different approaches to the challenge of optical coupling. The first contribution, by Gunderson and Wollack, describes the combination of ridged (scalar) horns. Ridged horns have a long heritage in space and the performance of ideal horns is well understood. These authors describe platelet fabrication technologies which can produce large arrays of such horns with near-ideal performance at acceptable cost. Most CMBpol mission concepts use focal planes with large arrays of superconducting bolometers at subkelvin temperatures. The use of ridged horns in such focal planes is likely to require a thermal break because arrays of horns may to be too heavy to cool to very low bolometer operating temperatures. Orthomode transducers (OMT), suitable for coupling ridged horns to bolometric detectors and HEMT amplifiers, are described in a separate paper by Wollack in these Proceedings.

The second contribution, by Bock uses a phased array of planar, subantennas for each pixel to synthesize a beam appropriate for coupling to the optics. In this aperture-synthesis approach, orthogonal arrays of polarization-sensitive subantennas feed microstrip outputs with orthogonal polarizations which couple naturally to microstrip filters and TES bolometers, All of these components are integrated on a single Si wafer. This approach has light weight and is very flexible. It can in principle produce coupling structures with any of several favorable properties. Interleaved antennas could even avoid the loss in mapping speed described above. In practice, however, there will be limits set by complexity that limit the number of these features that can be incorporated in a particular coupling structure. These issues, and the demands placed on fabrication by such high level integration, are being explored.

In the third contribution, by Lee, each pixel in a bolometric focal plane uses a broadband, planar, dual polarization, sinuous antenna, which is coupled to the optical beam by a hyperhemispherical Si lenselet. The microstrip outputs from the antenna can feed dual polarization diplexers, tripilexers, or multi band channelizers. The purpose of this approach is to

Journal of Physics: Conference Series **155** (2009) 012005 doi:10.1088/1742-6596/155/1/012005

use focal plane area efficiently by detecting several frequency bands in each optical pixel. Two TES bolometers will be used for each frequency band. If this approach proves successful, it could have a major impact on the design of CMBpol. Some aspects of this impact are described. The level of development if this approach, however, is lower than either of the approaches described above. In particular, the techniques for broad band antireflection coatings for optical elements such as a waveplate or a lense must be demonstrated. An approach to coating the Si lenselets is described. Neither of the new approaches to optical coupling being developed by Bock and Lee have a heritage in space. Many space specific issues, such as sensitivity to ionizing radiation, are yet to be explored.

Journal of Physics: Conference Series 155 (2009) 012005 doi:10.1088/1742-6596/155/1/012005

2 Millimeter-wave Corrugated Platelet Feeds

J Gundersen[1] and E Wollack[2]

[1]Department of Physics, University of Miami, Coral Gables, FL 33146

gunder@physics.miami.edu

[2]Observational Cosmology Laboratory, NASA/GSFC, Greenbelt, MD 20771

Edward.j.wollack@nasa.gov

Abstract. Feed horn arrays realized through platelet fabrication techniques represent a low cost, low risk approach to mass producing corrugated feed horns that maintain the superior performance of their electroformed equivalents. A brief review of the various approaches to fabricating these arrays is provided and their technical maturity is considered in comparison to their electroformed equivalents. The advantages and disadvantages of using corrugated platelet feed horns is provided, and an assessment of this approach's technology readiness level is established for its use with both modular coherent receiver systems as well as planar, bolometric arrays.

2.1 Introduction

Millimeter and submillimeter wave imaging systems for radio astronomy and remote sensing require high efficiency coupling structures. The use of such antenna elements in the design of focal plane arrays enables reductions in observing time and improvements in calibration accuracy, while maintaining desirable imaging properties. The realization of such low-loss transitions from free-space to the sensor requires a conversion in the impedance level and electromagnetic symmetry while achieving the radiometer's desired angular response, polarization, and frequency selectivity and maintaining an appropriate environment for optimal detector performance. The realization of such antenna elements in large focal plane arrays can potentially enable reductions in observing time and improvements in calibration accuracy while maintaining desirable imaging properties.

The conical corrugated horn is well known for its symmetrical Gaussian beam, low-side lobes, and cross polarization [1]. They have been the mainstay of optical coupling in cosmic microwave background (CMB) experiments. In particular, all of the experiments that have published detections of CMB polarization have used corrugated feed horns. These include DASI [2], CBI [3], CAPMAP [4], WMAP [5], Boomerang [6], and QUAD [7]. These experiments and their results provide ample evidence that a properly designed corrugated feed horn can meet the stringent systematic control associated with characterization of the CMB. Each of these experiments employed conventional techniques (either electroformed horns or directly machined horns) to fabricate individual feed horns that were then combined into arrays – none of which exceeded 50 elements. A difficulty with this approach arises when several hundred or a thousand elements is required as will likely be the case for a future CMBPol mission. In this case conventional machining techniques become impractical and/or very expensive.

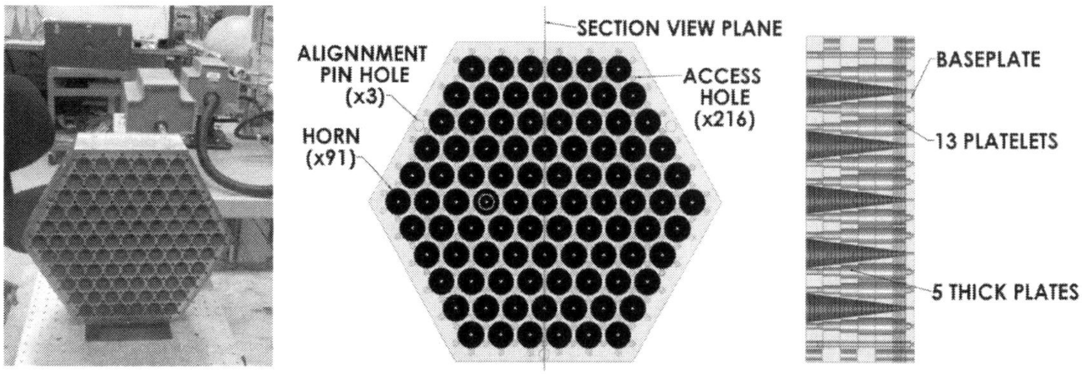

Figure 1. The left hand figure shows the 91-element W-band array that has been developed for QUIET as it was undergoing return loss measurements. The dimensions of the 20 kg array are 42.8 cm point to point, 37.0 cm across the flats, and 12.9 cm thick. The right hand figure shows that the platelet array consists of 5 thick plates each with 18 corrugations per horn hole, 13 platelets each consisting of a tooth and a groove, and a baseplate. The right hand figure also shows how the access holes open to light weighted pockets through the bulk of the structure.

2.2 Technical Description

A low cost alternative makes use of mass production to generate a pattern of holes on a large number of thin platelets that are subsequently bonded together to form a platelet array of corrugated feed horns. At centimeter to millimeter wavelengths this can be realized by direct machining bulk or sheets of material. In the submillimeter, chemical milling of the raw material is compatible with achieving the desired tolerances.

2.2.1 Design Features

The need for large numbers of corrugated feed horns was foreseen by [8] who were the first to develop platelet arrays of corrugated feed horns for remote sensing applications. These arrays consisted of ~100 thin sheets of zirconium copper with identical patterns of slightly different diameter holes etched into the plates. When these were diffusion bonded together, they formed a platelet array of corrugated feed horns. Recent variants on this approach have used different fabrication techniques (see below), but the overall design features have not changed drastically. In particular, the platelet array provides a flexible approach incorporating pocketing that can be used for either fluid cooling (important for [8]), light weighting, or to resolve complex interface issues as described in [9]. In addition, profiled corrugated feed horns such as those deployed in DASI can easily be implemented.

The largest departure from the fabrication described in [8] has been implemented in [9] who use a combination of thin platelets (each with a machined tooth and groove at the position of each horn hole) and series of thick plates each with 9-to-18 machined corrugations at the position of each horn hole. This design resulted in a significant reduction in setup time during machining, it minimized machining errors associated with the setup process, it significantly reduced the part count, and it eased the precision alignment and assembly process. Many of the design features that have been incorporated into platelet array assemblies are illustrated in Figure 1.

2.2.2 Fabrication Techniques

There are a number of techniques that have been employed to fabricate platelet arrays. These techniques use different materials, different material removal methods and different bonding techniques. These are summarized in Table 1. The material removal and to a lesser extent the choice of materials breaks depends on the frequency range. At low frequencies, computer

Journal of Physics: Conference Series 155 (2009) 012005 doi:10.1088/1742-6596/155/1/012005

numerical control (CNC) machining provides enough accuracy to define the holes. For fixed antenna gain, the horns are larger at lower frequencies and this drives the design toward lower density materials such as

Table 1. A summary of a number of platelet array fabrication techniques.

Reference	Material	Material Removal	Bonding	# of Elements	Frequency (GHz)
[8][10]	ZrCu +Al5052 Cu-plated SS#347	Etching	Diffusion Bonding	9, 16 16	90 626
[9]	Al6061	CNC	Diffusion Bonding	19 91	40 90
[11][12]	Copper	Etching	Diffusion Welding + Epoxy versions	1020	500
[13]	Brass	CNC	No bonding	31	100
[14]	Metalized Silicon	DRIE	TBD	TBD	150

aluminum. At higher frequencies, etching is necessary to precisely remove the material, and lower loss materials, such as copper, are preferred.

Coupling platelet feed horns to arrays of coherent receivers is straightforward since they can all reside at the same temperature, and the devices that follow the feeds, such as orthomode transducers (OMT) and septum polarizers, are waveguide based. Coupling platelet feed horns to an array of bolometric detectors is more challenging – see section 2.3. One of the challenges is to deal with the differential contraction between the metallic platelet array and the silicon used in bolometric arrays. One solution that is being pursued at the National Institute of Standards-Boulder [14], is to micromachined silicon platelets using Deep Reactive Ion Etch (DRIE) to enable easier coupling to arrays of bolometric detectors. Once the array structure is metalized, aligned, and fused, a low thermal mass structure with desired thermal and mass properties is envisioned. The resulting structure would also be thermally matched to silicon detector wafers, thus simplifying the alignment of the feeds to the detectors.

2.2.3 Coupling and Integration

For CMB polarization experiments, corrugated feeds are sometimes used with a waveguide polarization diplexer such as an OMT or a septum polarizer. For coherent receiver systems, these can be realized in waveguide assemblies and bolted onto the back of the platelet array. A more detailed description of these devices and how they would be used in both coherent and bolometric systems is provided in these proceedings [15]. For a bolometric array using a metallic platelet array, a thermal break would likely be necessary between the platelet array operating at 2-4 K and the bolometer array at 0.1-0.3 K. Introducing such a thermal break is challenging because it can cause unnecessary signal losses and it places cryogenic alignment between the horns and the detectors' waveguide inputs. One promising approach uses a photonic waveguide joint [16]. A dual-mode realization of this structure allows control over the power transport in the waveguide interface, a significant reduction in the necessary alignment tolerance, and a natural location for the detector and its supporting circuitry. This sensor configuration provides an environment which is compatible with the spaceborne thermal and ionizing radiation environments.

The final coupling occurs between dual mode waveguide (i.e. either rectangular or circular waveguide) and a planar structure – either a coherent receiver module or a bolometric detector element. Many approaches have been used to realize the transition between the waveguide and the detector. In the coherent case there are mature techniques to achieve this coupling and representative examples are provided in following references [17]. In the bolometric case, several coupling approaches have been explored and implemented [18]. Efforts to realize this coupling function via a planar OMT are arguably at a lower level of technical maturity; however, they are being investigated by several groups [15].

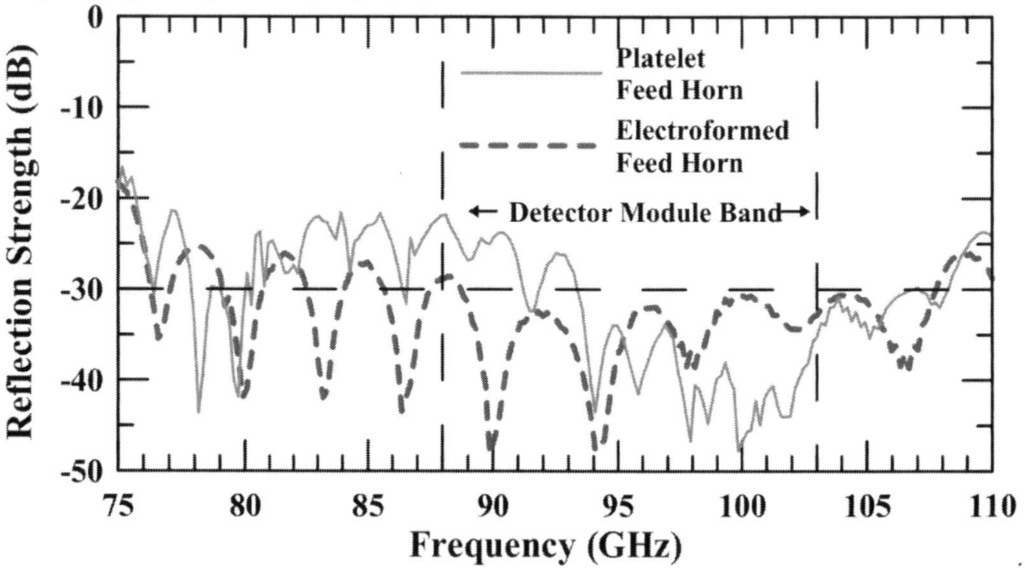

Figure 2. Return loss measurements from one of the horns in the QUIET 91-element W-band array is compared to an electroformed equivalent of the horn.

2.3 Technical Maturity

2.3.1 Performance Metrics

One measure of the technical maturity of platelet arrays is to compare their performance against theoretical models and/or electroformed versions of the platelet array feed horns. Most of the papers above that describe platelet feed horn array fabrication also summarize their observed performance characteristics. The quantities typically measured and reported include return loss, co- and cross-polar beam patterns and insertion loss.

Return Loss

Return loss is measured by connecting one port of a vector network analyzer to a horn with a circular to rectangular waveguide transition and then recording the S11 parameter while the horn is terminated by a piece of microwave absorber. Given this measurement approach the transition between the waveguide and the feed throat dominates the result. Return loss measurements are particularly sensitive to machining errors (e.g. [5]) and imperfect bonding or flanging which can be manifested as resonances (i.e., 'suck outs') in the return loss spectrum. Return loss values of 30 dB can be designed into corrugated feed horns over >25% bandwidths and realized in electroformed versions. Return loss measurements of platelet feed systems are typically close to but not identical to their electroformed counterparts. As an example, Figure 2 shows return loss values closer to 25 dB were observed in the lower third of the QUIET detector module's band (88-93 GHz), while from 93-103 GHz, the return loss was greater than 30 dB (see Figure 2). Even the 25 dB return loss is an acceptable level given that the septum polarizer used for QUIET's polarization diplexing is expected to dominate the return loss. Perhaps just as important, and not something that has been consistently reported in previous publications is that the return loss was very repeatable (typically +/- 1 dB for return losses <30 dB) from horn to horn for both the 19-element Q-band array and the 91 element W-band array.

Beam Patterns.

Beam pattern measurements are another diagnostic used to evaluate the performance of platelet arrays. All papers that have reported on platelet arrays have also reported some subset of E- and H- plane, co- and cross-polar beam measurements and compared them to either theoretical predictions and/or electroformed equivalents. The results are generally encouraging, although, as with the return loss measurements, there are cases where there are some noticeable

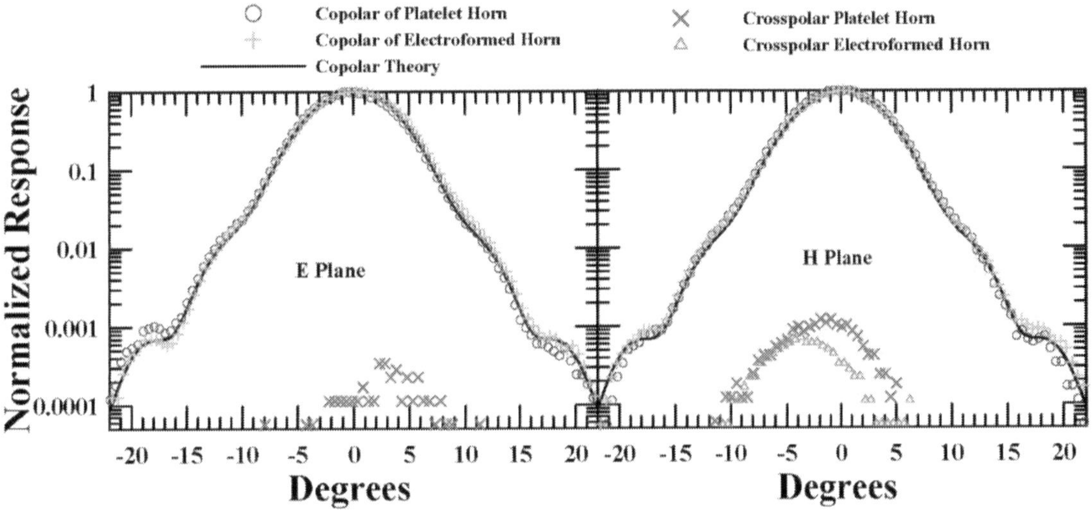

Figure 3. Co-polar and Cross-polar beam patterns in both E and H planes for one horn in the QUIET 19-element Q-band array are compared to measurements of an electroformed equivalent horn as well as to theoretical predictions. Theoretical predictions of the cross-polar levels as well as the electroformed horn's measured E plane were <-40 dB and are not shown.

deviations between the measured platelet response and the theoretical prediction or the electroformed equivalent. The pathologies associated with the platelet arrays that would give rise to these deviations have not been fully investigated, though in most cases the deviations are small enough so as not to cause significant deleterious effects.

The theoretical predictions of corrugated feed horn's beam patterns and return loss are extremely mature. One example is the software CCORHRN from YRS Associates which was used to design both the WMAP and QUIET feeds and has been experimentally verified down to the -80 dB level [5]. This agreement is possible due to the high degree of axial symmetry present in the structure studied. Finite element codes have been used to study the influence of a systematic reduction in axial symmetry and have found that the desired response is tolerant to adiabatic errors in the layup of the structure at level relevant to this application. Thus, in principle these pathologies could be investigated at a greater detail if necessitated by other design considerations. A measurement of particular relevance to CMB polarization experiments is the cross-polar response of the feed horns, and this is typically under reported in the above references. For an axially symmetric feed structure the cross-pol is a measure of asymmetry between the E- and H-plane angular responses. In the limit other modes are present; this simple interpretation can break down thus requiring knowledge or measurement.

Measured examples of beam pattern measurements from one of the horns on the QUIET 19-element Q-band array are shown in Figure 3. These illustrate that the co-polar beams match the theoretical predictions as well as the electroformed measurements very well down to the 30 dB

Journal of Physics: Conference Series **155** (2009) 012005 doi:10.1088/1742-6596/155/1/012005

level. The peak cross polar levels are predicted to be <-40 dB, while typical peak measured levels are in the -30 to -40 dB regime. These measured levels are higher than the theoretical predictions and are believed to be due to misalignments in the test apparatus. Measurements are currently being performed to confirm this.

Insertion loss

Insertion loss measurements are often difficult to perform because, for well manufactured horns with high electrical conductivity, the insertion loss would only manifest itself as a reduction in transmission efficiency (i.e., increase in system temperature). For applications at hand the influence of the insertion loss is expected to be subdominant compared to other potential sources of in-band emission. In cases where there is incomplete diffusion bonding, or as in the case of [13] where there is no bonding at all, increased insertion loss at cryogenic temperatures might be experienced due to the presence of gaps if insufficient platelet layer compression is realized. Crude measurements were performed at room temperature on the QUIET arrays by shorting the horn apertures with a metal plate during the room temperature return loss measurements, and typical values of <0.1 dB were observed. Further reductions would be anticipated upon cooling the feed array structure. Characterization of copper platelet waveguide sections at room and cryogenic temperatures indicate that loss within a factor of two of theoretical is achievable at millimeter wavelengths [11].

2.3.2 Readiness

In addition to the performance milestones discussed above, it is useful to establish the platelet array's stage of technology development. Up until very recently, all platelet arrays mentioned above were prototypes that were tested in the lab to various levels. Just recently (7/08) the first platelet array has moved out of the lab and into the field to begin taking data as part of the QUIET 19-element Q-band array. This has just been deployed to the Chajnantor Observatory on the Atacama plateau in Chile, and the 91-element W-band array will follow in early 2009.

2.4 Benefits and Disadvantages of Platelet Arrays

2.4.1 Benefits

1) Platelet arrays of corrugated feeds use a well established machining/fabrication processes for metallic structures which are compatible with reliable cryogenic end use.
2) Platelet arrays of corrugated feeds represent a viable technological path toward large number of detector elements in a focal plane.
3) Platelet arrays of corrugated feeds provide a well controlled and repeatable illumination pattern that enables desired return loss, beam shape, spill, and instrument polarization response.
4) Platelet arrays of corrugated feeds are compatible with high performance polarization diplexing and microstrip based filtering approaches and in addition they allow natural separation of IR and millimeter wave filtering needs.
5) The theoretical predictions of platelet array performance are very mature.

2.4.2 Disadvantages

1) Relatively high metal mass/volume used in convention approach required lightening to produce more optimal component for space borne application.
2) The simplest platelet array realization requires a flat focal surface. As in all antenna illumination approaches, the f/ and number of optimally performing detectors pixels are coupled in specification of the design. The necessary system sensitivity may push one toward a configuration with multiple optical assemblies in order to achieve the required number of pixels on the sky with the desired angular resolution.

3) The detector coupling structures are waveguide based. As a result, the packaging requirement can be more demanding than planar array based approaches
4) Depending upon system implementation, quasi-optical IR blocking filters may be required after feed illumination definition in order to achieve the desired thermal environment for the detector.

2.5 Technology Readiness Level (TRL)

The technical readiness of platelet feed array structures when used in conjunction with cryogenic coherent receiver systems is at level five or higher. For bolometric imaging systems the interface between the platelet array and the detector forces one to assign a lower level of readiness for the technology. Laboratory efforts in this area are underway and a TRL of ~3-to-4 is estimated for the approach at the present time. The resources required to push this technology to TRL five for this case will be dominated by detector and system level costs associated with fielding instruments which utilize and validate this technology.

Acknowledgements

The development of the QUIET 19 element Q-band array and the 91 element W-band array were supported by JPL and the NSF. William Imbriale, David Leibovitch, Joelle Cooperrider, Raul Monsalve and Yunior Savon were integral in the design and testing of the arrays. Todd Gaier, Suzanne Staggs and Glen Nixon provided assistance with return loss measurements.

References

[1] Clarricoats P J B and A.D. Olver A D 1984 *Corrugated Horns for Microwave Antennas* (Stevenage, UK: Peregrinus)
[2] Leitch E M, Pryke C, Halverson N W, Kovac J, Davidson G, LaRoque S, Schartman E, Yamasaki J, Carlstrom J E, Holzapfel W L, Dragovan M, Cartwright J K, Mason B S, Padin S, Pearson T J, Readhead A C S and Shepherd M C 2002 *Astrophys. J.* **568** 28
[3] Padin S, Shepherd M C, Cartwright J K, Keeney R G, Mason B S, Pearson T J, Readhead A C S, Schaal W A, Sievers J, Udomprasert P S, Yamasaki J K, Holzapfel W L, Carlstrom J E, Joy M, Myers S T and Otarola A 2002 *PASP* **114** 83-97
[4] Barkats D, Bischoff C, Farese P, Gaier T, Gundersen J O, Hedman M M, Hyatt L, McMahon J J, Samtleben D, Staggs S T, Stefanescu E, Vanderlinde K and Winstein B 2005 *Astrophys. J. Supp.* **159** 1
[5] Barnes C, Limon M, Page L, Bennett C, Bradley S, Halpern M, Hinshaw G, Jarosik N, Jones W, Kogut A, Meyer S, Motrunich O, Tucker G T, Wilkinson D and Wollack E 2002 *Astrophys. J. Supp.* **143** 567-76
[6] Jones W C, Bhatia R, Bock J J and Lange A E *Millimeter and Submillimeter Detectors for Astronomy: Proceedings of SPIE* vol 4855 ed T G Phillips and J Zmuidzinas (Bellingham, WA: SPIE) pp 227
[7] Hinderks J R, Ade P, Bock J, Bowden M, Brown M L, Cahill G, Carlstrom J E, Castro P G, Church S, Culverhouse T, Friedman R, Ganga K, Gear W K, Gupta S, Harris J, Haynes V, Kovac J, Kirby E, Lange A E, Leitch E, Mallie O E, Melhuish S, Murphy A, Orlando A, Schwarz R, O'Sullivan C, Piccirillo L, Pryke C, Rajguru N, Rusholme B, Taylor A N, Thompson K L, Tucker C, Wu E Y S and Zemcov M 2008 *Preprint astro-ph/0805.1990v1*
[8] Haas R W, Brest D, Mueggenburg H, Lang L and Heimlich D 1993, *Int. J. of Infrared and Millimeter Waves* **14**, 2289-93
[9] Gundersen J O, Imbriale W, Cooperrider J, Leibovitch D, Monsalve R and Savon Y 2008 Diffusion bonded corrugated platelet feedhorn arrays for mm-wave astronomy in preparation
[10] Haas R W 1995 *Multi-feed Systems for Radio Telescopes ASP Conference Series* vol 75 ed

Journal of Physics: Conference Series **155** (2009) 012005 doi:10.1088/1742-6596/155/1/012005

D T Emerson and J M Payne (San Francisco, CA: ASP) pp 99-105
[11] Wollack E J and Moseley S H 2005 Corrugated Platelet Feed Arrays for Millimeter Wave Imaging, online *Proc. of URSI National Radio Science Meeting* (Boulder, CO: URSI) p 398.
[12] Benford D J, Chervenak J A, Irwin K D, Moseley S H, Shafer R A, Staguhn J G and Wollack E 2003 *Millimeter and Submillimeter Detectors for Astronomy: Proc. of SPIE* vol 4855 ed T G Phillips and J Zmuidzinas, (Bellingham, WA: SPIE) pp 148-62
[13] Kangas M M, Ansmann M, Horgan B, Lemaster N, Leonardi R, Levy A, Lubin P, Marvil J, McCreary P and Villela T, 2005, *IEEE Ant. and Wireless Prop. Let.* **4** 245–8
[14] Irwin K 2008 private communication
[15] Wollack E 2008 Millimeter-wave Orthomode Transducers in these proceedings
[16] Hesler J 2001 *Microwave Symposium Digest IEEE MTT-S* **2** 783–6.
[17] Leong Y C and Weinreb S 1999 *IEEE MTT-S* 1435-8
Samoska L, Deal W R, Chattopadhyay G, Pukala D, Fung A, Gaier T, Soria M, Radisic V, Mei X, and Lai R 2008 *IEEE Trans. on Microwave Theory and Tech.* **56**, No. 6, 1380-8
[18] Yassin G, Withington S, Buffey M, Jacobs K and Wulff S 2000 *IEEE MTT* **48** 662
Glenn J, Chattopadhyay G, Edgington S F, Lange A E, Bock J J, Mauskopf P D and Lee A T 2002 *Applied Optics* **41** 136

3. Planar Antenna-Coupled Bolometers for CMB Polarimetry

James J. Bock

Jet Propulsion Laboratory

James.Bock@jpl.nasa.gov

Version 5, 19 August 2008

Abstract. Antenna-coupled detectors provide all the functions required of a CMB polarimeter, including beam formation, spectral band definition, and polarization analysis. Planar antennas are fully lithographed devices that do not require coupling optics, and thus readily scale to the production of large focal plane arrays. Antennas coupled to direct detectors such as transition-edge bolometers can realize background-limited sensitivity, and, in a single technology, cover the range of frequencies likely to be required by a future post-Planck satellite experiment for foreground monitoring and removal. We have successfully tested single devices at 150 GHz with the desired beam shapes, pass bands, and polarization properties with high optical efficiency, and are now developing focal plane arrays for the BICEP2 and SPIDER experiments.

3.1 Introduction

Antenna-coupled detectors provide polarization discrimination, beam formation and spectral band definition, and are thus a natural approach for CMB polarimetry. A planar antenna consists of a distributed arrangement of sub-antennas coherently summed with a combiner network. The response of the sub-antennas in coherent combination gives higher forward beam directivity than that of a single sub-antenna in isolation. The shape of the main beam is controlled through the size of the antenna, and the field amplitude and phase in each sub-antenna. This technique of beam formation eliminates the need for additional coupling optics, such as lenses or feedhorns. Planar antennas are a very flexible technology that can be used to produce highly tapered, directed, and/or overlapping beams with multiple frequency bands and the simultaneous analysis of multiple polarization parameters.

A network of sub-antennas with a combining network is shown in Fig. 4. The sub-antennas must have a minimum spacing to avoid generating grating lobes or surface waves, approximately given by the Nyquist condition for sampling a surface wave, $s < \lambda_{min}/2n$, where s is the spacing, λ_{min} is the shortest wavelength in the passband and n is the index of the substrate. A more exact analysis indicates the spacing can be somewhat relaxed to $\sim\lambda_{min}/1.2n$. The size of the antenna, effectively the number of sub-antennas commbined, is determined by the desired beam width, $\Delta\theta \approx \lambda/d$.

3.2 Advantages of Planar Antennas

Unlike 'cavity-coupled' bolometers, such as the devices used on Planck, where the absorber is approximately a wavelength in diameter, antenna-coupled bolometers use a resistive transmission line termination to absorb RF power, and this termination can be as small as lithographic techniques permit. Antenna-coupled detectors can operate over the range of frequencies anticipated for CMBPOL, 30 – 300 GHz [19], by scaling the

Journal of Physics: Conference Series **155** (2009) 012005 doi:10.1088/1742-6596/155/1/012005

antenna and filter design but leaving the detective element unchanged. This approach overcomes the limitation of bolometers with cavity-coupled absorbers, which have not been designed to operate at frequencies lower than 70 GHz. While it is possible this technology could be pushed to still lower frequencies, the lowest frequency Planck devices began to show both lower mechanical yield and slower time constants due to their large physical size.

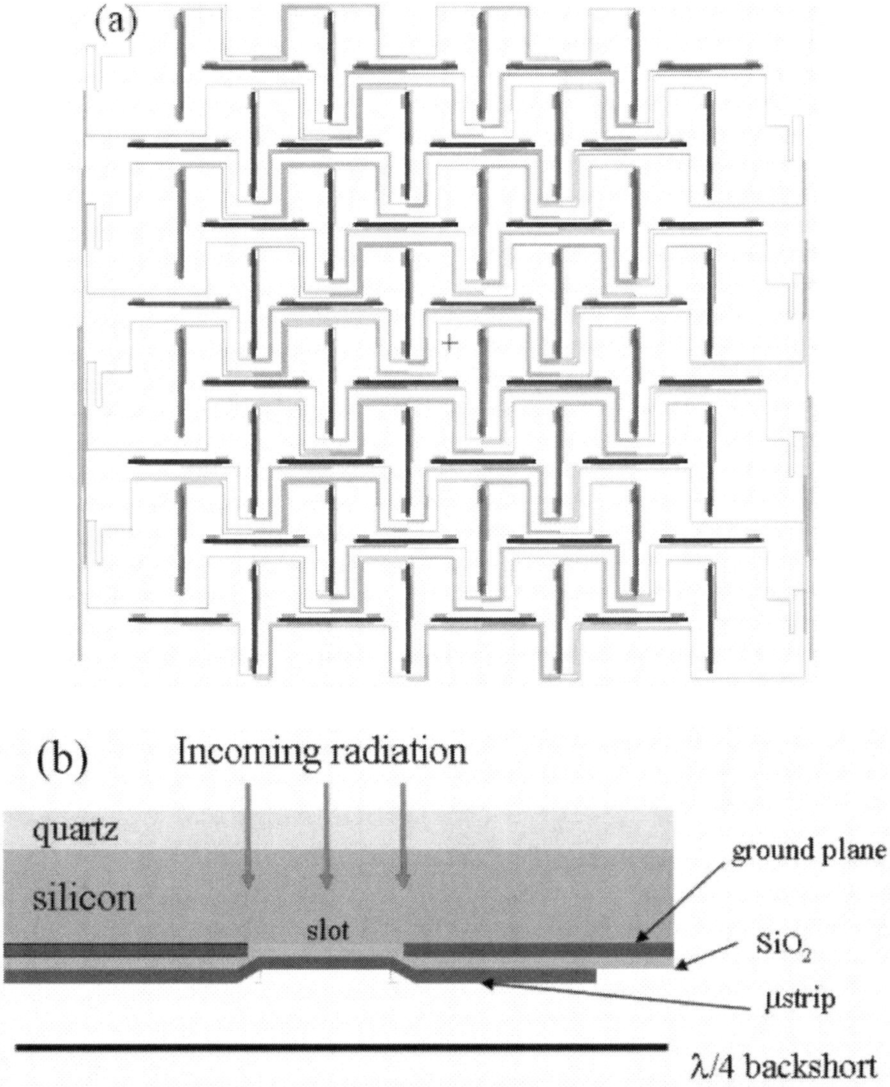

Fig. 4. Schematic of a planar antenna (a). The sub-antennas are slots (red and blue lines) combined by a microstrip summing network (grey lines). There are two antennas in this configuration which are closely interleaved, one in vertical polarization and one in horizontal polarization, and the slots and summing networks are independent. Actual antennas combine larger numbers of sub-antennas (see Fig. 5c) than shown in this illustration. The antenna is illuminated through the silicon substrate (b), with microstrip taps coupling to the slot antennas. A λ/4n layer of fused quartz serves as an anti-reflection coating. A reflective λ/4 backshort is used to improve the optical efficiency, preventing radiative loss to free space.

Planar antennas are entirely lithographed and thus naturally lend themselves to the production of large focal plane arrays. Planar antennas do not require coupling optics such as lenses or feedhorns, but do require a flat anti-reflection coating. This coating can be quite simple, e.g. for devices with 30 % bandwidth a single $\sqrt{n}$ coating suffices, and, for devices made on Si, a tuned wafer of fused quartz is a highly efficient coating. Compared with lens-coupled antennas [20][21], planar antennas eliminate the alignment between the lenses with the exciting antenna, and the potentially difficult problem of anti-reflection coating the highly curved lens surface. Compared with feedhorn-coupled antennas [22], planar antennas eliminate the mass of the feedhorns. The mass of such structures at sub-K temperatures drives the design of the mechanical suspension and the sub-K cooling system, and can have adverse systems implications.

Planar antennas provide an elegant means for controlling the beam pattern [23]. With a suitable summing network, the amplitude and phase of the electric field can be chosen at every sub-antenna. For example, by changing the phase across the sub-antennas, the beam direction can be controlled. In particular the beam can be steered away from normal incidence, while keeping the focal plane flat, so that planar antennas can be used with non-telecentric optics. By varying the amplitude of the electric field over the sub-antennas, the beam pattern can be tapered to reduce spillover off the telescope optics. As the electric field can be controlled exactly at every sub-antenna, the illumination pattern from a planar antenna can be highly optimized. This degree of control is not possible with lens coupled antennas. Planar antennas can be designed to eliminate differential beam ellipticity, because the x-y beam widths are determined by the x-y field distribution of the sub-antennas, and only weakly by the radiation pattern of the sub-antennas themselves.

A future CMB polarization mission will require efficient use of the available field of view provided by the optics. Planar antennas offer several advantages for increased packing density. Like lens-coupled antennas, planar antennas can be operated in multiple polarizations simultaneously, or in multiple bands. The antennas are currently designed to extract a single Stokes polarization parameter (Q or U) by differencing 2 matched detectors and antennas, providing common-mode rejection of temperature anisotropy, unpolarized emission from the optics, and focal plane temperature fluctuations. Multi-color designs are also possible, although this brings the additional difficulty of anti-reflection coating over a wide bandwidth and controlling the beams in multiple bands. We have demonstrated a design that realizes more than an octave of bandwidth in one polarization [24].

In any focal plane, configured with either antennas or feeds, there is a tradeoff between the density of detectors and the illumination pattern on the primary aperture. We have compared the mapping speed of various focal plane architectures assuming the available $A\Omega$ product of the available focal plane is held constant [25]. A design with higher edge taper on the primary necessitates a larger antenna (or feed). Focal planes with low spillover, characteristic of CMB instruments, suffer an overall loss in mapping speed compared to bare detector focal plane arrays, which detect all the incident photons but provide no beam control. An aggressive design with ~10 % edge spillover, typical of $2f\lambda$ antennas or feeds, suffers an overall loss in speed of a factor of ~3 compared with an ideal bare-detector focal plane array. More conservative designs with lower spillover have even larger losses in speed due to the fact that the pixels must be larger. This factor

Journal of Physics: Conference Series **155** (2009) 012005 doi:10.1088/1742-6596/155/1/012005

can be recovered using a focal plane array of planar antennas with overlapping field distributions [26]. While the design of such a focal plane array will be involved, it is possible to simultaneously control the illumination on the primary aperture, while fully realizing the ideal mapping speed [27].

3.3 State of Development

We have developed focal plane arrays of planar-antenna detectors [28] as shown in Fig. 5. The device has two interleaved antennas detecting vertical and horizontal polarization, terminated in two TES bolometers. The sum and difference of the two signals produces Stokes Q and I. The sub-antennas are slot antennas in a superconducting Nb ground plane, and the antenna is backside illuminated through the substrate Si. A single $\lambda/4n$ layer of fused quartz is placed on the backside Si for anti-reflection coating. The slot arrangement places the RF components of the detector (summing network and stripline filters) behind the ground plane so they cannot couple to incident radiation.

The slots are interleaved, as shown in Fig. 4, to achieve the required sub-antenna spacing. Each slot is fed off center with two microstrip taps, at a location where the antenna has a suitable impedance to conveniently match to microstrip. Each tap is combined with a rectangular capacitor to tune out the reactance of the antenna. The sub-antennas are combined by a summing network, which in the time reversed sense couples equal amplitude and phase to each sub-antenna. This arrangement gives a beam pattern resembling a two-dimensional sinc function.

Fig. 5. (a) Focal plane array of planar antenna-coupled bolometers produced on a 100 mm wafer consisting of 64 antennas each with two TES bolometers measuring vertical and horizontal polarization. (b) A focal plane utilizing 4 such arrays with a total of 512 TES bolometers. The focal plane arrays are read out with time-domain SQUID multiplexers, mounted at the top and bottom of the circuit board. Each planar antenna (c) is a dual polarization device with the sub-slot driven with identical phase and amplitude by a combining network. The antenna is designed to produce equivalent beam widths along the co-polar and cross-polar direction, eliminating the false polarization signal caused by differential ellipticity, and identical beam centers in the two polarizations, eliminating the false polarization signal caused by differential pointing [29][30]. The summing network combines the RF signal onto a single transmission line which then passes through a lumped-element RF filter (d) and then to a thermally-isolated termination resistor (e) where the thermal rise from dissipated RF power is detected by a TES bolometer.

These antennas were designed to be coupled to f/2 optics with a cold absorbing aperture stop. The antennas are 8 mm on a side and operate at 150 GHz, giving 2fλ pixels at f/2. We chose to have an aggressive spillover in order to use smaller pixels with a higher packing density. The aperture is approximately near the first beam minimum, and the secondary lobes fall onto the aperture stop giving a power spillover of

~10 %. Tapering the beam produces little improvement for an antenna of this size. Larger antennas with tapering would provide better beam control, but at the price of reduced packing density.

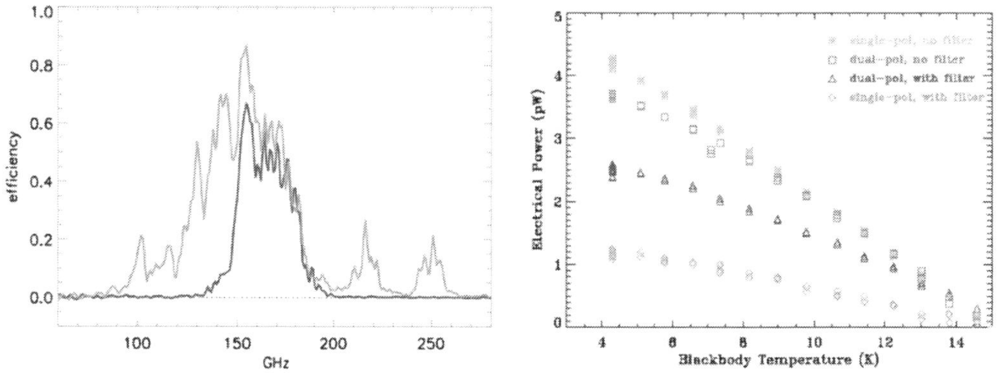

Fig. 6. (Left) Measured spectral response of an unfiltered (orange) and filtered (blue) antenna, including the quartz anti-reflection coating, antenna, combiner network, and meander absorber but not including the λ/4 backshort. The response curves are absolutely normalized by measuring the response to a cryogenic blackbody source (Right) so that the measured response convolved with the blackbody function with single-mode coupling in one polarization produces matches the measured power from the blackbody source. This efficiency includes the filter, absorber, antenna, anti-reflection coating, and (absorbing) backshort.

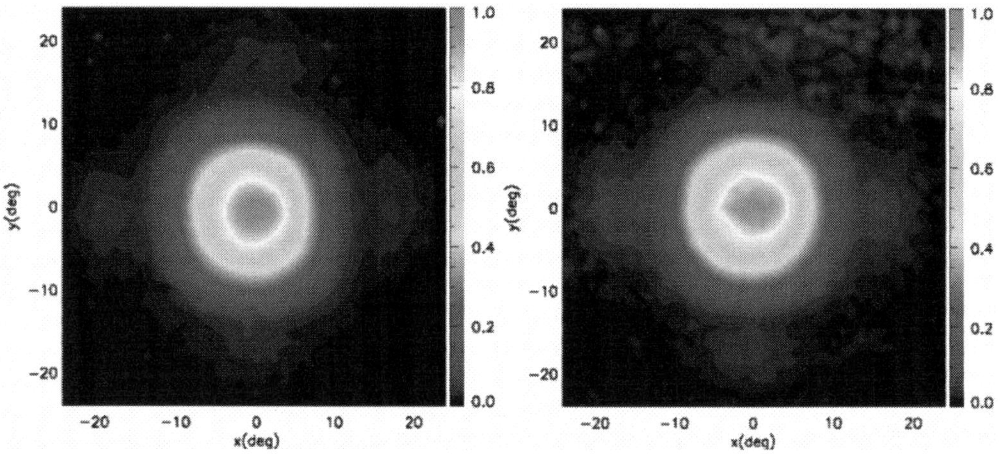

Fig.7. Beam patterns of a filtered dual-polarization antenna in vertical and horizontal polarizations measured with a chopped thermal source. The original of the small deviations seen in the above panels are not clear, though we have seen evidence that reflections in the testbed windows and filters can effects at this level.

Measurements of full planar antenna-coupled TES bolometers are shown in Figs. 6 and 7. The response was measured looking out through a low-pass filter stack and window into a Fourier-transform spectrometer. The spectral bandwidth of the antenna is $\Delta\nu/\nu = 35$ %. The efficiency shown in Fig. 6 was normalized by measuring the response to a cryogenic blackbody source. The devices show high optical efficiency, typically 50 – 80 %, measured using an absorbing backshort. This efficiency is notably improved over that of corrugated feeds combined with cavity-coupled polarization sensitive bolometers [31][32], typically 20 – 40 %. Using a $\lambda/4$ reflecting backshort will recover the ~10 % loss that the antennas radiate into the vacuum side of the device [33]. The beam patterns, measured with a chopped thermal source and shown in Fig. 7, are highly symmetric and well matched. The devices clearly show the expected characteristic two-dimensional sinc pattern and secondary sidelobes. The cross-polarized response, measured with a chopped source and a rotating polarizing grid, is < 3 %.

Breaks or shorts in the feed network can cause imperfections in the beam patterns. The effect of a single missing sub-slot in the feed network shown in Fig. 5c is fairly subtle, causing a deviation from the expected pattern at the -28 dB level, which is not measurable with our existing test apparatus. Of course more missing slots cause larger deviations, and the deviations depend on the location of a defect in the feed network. We carried out electrical continuity measurements of long Nb/SiO2/Nb test striplines with tapered sidewalls that indicate defects in freshly deposited and bias-sputtered dielectric are basically negligible. However, in fully processed wafers, we have noticed occasional shorts between the wiring layer and the ground plane, apparently due to pinholes in the dielectric layer developing as a result of the additional processing steps required to form the TES and absorber. We are currently testing several techniques to reduce defects to a negligible level for arrays.

3.4 Future Developments

With single-element 150 GHz planar antennas successfully demonstrated, we are now in the processes developing focal plane arrays. Testing of these focal plane arrays will be used for process control at wafer level and wafer to wafer that can affect the optical performance, e.g. fabrication errors (shorts and breaks) in the combining network and variability (thickness, width and index) in the microstrip dielectric. The next step is to expand the frequency range of operation. We are currently testing designs for 100 GHz, and ultimately plan to demonstrate devices in bands ranging from 30 – 300 GHz.

A simple modification to the antennas allows each pixel to sense Stokes Q and U simultaneously. This may provide improved systematic error control, since otherwise Stokes Q and U must be extracted either by using different detector pairs, or using the same detector after a boresight rotation and corresponding time interval. To measure Q and U, two antenna output lines E_x and E_y may be first power divided and half of each line routed to two detectors producing $E_x^2/2$ and $E_y^2/2$. The remaining half can be routed though a broadband 180° hybrid followed by two detectors, producing $(E_x+E_y)^2/4$ and $(E_x-E_y)^2/4$. The difference between each detector pair produces Q and U respectively. If the detectors are photon noise limited, there is no sensitivity penalty in such an arrangement. Two balanced hybrids can be used in place of one hybrid to minimize differential spectral features from the hybrid.

Tailoring the field amplitude and phase distribution over the antenna can provide improved sidelobe control, an important development for optical systems without a cold stop, e.g. the Planck telescope or crossed-Dragone optics. Because the field distribution over the antenna may be designed exactly, planar antennas are an ideal technology for such precise beam control. With a focal plane array of antennas, the field distribution may be shared between multiple detectors to produce overlapped beams to simultaneously provide beam formation with maximum mapping speed, though this will require the development of complex and dense feed networks with crossovers.

Planar antennas are a flexible technology. Some attributes represent enabling technology for CMB polarimetry from space, while other capabilities are merely advantageous. We have listed a set of milestones to demonstrate readiness (TRL ~ 6) for space, and list separately capabilities that provide additional advantageous that would benefit CMBPOL. The milestones described below are what we consider to be the necessary and minimum demonstration of technical readiness. Where the capability is likely to interact with a full system, we recommend a demonstration at instrument level. Where the required performance can be fully characterized in the laboratory we recommend the development of a single pixel, and assume that space-qualified focal plane arrays developed under the auspices of the future CMBPOL mission can be based on this demonstration. Full focal planes (see Fig. 5b) intended for the BICEP2 and SPIDER experiments are currently in development to fulfil milestone #2. These would provide an end-to-end test in a representative instrument, demonstrating suitability for a space-borne densely-packed focal plane appropriate for an optical system with a cold aperture stop.

Table 2. Milestones Required to Demonstrate Technology Readiness for Space

Capability	**Advantage for CMBPOL**	**Technical Challenges**	**Milestone**
Attributes Necessary for CMBPOL			
Optical coupling	Beam formation, polarization analysis, band definition	Antenna and filter design, RF properties and losses	Pixel demonstration (completed at 100 and 150 GHz)
Focal plane arrays	System sensitivity	Process uniformity and reliability	Field focal plane arrays in a CMB receiver
Frequency coverage	Foreground removal	Antenna and filter design	Pixel demonstrations from 30 to 300 GHz
Attributes Necessary for Some CMBPOL Mission Designs			
Highly tapered beams	Sidelobe control for optics without a cold stop	Antenna and combining network design	Pixel demonstration in a single band
Simultaneous Stokes I, Q and U	Systematics control, depends on scan strategy	Hybrid design and spectral band matching	Pixel demonstration in a single band
Attributes Advantageous for CMBPOL			
Directed beams	Use of non-telecentric optical designs	Antenna and combining network design	Pixel demonstration in a single band
Multiple frequency bands per antenna	System sensitivity	Diplexer design Dual-polarization design	Pixel demonstration in multiple bands
Overlapping antennas	System sensitivity	Antenna and combining network design, transmission line density and crossovers	Focal plane array demonstration in a single band
Polarization modulation	Noise stability	Active device to switch polarization states, RF design	Field focal plane arrays in CMB receiver

References

[19] R. Weiss *et al.* 2006, "Task Force on Cosmic Microwave Background Research", astro-ph 0604101.

[20] M.J. Meyers *et al.* 2008, "Antenna-Coupled Bolometer Arrays for Measurement of the Cosmic Microwave Background Polarization", J. Low Temp. Phys. 151, 464.

[21] J. Zmuidzinas and H.G. LeDuc 1992, "Quasi-Optical Slot Antenna SIS Mixers", IEEE Trans. Micro. Th. Tech. 40, 1797.

[22] T. Stevenson et al. 2008, "Cosmic Microwave Background Polarization Detector with High Efficiency, Broad Bandwidth, and Highly Symmetric Coupling to Transition Edge Sensor Bolometers", J. Low Temp. Phys. 151, 471.
[23] P. Day *et al.* 2008, "Design of slot-array antennas for applications in astronomy", in prep..
[24] J. Schlaerth *et al.* 2008, "A Millimeter and Submillimeter Kinetic Indutance Detector Camera", J. Low Temp. Phys. 151, 684.
[25] M.J. Griffin, J.J. Bock and W.K. Gear 2002, "Relative Performance of Filled and Feedhorn-Coupled Focal-Plane Architectures", Applied Optics 41, 6543.
[26] S. Stein 1962, "On Cross Coupling in Multiple-Beam Antennas", IEEE Trans. Ant. Prop. 10, 548.
[27] J.F. Johansson 1989, "Stein's Limit for Finite and Infinite Multi-Beam Arrays", Ant. Prop. Soc. Int. Symp., 713.
[28] Kuo et al. 2008, "Antenna-Coupled TES Bolometer Arrays for CMB Polarimetery", SPIE, in press.
[29] W. Hu, M.M. Hedman, and M. Zaldarriaga 2003, "Benchmark Parameters for CMB Polarization Experiments", Ph. Rev. D. 67, 3004.
[30] M. Shimon et al., 2007, "CMB Polarization Systematics Due to Beam Asymmetry: Impact on Inflationary Science", arXiv 0709.1513.
[31] J.M. Lamarre *et al.* 2003, "The Planck High Frequency Instrument, a 3rd Generation CMB Experiment and a Full Sky Submillimeter Survey", New Astro. Rev. 47, 1017.
[32] J. Hinderks *et al.* 2008, "QUaD: A High-Resolution Cosmic Microwave Background Polarimeter", arXiv 0805.1990.
[33] G. Chattopadhyay and J. Zmuidzinas 1998, "A Dual-Polarized Slot Antenna for Millimeter Waves", IEEE Trans. Ant. Prop. 46, 736.

Journal of Physics: Conference Series 155 (2009) 012005 doi:10.1088/1742-6596/155/1/012005

4. Multi-band Dual-Polarization Lens-coupled Planar Antennas for Bolometric CMB Polarimetry

Adrian T. Lee

Department of Physics, University of California, Berkeley CA 94720
Physics Division, Lawrence Berkeley National Laboratory, Berkeley, CA 94720

Email: Adrian.Lee@berkeley.edu

Abstract. Len-coupled planar antennas provide a means to efficiently utilize the focal-plane area of a CMB polarization telescope. Our current pixel designs sense both linear polarizations and multiple photometric bands with an overall bandwidth ratio of 3:1. Planar RF channelizing filters distribute power to two bolometers for each frequency band, one for each polarization. Compared to a focal-plane of single-band pixels, the use of these multi-band pixels allow some combination of lower size/mass, larger frequency range, more frequency bands, and higher overall sensitivity. The detector wafer is fully lithographed, and the contacting hemispherical lenses that are required for each pixel are positioned using a second lithographically defined wafer. We have developed a multi-layer broad-band antireflection coating for the lenses. We have tested a superconducting prototype with two polarizations and frequency bands at 150 and 220 GHz, and we have performed accurate beam pattern measurements using scale models at room temperature. To demonstrate this technology at NASA TRL-5, a laboratory demonstration of a full-featured pixel including beam patterns, optical efficiency, band shape, and polarization performance is required, followed by the use of such a pixel in a sub-orbital experiment. Finally, a sketch of a mission design using a single 30-centimeter-diameter aperture and two sizes of multichroic pixel is presented.

4.1 Introduction

A dedicated CMB polarization mission, such as CMBPOL, will require high overall sensitivity and a broad frequency range with 5-10 or more photometric frequency bands to characterize and subtract foregrounds. Since bolometers operated at temperature < 250 mK are near the sensitivity floor set by photon statistics, the overall sensitivity will be given by the number of detectors, assuming that each detector detects a single-mode to get the highest possible angular resolution. In the current EPIC-JPL concepts, for example, a combination of sufficient frequency bands and a high detector count are achieved using six 30 cm aperture telescopes with seven total bands or one 1.5-3 meter aperture with eight bands. Total optical throughput, and therefore the possible number of single-mode detectors, increases with aperture size in addition to increasing angular resolution, but obviously at the cost of size and weight.

Given the need for a high product of band number multiplied by detector number, it is very advantageous to have a pixel that can detect more than one band simultaneously. We are developing a pixel architecture that can sense two linear polarizations and multiple photometric frequency bands simultaneously. Such a design can greatly reduce the mass of the experiment while holding the detector and band number constant. For example, it may be possible to achieve the required total sensitivity and band number with a single 30-50 cm diameter reflective or refractive telescope.

4.2 Lens-Coupled Planar Antennas

In general, planar antennas use elements that have a length comparable to the wavelength of the radiation detected. For ``resonant'' antennas, the length of the antenna is often 1/4, 1/2, or 1 λ. For log-periodic antennas, which will be discussed below, the antenna is made up of ``arms'' that vary in length such that some part of the antenna is resonant over a broad range of wavelengths. An antenna that is comparable in length to the wavelength will have a very broad antenna pattern, since the antenna pattern is essentially set by diffraction and $\theta_{beam} \sim \lambda/length \sim 1\ radian$. To increase the directivity of an antenna, many wavelength elements can combined into a ``phased array'' or the antenna can be placed behind a lens in direct contact with the antenna. The presence of a high dielectric constant silicon lens means that 90% of the power is received from the front. Lens-coupled planar antennas were developed by the antenna-engineering community and they are commonly used there. In astrophysics, they have been used for coupling to SIS mixers, for example, by the Zmudzinas group [34].

Before working on the multichroic antenna, we developed an antenna-coupled bolometer using the well characterized double-slot dipole design [34]. We integrated the antenna with band-defining planar filters and Transition-Edge Sensor (TES) bolometers [35]. We have been able to demonstrate ~ 35% receiver optical efficiency, a band-pass consistent with design, and an upper limit on main-beam cross-pol of 2%. The optical efficiency would improve to > 40% by adding an antireflection coating on the contacting silicon lens. A focal plane of 644 such pixels (1288 bolometers) will be fielded on the POLARBEAR experiment in 2009/2010.

4.3 Log-Periodic Antennas

Planar dipole antennas have less than 40% fractional bandwidth over which the impedance match offers better than -10dB return loss [34]. The bandwidth where the beam has usable properties is often smaller. Log periodic antennas offer a much wider bandwidth because they are self-similar. They are composed of repeating geometrically-similar resonant sections known as cells that differ in size from their neighbors by a common factor τ, typically between 1 and 2. As a result, the antenna impedance and beam-properties are identical for frequencies whose logarithm differ by ln(τ) and vary marginally for frequencies within a log-period [36]. The bandwidth of these antennas is only limited by the size of the smallest and largest log-periodic cells.

4.4 Lens-Coupled Sinuous Antenna

The sinuous antenna (Figure 8) is a log-periodic antenna that has been known for useful polarization properties since DuHamel invented it in the 1980s [37]. The edges of the switch-backing arms follow the equation [38]:

$$\varphi = \alpha \sin (\pi \ln(r/R_0)/\ln \tau) \pm \delta \quad \text{for } R_0 < r < R_0\, \tau^N \qquad (1)$$

Each of the four arms snake through an angle of $\pm\alpha$ every rescaling of τ^2. Typical values of α are between 30° and 70°, while typical values of τ are between 1.2 and 1.5. The standard log-periodic antenna has a radial metal arm with azimuthal "teeth" connected to the arm. In the sinuous, by contrast, the antenna is formed from a continuous winding structure.

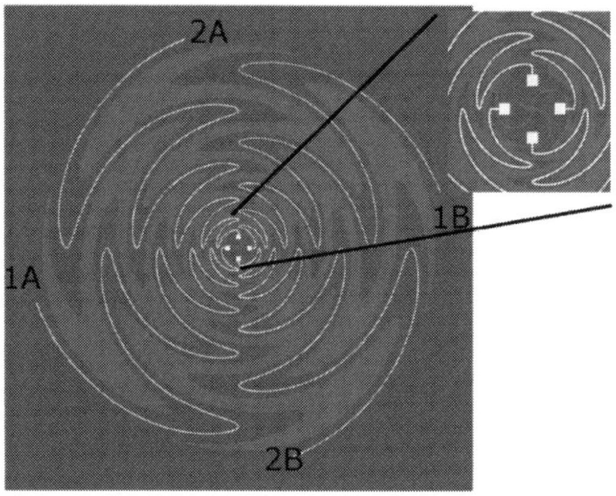

Figure 8. Drawing of a sinuous antenna. This antenna is designed to be placed in contact with a half-space of silicon and to be driven with oppositely phased currents on the microstrip lines (yellow) that couple to the slots (gray) in the center. The polarization of this antenna rotates ±4.5° as a slowly varying function of frequency.

4.5 Telescope Coupling and Mapping Speed

The antenna gain of the lens-coupled sinuous antenna varies with frequency. In the reverse time sense, the entire lens can be considered a constant-size transmitting surface for all frequencies with diffraction then determining the antenna gain. As the antenna gain varies, so will the fraction of power that "spills over" the primary aperture (again in the reverse time sense), and therefore the lens-coupled sinuous antenna will require an aperture stop at sub-Kelvin temperature to avoid sensitivity degradation due excess loading and photon-noise from the aperture stop. A trade-off analysis of pixel NET versus antenna-gain has been done by Griffin et al. for pixels with a single-frequency band [39], and it can be applied to the multichroic pixel by calculating the sensitivity at each frequency band for a fixed lens size. Assuming an aperture stop temperature of 500mK, such that power from the sky is 1000 times higher than power from the aperture stop, the sensitivity of a 3:1 ratio bandwidth pixel is within 15% of an optimized single-frequency array at all frequencies. A further improvement can be made by phase arraying of the pixels, including lenses, where the number of pixels that are phased together scales with frequency such that the antenna gain stays roughly constant.

4.6 Measured Beams of Scale Models

After detailed electromagnetic simulations, we built and tested a 3-12 GHz scale model. The availability of low-cost microwave hybrids, standard gain horns, and network analyzers allow for far easier testing than would be possible with mm-wavelength antennas.

The tested antenna was a wire (non-slot) version instrumented with chip diode detectors mounted across opposite arms in the center. We used a scale model of the

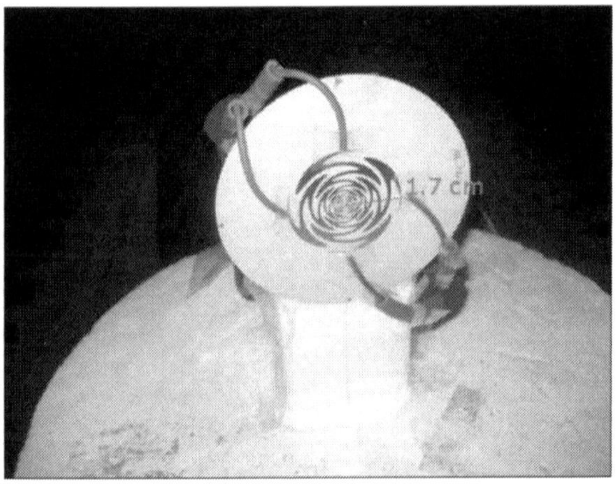

Figure 9. A 3-12GHz scale model center-fed with diodes for lens-coupled measurements was fabricated on 25 mil Roger's 3010 duroid. The antenna size is indicated in the photograph.

contacting silicon lens made from Emmerson-Cuming Microwave Eccostock HiK 12 material, chosen because its dielectric constant of $\varepsilon = 12$ closely matches that of silicon. The lens consisted of a 6 inch diameter hemisphere and a 1.5 inch extension block.

Beam maps of this diode-fed antenna were measured in an anechoic chamber and are shown in Fig. 10. The beam maps are largely Gaussian in shape, symmetric, and have sidelobe levels of less then -20 dB. In simulation, these antennas exhibit low cross-polarization, but the axis of polarization slowly rotates as frequency is varied. The large difference in power received by orthogonal axes in the scale-model beam maps demonstrates polarization properties that are consistent with simulations that show $\pm 5^\circ$ polarization rotation and cross-polarization well below the -20 dB level.

Polarization rotation has two effects on a CMB polarization measurement. The main effect is the average polarization rotation angle will be slightly different for each frequency band. This effect can be mitigated by doing a polarization calibration of the entire experiment for each frequency band. The second effect is depolarization, or signal loss, if there is significant polarization rotation within one frequency band. For our current design, a $\pm 5^\circ$ polarization rotation, the depolarization is at the 10^{-3} level for the worst case band placement at a frequency where the polarization rotation has a maximum slope with respect to frequency. This factor depends on the emission spectrum of the source, but the factor is sufficiently small that it should not cause any significant systematic error.

4.7 Superconducting Prototype

We have built and tested a superconducting millimeter-wave prototype pixel as shown in the photograph of Fig 11(a). This device was described in O'brient et al. [40]. A sinuous antenna with bandwidth from 90 to 270 GHz was connected to a simple diplexing RF filer where two bandpass filters are simply connected to the output of the antenna. The filters are optimized to work well as a diplexing pair using the MMICAD microwave design program.

Frequency response for the two bands was measured and is shown in Fig. 11(b). The two measured bands are centered close to the design frequency. The efficiency of this

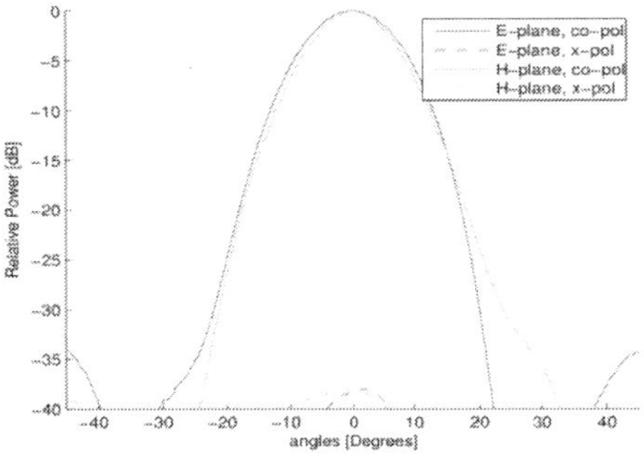

(a) 5 GHz Measured Cuts

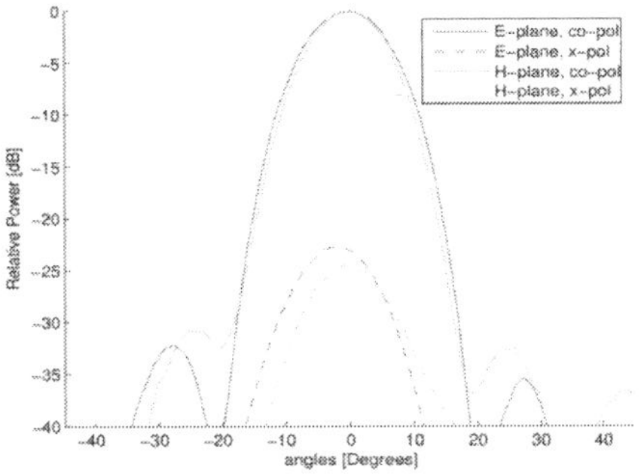

(b) 6 GHz Measured Cuts

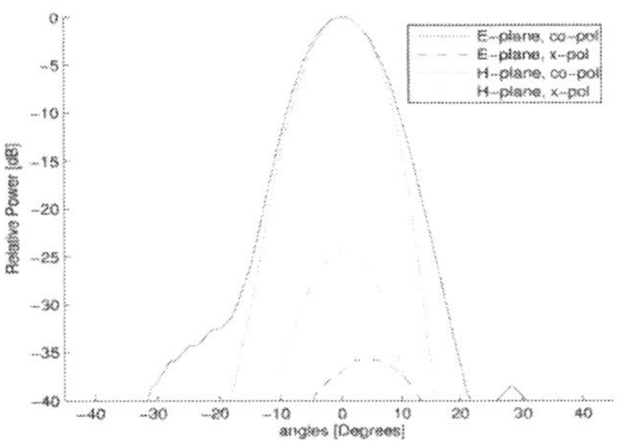

(c) 8 GHz Measured Cuts

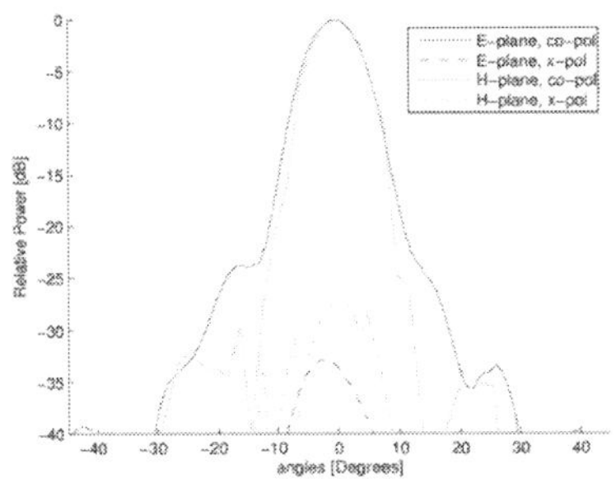

(d) 9 GHz Measured Cuts

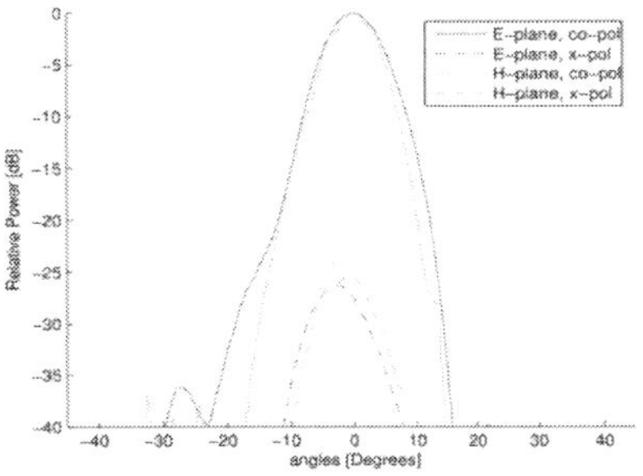

(e) 11 GHz Measured Cuts

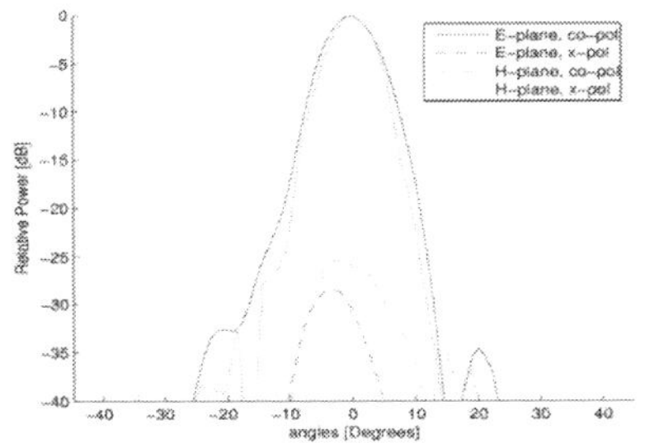

(f) 12 GHz Measured Cuts

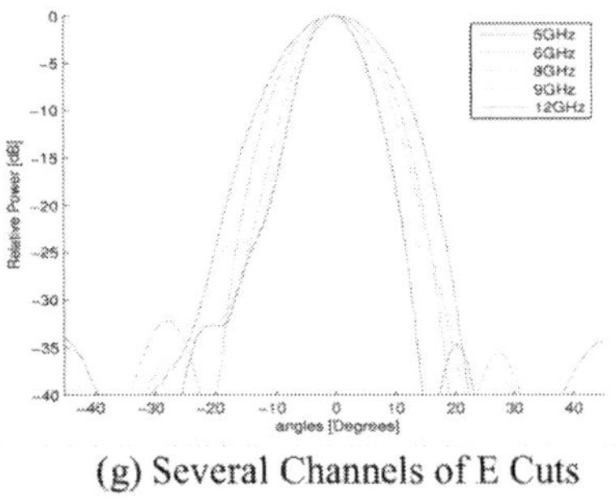

(g) Several Channels of E Cuts

Figure 10. Measured Beams from the diode-fed lens-coupled sinuous antenna. First sidelobes are below -20 dB at all frequencies. No attempt was made to adjust the antenna with frequency to keep up with the polarization rotation, so these are not proper E and H cuts and some of the cross-polarized power is from misalignment. Despite this, the off-axis polarization seen in figs. (a) through (f) is very low. The magnitude of this cross-polarization is consistent with a rotation of $\pm 5^{\circ}$. The beams also narrow as frequency increases due to the contacting lens as seen in (g) where several co-polarized E-cuts are plotted together.

Journal of Physics: Conference Series **155** (2009) 012005 doi:10.1088/1742-6596/155/1/012005

pixel was low due to an avoidable design fault that gave a spurious coupling between the antenna feed lines and the antenna. Our subsequent designs have eliminated this unwanted coupling.

The diplexer design shown can be extended to a triplexer or quadplexer, but it is not appropriate for a larger number of bands. We have designed a built a scale model of a log-periodic RF channelizer that has eight frequency bands in an overall bandwidth of a decade as shown in Fig. 14 [41,42]. In this topology, the backbone and arms contain a series of low-pass and high-pass filters that are scaled in a log-periodic series. The placement, width, and isolation of the bands are adjustable parameters in the general design. The bands must be contiguous, but unwanted bands can be terminated in a passive resistor.

4.8 Antireflection Coating

A key technology for the lens-coupled sinuous antenna is a broadband antireflection coating for the silicon lens. We have developed a four layer coating that uses three layers of a loaded resin (Rogers Corp. TMM) which is sold in several refractive indices and a final porous Teflon (Zitex) layer. We have achieved 95% transmission over a 3:1 bandwidth ratio, and the performance of this coating is the main limitation on the overall bandwidth of our designs.

Applications of this technology on curved surface are in development by thermosetting TMM on a curved form. At present, the TMM broadband coatings applied to curved surfaces are at TRL 2. In order to reach TRL 5, the following are necessary: Procedure for applying Zitex, or another low index layer, to a curved surface; Electromagnetic simulation of the effects of a layered AR coating on beam shape; Fabrication and optical testing of a prototype, including effects on beam shape; and technology scaling to efficiently produce many lenslets.

4.9 Low-mass Mission Concept

In this section, a sketch of a mission design using multichroic pixels is presented. For one type of telescope design, focal plane area and telescope throughput scales linearly with aperture area. Even when different telescope designs are included, there is a general trend of increasing throughput with telescope size. Therefore, if a minimum throughput is required for sensitivity, it will drive a CMB polarization mission design toward one large aperture or multiple smaller apertures. The JPL-EPIC designs reflect this tradeoff. EPIC-LC has six 30-centimeter diameter apertures and EPIC-comprehensive has one 2.5-m aperture.

The mission design presented here uses a single 30-centimeter-diameter aperture cooled to 2K (see figure 14(a)). Two types of pixel are used. A low-frequency pixel with bands centered at 30, 40, 50, 65, and 85 GHz and a high-frequency pixel with bands centered at 110, 145, 190, 242, and 314 GHz (see figure 15). Since each pixel covers a factor three in total bandwidth range, the spillover efficiency and aperture efficiency will vary with frequency for a single pixel. The pixel diameter is chosen with a compromise between edge taper on the Lyot stop (which determines the sidelobe level) at the lowest frequencies and mapping speed at the highest frequencies. For the design presented here, the overall middle of the frequency range was emphasized such that edge taper at the 30

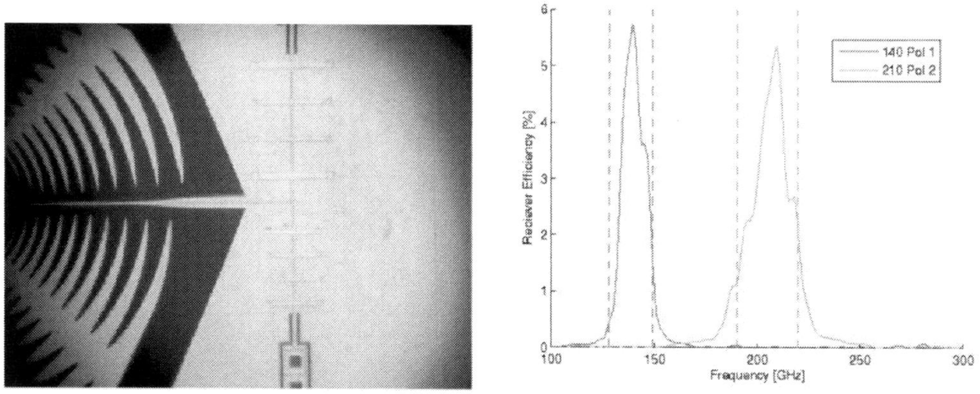

(a) Photograph of a chip (b) FTS Measured frequency

Figure 11. (a) A photograph of a superconducting mm-wave sinuous antenna with RF diplexing filter that gives two frequency bands centered at 140 and 220 GHz. (b) End-to-end receiver measured passband response for the pixel shown in Fig 11(a) using an FTS. The efficiency of this pixel was low due to an avoidable design fault that gave a spurious coupling between the antenna feed lines and the antenna. Our subsequent designs have eliminated this unwanted coupling.

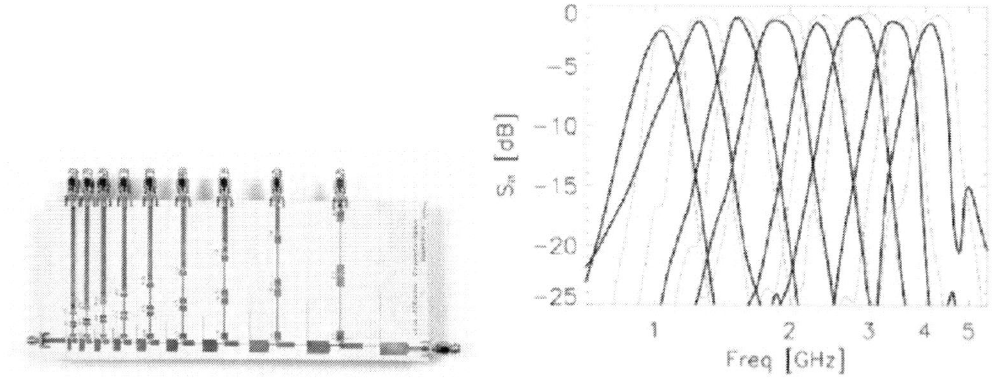

(a) Photograph of scale model (b) Measured frequency response

Fig 12 (a) A photograph of a microwave scale model of an eight channel RF channelizer. The backbone and arms contain a series of low-pass and high-pass filters that are scaled in a log-periodic series. (b) Measured passband response for the model shown in panel (a). The dashed lines show the simulation of this circuit and the solid lines show measurements for this scale model. The apparent loss at the peaks of the passbands is largely due to power sharing between bands rather than dissipative loss.

GHz end and the mapping speed at the 314 GHz end are compromised. The overall sensitivity is comparable to the JPL-EPIC concepts.

4.10 Future work

This section describes the work needed to bring this technology to NASA TRL-5 such that it could be proposed for a CMBPOL mission. The minimum implementation of the lens-coupled sinuous antenna would use a "manifold" channelizer where up to four bandpass filters are simply connected to each of the two polarized output ports of the sinuous antenna as was done with our demontrated diplexer circuit. We need to improve

Journal of Physics: Conference Series **155** (2009) 012005 doi:10.1088/1742-6596/155/1/012005

upon the working prototype that we have already built and test it cryogenically for antenna patterns, polarization properties, band-definition, and optical efficiency. As

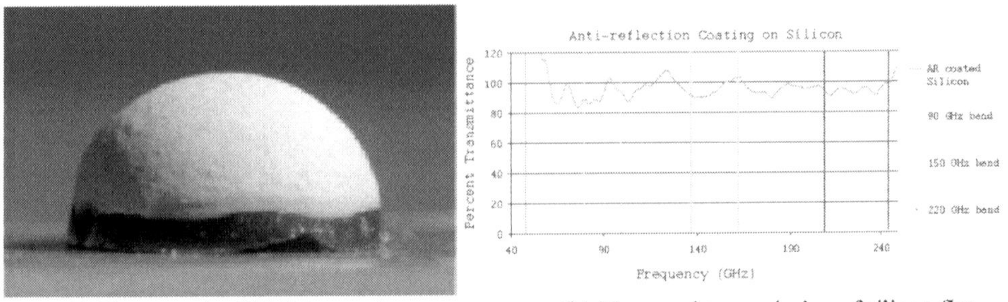

(a) Photograph of TMM coated hemisphere

(b) Measured transmission of silicon flat with two sided coating

Fig 13 (a) A photograph of a 5 mm diameter silicon hemisphere with single-layer TMM coating. This sample was cycled to immersed in liquid nitrogen with no damage to the coating. We have also fabricated multi-layer coatings that have been ground to the required thickness between coats. (b) Measured transmission for a flat piece of silicon with four layer coating. The flat has an antireflection coating on both sides, and therefore a lens with only a single coating would have an average of 95% transmission.

mentioned, we have to test our broadband antireflection coating for the silicon lens in situ. Once these tests have been successfully achieved, the next step would be to demonstrate the design in a sub-orbital experiment e.g. POLARBEAR, EBEX, or SPT.

A more capable implementation of the lens-coupled sinuous antenna would exchange the simple manifold RF channelizer with a log-periodic RF channelizer. In this case, the number and placement of the frequency bands is more flexible and the entire bandwidth of the antenna can be used. Again, this step would have to be tested first in the lab using superconducting mm-wave pixels and then in a sub-orbital experiment.

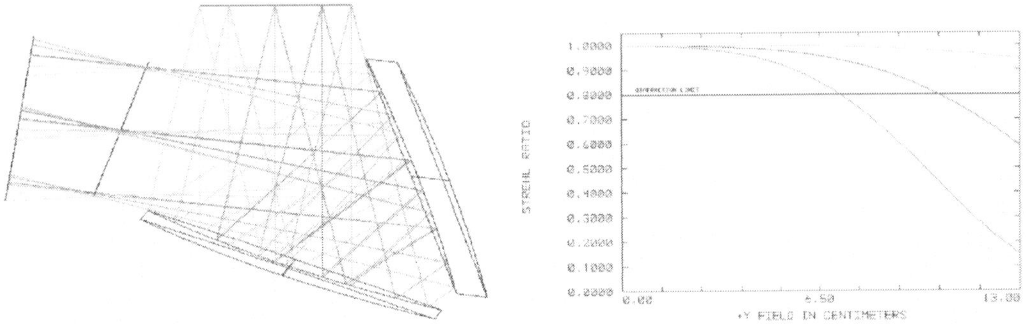

(a) 30 cm crossed Dragone telescope

(b) Strehl ratio for 30 cm crossed Dragone

Fig 14 (a) ZEMAX geometric optics simulation of a 30-centimeter-diameter aperture crossed Dragone telescope. An aperture stop is included at the entrance of the telescope. The focal plane is limited to 26 centimeters in height along the symmetry axis of the telescope due to vignetting. The entire telescope would be cooled to 2K. This design was produced by Huan Tran. (b) Calculated Strehl ratios for 850, 450, and 150 GHz, with 150 GHz line having the highest Strehl ratio in the figure. The assumed focal plane uses the central half of the focal plane area for 110-314 GHz and the outer half of the area for 30-85 GHz. This allocation emphasizes the 110-314 GHz range.

Another future development would be to expand the overall bandwidth beyond our current design with a 3:1 ratio. Our design is currently limited by the antireflection coating on the lens, and therefore we would have to develop a coating with a larger overall bandwidth and develop a practical method for its manufacture.

Finally, a mature design would need to take system level issues into account such as the characteristics of the front-end optical filtering chain, array temperature control and stability, and robustness against ionizing radiation, surface charging, and radio-frequency interference.

Acknowledgements

This paper summarizes work that has been previously reported largely by O'brient *et al.* [40]. The measurements reproduced here were done by Roger O'brient of Berkeley and Jen Edwards of UCSD.

References

[34] Chattopadhyay, G. and Zmuidzinas, J., "A Dual-Polarized Slot Antenna for Millimeter Waves," *IEEE Transaction on Antennas and Propagation*, Vol. **47**, No. 5, 736 (1998).
[35] Myers, M. Ade, P.A.R., Engargiola, G. Holzapfel, W.L. Lee, A.T., O'Brient, R. Richards, P., Smith, A. Spieler, H., and Tran, H., "An antenna-coupled bolometer with an integrated microstrip bandpass filter," *Applied Physics Letters*, **86**, 114103 (2005).
[36] Rumsey, V. [*Frequency independent antennas*]. New York Academic Press (1966).
[37] Saini, K., Bradley, R., "The Sinuous Antenna- A Dual Polarized Element for Wideband Phased Array Feed Application," Electronic Division Internal Memo N0.301. National Radio Astronomy Observatory (1996).
[38] DuHamel, R. "Dual polarized sinuous antennas," US Patent No. 4658262 (1987).
[39] Griffin, M., Bock, J., Gear, W., "Relative performance of filled and feedhorn-coupled focal-plane architectures" *Applied Optics* **41** 6543-6554 (2002).
[40] O'Brient, R., Ade, P., Arnold, K., Engargiola, G., Holzapfel, W., Lee, A.T., Myers, M., Meng, X., Quealy, E., Richards, P., Spieler, H., Tran, H., "A Multi-Band Dual-Polarized Antenna-Coupled TES Bolometer," *Journal of Low Temperature Physics* **151** Vol. 1 (2008).
[41] Engargiola, G., Holzapfel, W., Lee, A.T., Myers, M., O'Brient, R., Richards, P., Tran, H., Spieler, H., "Planar Channelized Log-Periodic Antenna," *IEEE 2005 International Symposium on Microwave, Antenna Propagation and EMC Technologies for Wireless Communications* **1** pp. 306-9 (2005).
[42] Engargiola, G., "Channelized log-periodic antenna with matched coupling," US Patent No. 7109821 (2006).

Band	Lens Diam.	D/(f * λ)	Resol.	#pix	#bolo	Spill. Eff	taper (dB)	Load (pW)	NEP_phon	NEP_poiss	NEP_total	NET_RJ	NET_CMB	NET_total
30	25	1	124	42	84	0.45	-3	0.22	3.66E-18	2.95E-18	4.70E-18	74	75.4	8.21
40	25	1.33	95	42	84	0.55	-4	0.28	4.10E-18	3.82E-18	5.60E-18	54	56.2	6.11
50	25	1.67	79	42	84	0.8	-6	0.36	4.70E-18	4.90E-18	6.79E-18	36	38.3	4.17
65	25	2.17	64	42	84	0.92	-10	0.42	5.09E-18	6.05E-18	7.91E-18	28	31.3	3.4
85	25	2.83	53	42	84	0.97	-16	0.45	5.23E-18	7.12E-18	8.83E-18	23	27.4	2.98
110	10	1.47	35	243	486	0.75	-5	0.34	5.23E-18	7.12E-18	8.83E-18	23	27.4	1.24
145	10	1.93	27	243	486	0.85	-8	0.32	4.40E-18	7.80E-18	8.97E-18	15	26.2	1.19
190	10	2.53	23	243	486	0.95	-14	0.24	3.68E-18	7.49E-18	8.35E-18	11	26.3	1.19
242	10	3.23	19	243	486	0.99	-23	0.15	3.05E-18	7.01E-18	7.65E-18	7	26.1	1.18
314	10	4.19	18	243	486	1	-38	0.07	2.07E-18	5.40E-18	5.79E-18	4	32.2	1.46
Total				286	2852						All frequencies ->		Array NET	0.53
											50 to 190 GHz ->		Array NET	0.66
												Arr. NET*rt2		0.93

Table 3. Table of design parameters and sensitivity for a 30 centimeter crossed Dragone telescope using multichroic pixels with five frequency bands per pixel. The columns with units are: Central frequency of the band in GHz, the lens diameter in millimeters, the ratio of diameter to the product of f λ, the number of pixels with the given band, the number of bolometers with the given band, the spillover efficiency through the Lyot stop, the edge taper in dB, the loading in pW, the Noise Equivalent Power (NEP) in W/√Hz, the photon noise NEP, the total NEP, the Rayleigh-Jeans Noise Equivalent Temperature (NET) in μK√s, the CMB NET, and the total NET for all the bolometers at that frequency. The assumptions in this model are: focal ratio at feed = 2.5, diameter of focal plane = 26 cm, the diameter of the focal plane region used for the ``high-frequency" pixels = 18 cm, the maximum theoretical area efficiency for a hexagonal close packed array = 90%, the efficiency assumed here = 75%, the telescope temperature = 2 K, the bolometer stage temperature = 100 mK, the bolometer temperature rise = 100 mK, optical efficiency to sky = 40%, optical efficiency to telescope = 30%, optical efficiency to 100mK = 30%, the fractional bandwidth for all bands = 30%, the ratio of saturation power to optical power = 4, and index of temperature dependence of the bolometer's weak link = 1.

The edge taper determines the sidelobe level of the telescope, and it is therefore important to keep this number low. However, given the wide bandwidth of these pixels, a compromise of edge taper at the low-frequency end of the pixel has to be made with overall mapping speed for a fixed focal plane area at the high-frequency end. In this example, the low-frequency pixels are weighed toward efficiency at the higher frequencies and the high-frequency pixels are weighted toward good edge taper at the low-frequency end.

Journal of Physics: Conference Series **155** (2009) 012005 doi:10.1088/1742-6596/155/1/012005

5. Conclusion

We do not anticipate that the optical coupling scheme will be a source of high technological risk for a CMB polarization space mission. The traditional approach using corrugated feeds can certainly be made to work well for single band pixels. Lens-coupled antennas and planar antennas are developing rapidly and may offer advantages in focal plane mass, system sensitivity, and complexity. The tradeoffs between these different approaches can be difficult to assess because they interact strongly with the system design. Demonstrations in ground-based and sub-orbital experiments are thus a critical step in assessing readiness for space.

Polarization Modulators for CMBPol

P A R Ade[1], David T Chuss[2], Shaul Hanany[3], V Haynes[4], Brian G. Keating[5], A. Kogut[6], John E. Ruhl[7], G Pisano[4], G Savini[1], and E J Wollack[8]

[1] Cardiff University, School of Physics and Astronomy, Queens Buildings, The Parade, Cardiff, CF24 3AA, UK
[2] NASA Goddard Space Flight Center, Code 665, Greenbelt, MD, 20771 USA
[3] School of Physics and Astronomy, University of Minnesota/Twin Cities, Minneapolis, MN, 55455, USA
[4] The University of Manchester, School of Physics and Astronomy - Alan Turing Building, Upper Brooke street, Manchester, M13 4PL, UK
[5] Department of Physics, University of California, San Diego, La Jolla, CA 92093-0424 USA
[6] Code 665 Goddard Space Flight Center, Greenbelt, MD 20771
[7] Physics Department, Case Western Reserve University, Cleveland, OH, 44106 USA
[8] Observational Cosmology Laboratory, NASA/GSFC, Greenbelt, MD 20771

Abstract.
We review a number of technologies that are candidates for active polarization modulators for CMBPol. The technologies are appropriate for instruments that use bolometric detectors and include birefringent crystal-based and metal-mesh-based half-wave plates, variable phase polarization modulator, Faraday rotator, and photolithographed modulators. We also give a current account of the status of millimeter-wave orthomode transducers.

1. Introduction

Shaul Hanany
School of Physics and Astronomy, University of Minnesota/Twin Cities, Minneapolis, MN, 55455, USA

E-mail: `hanany@physics.umn.edu`

The experimental technique of temporal modulation of amplitude, frequency or phase has long been a powerful technique to separate signals from sources of noise or from sources of systematic errors. In a generic amplitude modulation application the signal is modulated at a frequency f and then demodulated using phase-sensitive detection (either in hardware or in software) with a bandwidth Δf around the center frequency. Rejection of noise, or equivalently improved sensitivity, comes from moving the signal to high frequencies, away from typical $1/f$ noise sources in the experiment. Rejection of systematic errors comes from moving the signal to frequencies that do not contain contributions from specific systematic effects.

The cosmic microwave background's B-mode polarization is expected to have an RMS magnitude that is smaller than 100 nK. This magnitude is at least a factor of $\sim$10 smaller than the E-mode signal, a factor of $\sim$100 smaller than the temperature anisotropy, and a factor of $\sim 10^7$ smaller than the temperature of the CMB itself. The smallness of the signal requires high sensitivity and exquisite control of polarimetric systematics. Temporal modulation of the incident polarization can provide improved sensitivity and rejection of systematics errors, hence the interest in polarization modulation for future CMB polarization experiments, as well as for CMBPol.

This paper discusses a number of polarization modulation techniques that are technology candidates for CMBPol. Since the TFCR report [1] has endorsed a bolometric detector technology for the satellite, this paper will concentrate only on polarization modulation techniques for bolometric experiments. It is broadly expected that the scan pattern of CMBPol will be optimized for polarization measurements.

There is no evidence yet that it is imperative that CMBPol will have an active polarization modulator. A number of experiments have already made detections of the E-mode polarization without employing such modulator [2, 3]. The term 'active' is used here to denote a specific hardware implementation inside the instrument that is meant to switch polarization states, as opposed to modulation only through the scan pattern across the sky. MAXIPOL [4] was the first experiment to measure the polarization of the CMB with an active modulator, which was a combination of a rotating half wave plate and a stationary grid. The lessons learned from these experiments were discussed in a separate workshop dedicated to discussion of systematic errors from polarization experiments. Our goal in this paper is to survey candidate active polarization modulation approaches from a technical point of view. The question of whether such modulation is necessary for CMBPol is outside the scope of this paper.

It is useful to enumerate the properties of an ideal polarization modulator. The different technologies that will be described in subsequent chapters address some of these properties well, while falling short with others. This is the inevitable trade-off that the experimental designer must consider.

- An ideal modulator should be the first element in the optical train of the instrument. In this way it modulates only the incident polarization from sky sources; not any spurious polarization that is created by the instrument.
- An ideal modulator should modulate the polarization on time scales that are fast compared to 1/f noise contributions to the measurement.
- An ideal modulator must have high modulation efficiency. In other words, an ideal modulator should not decrease the level of polarization of the incident signal.
- Almost all modulation techniques give rise to spurious signals that are synchronous with the modulation. Such spurious signals can be induced by a number of sources including mechanical vibrations, EMI, scattering, or modulation of transmission, to name just a few. An ideal modulator addresses potential sources of systematic effects, while not inducing systematic effects of its own.
- An ideal modulator is easy to fabricate, easy to implement mechanically, electrically and cryogenically, it does not produce significant power dissipation, and has low mass.

The technologies discussed in subsequent chapters were selected out of a wider array of possibilities because they are currently or have recently been actively pursued by a number of groups carrying out CMB polarization experiments. Section 2 discusses the use of a birefringent half wave plate made of crystalline material. The technique has substantial heritage in astronomy and has already been used in an experiment that reported CMB measurements [4]. Sections 3 and 4 discuss technologies that also have past heritage in millimeter-wave astronomy. However, no detection of CMB signals is yet reported with either. Both technologies are integrated into experiments that have recently searched, or will search in the near future, for the polarization of the CMB. In Section 5 the authors present a novel approach whereby standard metal-mesh filtering technology is extended to form a metal-mesh-based birefringent half-wave plate. A prototype has been constructed and there are plans to make the first astronomical observations in the near future. Section 6 reviews the status of polarization modulators that are fabricated photolithographically on a silicon wafer. This technology is in relative infancy. There are no instruments yet operating with photolithographic modulators. The last section in this compilation discusses the status of ortho-mode transducers. Although these are not polarization modulators, we include the section because the review of the technology might be useful for designers of future polarization instruments.

All-in-all the goal of this paper to inform the choice of a polarization modulator for a future CMBPol mission.

Journal of Physics: Conference Series **155** (2009) 012006 doi:10.1088/1742-6596/155/1/012006

2. Half Wave Plate Modulator

John E Ruhl

Physics Department, Case Western Reserve University, Cleveland, OH, 44106 USA

E-mail: ruhl@case.edu

Abstract. Polarization modulators that are based on the introduction of a quasioptical phase delay between two orthogonal linear polarizations are reviewed. The general principle behind the device and the application to the Beyond Einstein Inflation Probe are addressed.

2.1. Introduction

The birefringent halfwave plate (HWP) polarization modulator, while new to the CMB field, has been used for decades at shorter wavelengths. The basic idea is simple; a polarized plane wave passing through a HWP will have its polarization orientation rotated by $\Delta\psi = 2\theta$, where θ is the angle between the incident polarization and one of the optical crystal axes. Rotating a HWP in front of a linearly polarized detector rotates the polarization sensitivity of that detector on the sky, nominally without affecting its beam pattern. This virtue, rotating polarization sensitivity without rotating a potentially asymmetric beam, is the main reason many upcoming CMB polarimeters (including ACT, ABS, Bicep2/Spud, Clover, EBEX, Polarbear, and Spider) will use birefringent sapphire HWP polarization modulators.

Maxipol [5, 6] is the only experiment so far to publish CMB observations using a HWP polarization modulator. With a single night of data from a balloon platform, the sensitivity was insufficient to probe B-modes, but was a good demonstration of using a HWP modulator to measure polarization on the sky.

The future use of HWPs for B-mode polarization observations will test a variety of designs and strategies for using these devices. Here we review some of the issues associated with using HWPs, and the ongoing work that will test these devices in future CMB polarization observations.

2.2. Major design and observing choices

There are three major design choices that each HWP implementation must make: bandwidth, operation temperature, and rotation style (continuous or stepped).

The optical design of the HWP is driven by desired bandwidth. Some experiments will use separate optical paths for each observing band, with a separate HWP optimized for each. For these, the required fractional bandwidth is of order 25%. Other experiments put the HWP in an optical path that is shared by several bands; in this case, the required fractional bandwidth is much larger, of order unity. In practice, the single-band design can be simpler, using only a single-layer birefringent crystal and a single-layer anti-reflection coating. Multi-band designs require a stack of multiple birefringent plates with multilayer anti-reflection coatings. Of course all optical designs must also respect the upper limit on available diameter of materials used to construct the HWP; birefringent sapphire is available in diameters up to at least 330mm.

Most upcoming experiments are baselining cryogenic HWPs, which ensures very low thermal emission from the HWP in the observing bands and minimizes signals from reflections. However, even at 300 K the in-band emission of a single-plate sapphire HWP design is only about 2 K at 150 GHz, so ground-based experiments may be able to use ambient temperature HWPs. ACT has investigated such ambient-temperature operation [7]. The mm-wave loss in sapphire drops dramatically with temperature, so loading is unlikely to be an issue for any instrument (including CMBPol) using a cryogenic HWP.

Journal of Physics: Conference Series **155** (2009) 012006 doi:10.1088/1742-6596/155/1/012006

HWPs can be used to separate beam asymmetries from polarization signals by occasional stepping of the HWP orientation or by continuously rotating it. The choice between these two strategies is intimately linked with the observing scan pattern and detector noise properties. This choice drives the mechanical design; stepper motors and conventional bearings are a good match to occasional stepping, while low heat dissipation requirements at cryogenic temperatures makes a passive magnetic bearing attractive, though perhaps not essential, for continuous rotation.

Continuous rotation comes in two flavors, which color the way one views the optical design criteria. The simplest use of a continuous rotation is to ensure the HWP modulates the polarization (at four times the HWP rotation rate f) very fast compared to rate at which the instrument's beam response is scanned across the sky. For such fast modulation, one can just lock in to the $4f$ component of the timestream and ignore other harmonics. Alternatively, one can scan across the sky fast compared to $4f$; in terms of analysis, this is much like a very rapid stepping of the waveplate.

2.3. Optical Designs and characterization

The coupling from a given input polarized signal to a polarized detector, as a function of photon frequency and HWP angle, can be calculated using E and H field boundary matching at each dielectric interface. An unpolarized version of this matching is worked out in matrix formulation in [8], while a polarized version is discussed in [9]. It is worth considering a few simple single-plate cases, however, to build intuition.

Case 1: When the input polarization is aligned with one of the crystal axes and the detector polarization sensitivity, the polarization is not rotated by the HWP at *any* photon frequency ν. Reflections at the various dielectric interfaces, however, will affect the transmission amplitude as a function of ν.

Case 2: Rotate the detector polarization sensitivity by 90° from Case 1. Again, the HWP does not rotate the input polarization at any ν; since the sky signal and detector sensitivity (assumed to have no crosspolar response) are crossed, there is no detector response at *any* ν. This is independent of reflections, since zero times anything is zero.

Case 3: Start with Case 1, but rotate the HWP by $\theta = 45°$. At the design frequency, the input polarization is rotated by exactly $\gamma = 2\theta = 90°$, so the detector response is zero. However, at other photon frequencies γ is not exactly 90°, so there is some detector response. This means that when the sky signal and detector sensitivity are aligned, rotating the HWP takes the band averaged signal from some large value (Case 1) to some small but non-zero value (Case 3).

Case 4: Start with Case 2, but rotate the HWP by $\theta = 45°$. At the design frequency, the input polarization is rotated by exactly $\gamma = 2\theta = 90°$, so the detector response is maximized. At nearby photon frequencies γ is not exactly 90°, so detector response is not maximized but it will still be large. This means that when the sky signal and detector sensitivity are orthogonal, rotating the HWP takes the band averaged signal from zero (Case 2) to some large value (Case 4).

Full calculations of the coupling as a function of HWP orientation angle, for normal incidence and ideal materials, generally give signals that contain a DC level, a second harmonic of the HWP rotation frequency f, and the expected signal at $4f$, the amplitudes of which are a function of photon frequency. The $2f$ signal is the result of reflections that depend on how the input signal is aligned with the crystalline axes. A figure of merit called the "polarization modulation efficiency", particularly relevant to experiments using rapidly rotating HWPs, is unfortunately defined in different ways by different authors so we will avoid that terminology here.

The importance of various optical characteristics are driven by the desired bandwidth and the mode of operation (stepped or slow rotation vs. fast rotation). In the fast-rotation case, the $4f$ signal amplitude and phase (as a function of ν across the band) are the major parameters of interest. In the slow-rotation and stepped modes, one needs to consider the spectral response as a function of input (sky) polarization angle, for each orientation of detector sensitivity and HWP rotation angle that will be used.

2.3.1. Single-plate designs A HWP constructed from a single slab of birefringent sapphire with a single-layer quarter-wave antireflection coating has a very large and uniform $4f$ signal across a 25% fractional bandwidth (top panel of Figure 1). The $2f$ signal, due to differential reflections when the input signal is aligned with the two crystal axes, is small; the DC signal is nearly the same as the $4f$ signal, indicating that the signal minima (dominated by DC + $4f$) are nearly zero. Note the absence of a $2f$ signal in the perpendicular case (Case #2 above); for the same reason, the signal minima area exactly zero, which means that the DC offset is equal to the $4f$ signal amplitude.

Several groups are building single-band, single-plate HWP systems that will be used in upcoming experiments. Our group at Case is constructing sets for Bicep2/SPUD and Spider, centered at 90, 145, 220, and 270GHz, each with about 25% fractional bandwidth. ACT, ABS, and Polarbear plan to use single-band, single-plate designs at 150 GHz in their first incarnations [7, 10, 11].

As Figure 1 illustrates, the HWP introduces features into the $4f$ amplitude bandpass of the system. These $4f$ features are independent of detector orientation relative to the sky signal, which means this can be treated just like another multiplicative effect defining the detector bandpass - there is no extra complication.

However, the $2f$ and DC features are not independent of detector orientation. For slow continuous HWP rotation, or stepped rotation, these effects must be accounted for. Figure 2 shows how this same HWP design affects sky-to-detector coupling depending on the relative angles of the sky signal, detector orientation, and HWP, for two orientations of sky signal and detector (coaligned and crossed) and four HWP orientations. The important thing to note is that in general a crossed detector pair with otherwise identical bands will have different frequency response to the same polarized sky signal. This is an extra complication that must be accounted for in the data analysis.

2.3.2. Multi-plate designs Multiple birefringent plates can be combined to significantly increase the useful bandwidth of the HWP [12, 13, 14]; these must be combined with wide-bandwidth anti-reflection coating to get good overall performance over a wide range of photon frequencies.

As an example, the calculated performance of representative a 3-plate design with a 5-layer AR coat is shown in Figure 3; with the $4f$ amplitude nearer to unity over a much larger range of photon frequencies than the single-plate design, this system enables optical designs where multiple bands share a single HWP.

Several groups are implementing multi-plate designs; Clover and EBEX are baselining them for their first observations, while other groups are considering them for use when they implement multiple bands in the future.

2.3.3. Characterization While many experiments are planning to implement HWPs into their B-mode CMB observing instruments in the near future, there has been only a relatively small amount of detailed optical testing published so far.

The only published CMB observations with a HWP were done by Maxipol [6]. That system used a single birefringent sapphire plate and single-layer AR coat, optimized for its 150 GHz observing band. Significant offsets at the HWP rotation frequency and harmonics were caused by the drive system [15], which incorporated a G10 shaft attached to the center of the HWP and

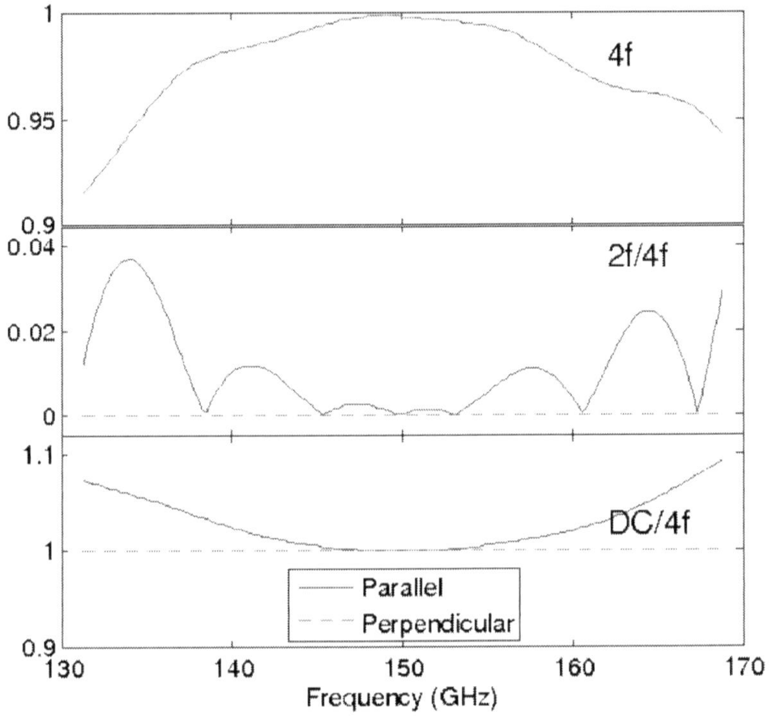

Figure 1. Calculated transmission signals from a completely polarized input, coupling into detectors that are sensitive to polarization parallel to (solid blue line) and perpendicular to (dashed red line) the input polarization. The top panel shows the signal that is modulated four times per HWP rotation ($4f$ component), which is the same for both plotted cases; the middle shows the ratio of the $2f$ to $4f$ (which is zero for the perpendicular case), and the bottom panel shows the ratio of the DC to $4f$, which is unity for the perpendicular case. The plotted points cover a 25% fractional bandwidth centered on 150GHz, using a single lossless sapphire waveplate ($n_s = 3.4$, $n_f = 3.07$) with an "ideal" quarter-wave AR coat that has an index $n_{AR} = \sqrt{(n_s + n_f)/2}$.

therefore within the optical path. Despite that issue (which future implementations of HWPs for CMB observations are all avoiding), the optical performance was sufficiently good to reconstruct sky maps at the detector sensitivity limit with no obvious systematics.

The ACT team [7] tested an ambient temperature HWP (again, single sapphire plate with a single layer AR coat) briefly installed on the CCAM instrument in Chile. Only a small amount of data was taken, but the optical performance was again promising.

Laboratory optical testing has included very narrow-band measurements using coherent sources and broadband FTS testing [13, 14]. To a large extent, this work has confirmed expectations, and has not yet raised any serious unexpected issues.

However, these tests have not yet probed all the relevant questions, such as whether spurious signals (eg from reflections) will be stable enough to avoid signal contamination, whether non-uniformities will induce systematics or degrade performance, and whether the frequency response can be characterized well enough to avoid limiting foreground subtraction. The importance of these issues is as much a function of the rest of the optical and detector system, and of the scan strategy, as they are of the HWP. We will learn an enormous amount from upcoming

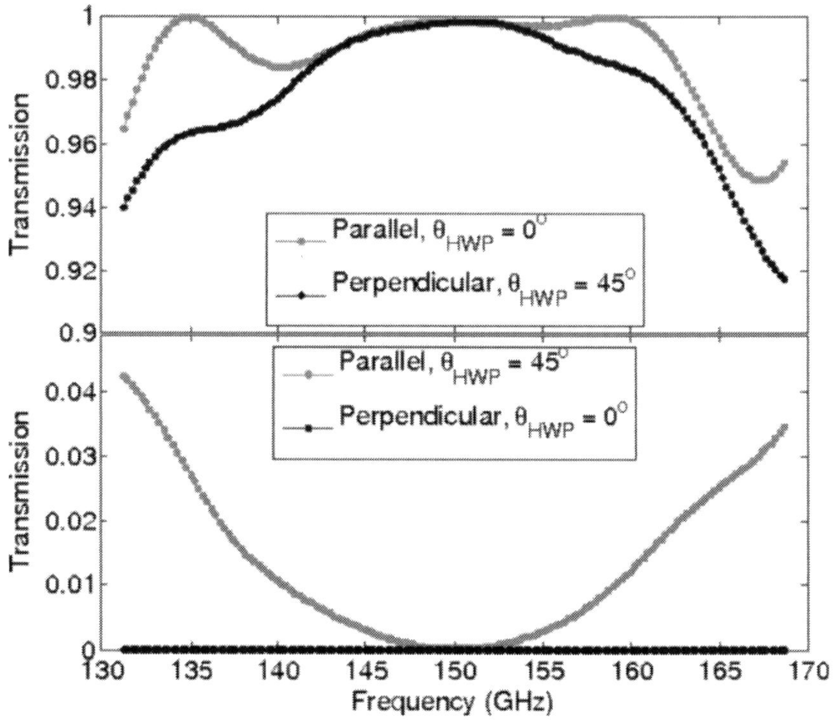

Figure 2. Calculated transmission as a function of photon frequency for detectors parallel (black symbols and lines) and perpendicular (red symbols and lines) to the incident polarization direction, for HWP orientations that nominally maximize (upper panel) and minimize (lower panel) those couplings. The HWP characteristics are the same as for Figure 1.

observations.

2.3.4. Cryogenic rotation schemes To be useful, a HWP must be rotated (continuously or stepped) during or between observations. If the waveplate is used at 300K (as is being investigated by the ACT team), such rotation and encoding of the HWP angular position is straightforward. Rotating at cryogenic temperatures, and knowing the HWP angular position to the required level of $\approx 0.1^\circ$, is another technical challenge that must be dealt with.

2.3.5. Stepped rotation Cryogenic stepped rotation is conceptually straightforward, with the caveat that linear thermal contractions must be handled. Several groups have used stepper motors at 4K to drive such motion; options range from commercially available cryo-vacuum stepper motors to simple modifications of motors built for room temperature operation. Incremental cryogenic encoders with the required precision have been constructed using cryogenic lasers, LED's, and photodiodes, and tested in the lab.

The large optical throughput of CMB polarimeters is driving a desire for larger diameter HWP's than have been used before - as large as $\sim$ 30cm clear diameter. These large diameters increase the likelihood of failure of the large diameter ball bearing assemblies (with I.D. equal to or greater than the clear aperture) used by previous designs [16] as they thermally contract. We have built a design that avoids this problem by using three small rollers rather than one large bearing, which minimizes the likelihood of binding due to thermal contraction mismatches.

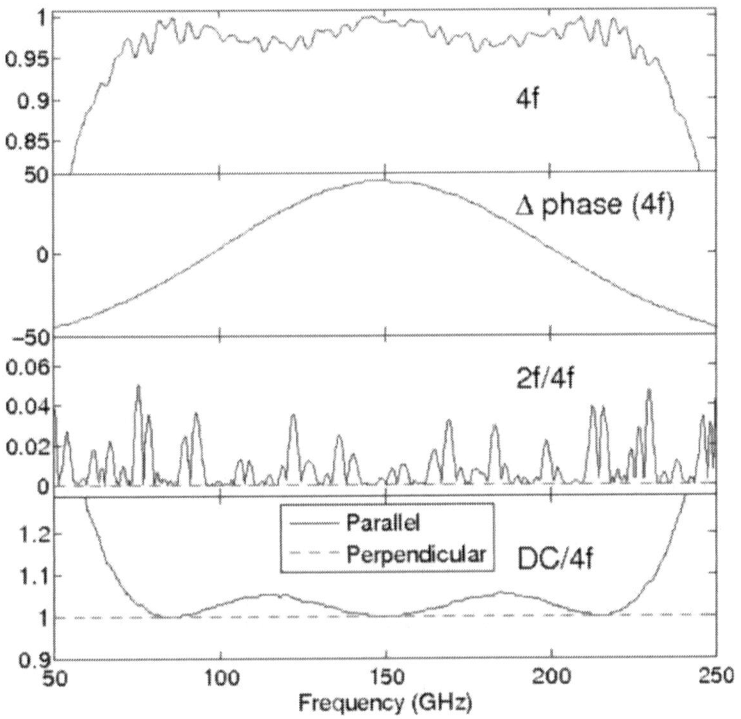

Figure 3. Similar to Figure 1, but for an example HWP system constructed from three birefringent sapphire plates (the middle layer is rotated by 50° with respect to the outer two layers, in this case) with a 5-layer AR coat. Here we have added another panel to the plot to show the phase difference of the $4f$ signal as a function of photon frequency, which limits the useful width of individual bands using this HWP; that is, the HWP system is still very good for multiple (narrow) bands within the frequency limits shown, but is loses efficiency for very wide bands. The reflections at the internal and external interfaces create a rich set of features in the $4f$ and $2f$ amplitudes as a function of photon frequency.

2.3.6. Continuous rotation Cryogenic continuous rotation raises the issues of microphonic noise and the generation of excessive heat at those temperatures. Passive superconducting magnetic bearings (SMBs) [17] have very low friction, and therefore minimize both microphonics and heating. Polarbear, however, will be using a standard ball-bearing at 50K, where the heat loading is less of an issue [11].

Several drives have been developed for cryogenic, continuously rotating HWPs; EBEX will be using a magnetically levitated HWP, driven by a brushless DC motor outside the cryostat, coupled to the HWP by a shaft and a belt [18, 19]. Polarbear is baselining a similar drive, but using standard ball bearings at the 50K HWP [11]. Here at Case, we have also prototyped and tested an integrated synchronous AC motor built into the rotor, riding on a SMB.

Knowledge of the rotation angle is of course key; as with stepped systems, encoders based on cryogenic lasers, LED's and photodiodes should provide the required precision.

2.4. Future Development Work

A large number of of upcoming CMB polarization experiments (eg ABS, ACT, Bicep2/SPUD, Clover, EBEX, Polarbear, and Spider) are planning to use birefringent crystal HWPs. Calculations and preliminary testing indicate that HWPs should work well for these observations.

Journal of Physics: Conference Series 155 (2009) 012006 doi:10.1088/1742-6596/155/1/012006

These experiments will test a wide variety of optical designs and rotation schemes, exactly as is needed to explore the usefulness of HWP polarization modulators for CMBPol.

After these upcoming experiments are done, the various HWP implementations used would be at TRL 7, defined as "System prototyping demonstration in an operational environment (ground or space)." Right now they are around TRL 4, "Component/subsystem validation in laboratory environment". The real question for CMBPol is whether it will be using a HWP very similar to those implemented in the upcoming generation of experiments, or will require modifications; given the newness of HWPs in CMB instruments and the differences between the suborbital and orbital environments, the need for some evolution is quite likely. For this reason, it would be good to support efforts to further develop HWP concepts that specifically target issues important for their actual use in CMBPol, above and beyond the existing suborbital experiments.

3. Faraday Rotation Modulators for mm-Wavelength Polarimetry

Brian G Keating

Department of Physics, University of California, San Diego, La Jolla, CA 92093-0424 USA

E-mail: bkeating@ucsd.edu

Abstract. The design and performance of wide bandwidth linear polarization modulators based on the Faraday effect in corrugated cylindrical waveguide is described. The device can be used as a polarization modulator for millimeter-wave and microwave polarimetry and is functionally similar to a birefringent crystal waveplate, but with no moving parts. The Faraday rotation modulator's intrinsically single-moded design results in wide-bandwidth performance with low-loss, and is capable of polarization modulation up to $\sim$ 10 kHz. The Faraday Rotation Modulator is capable of operation at cryogenic temperatures and is scalable for use up to electromagnetic frequencies of 300 GHz. Faraday rotation modulators operating at 100 and 150 GHz, with 30% fractional bandwidths in each band have been used to make astronomical observations of polarized galactic emission.

3.1. Introduction and Motivation

Measurements of the polarization of the cosmic microwave background (CMB) have the promise to revolutionize our understanding of the early universe. Unlike the temperature anisotropy of the CMB, which has been measured to exquisite precision over a wide range of angular scales, the polarization of the CMB has only recently been detected and remains relatively unexplored. To detect the faint CMB polarization it will also be necessary to detect and remove polarized galactic emission from interstellar dust and synchrotron sources which can be accomplished if polarimetry is done in multiple millimeter-wave bands.

Polarimeters often employ mechanisms to modulate the plane of polarization of the incident radiation field about the optical axis. In conjunction with an analyzer (to decompose the radiation into orthogonal polarization states), a modulator is used to exchange the polarized intensity between the two detectors (or two orientations of a single detector). If the modulation is done rapidly enough, this technique mitigates the effects of detector gain instability. If the detector's gain instability systematics are mitigated by other means, e.g. spatial scanning, then the FRM can be used in a "point-and-integrate" mode where it is used solely to modulate polarization. This can simplify either the optics, detectors (allowing for only one linear polarization state to be detected) or the telescope's mount. In such situations, having both FRM and scanning modulation, allow the tasks of beam systematics mitigation and detector fluctuation systematics mitigation to be separated.

3.1.1. Polarization modulation Traditionally, polarization modulation for millimeter-wave polarimeters was accomplished by physical rotation of the polarimeter about the optical axis [20, 21], mechanical rotation of a waveguide polarizer [22], rotation of a wire grid polarizer [23], or rotation of a birefringent half-waveplate [24, 25]. In many CMB experiments, polarization modulation is often effected by parallactic angle rotation ("sky rotation") with respect to the instrument coordinate system.

Classical polarization modulation mechanisms often employ bulky and complex mechanisms which are subject to failure with repeated use. The modulators described in this paper represent a novel approach to the technology of CMB polarization modulation. Early CMB polarimeters (including Penzias and Wilson's, which *was* polarization-sensitive) used a combination of sky-rotation with rotation of the entire telescope to modulate CMB polarization. These experiments

[26, 27] rotated hundreds or thousands of kilograms, were susceptible to vibration induced microphonic noise, and were limited mechanically to modulation rates < 0.1 Hz.

The next polarization modulation innovation was a birefringent half-waveplate: a single crystal of anisotropic dielectric (typically quartz or sapphire) that phase-delays one of the two linear polarizations [23, 5]. While the fragile ~ 1 kg, cryogenically-cooled crystal *can* be rotated at ~ 10 Hz with lower-vibration than rotation of the entire telescope, such a mechanism is prone to failure since bearing operation is a severe challenge at cryogenic temperatures. And since bolometers are sensitive to power dissipation at the 10^{-17}W level, even minute mechanical vibrations produced by the bearings are intolerable.

The Faraday Rotation Modulator[1] (FRM) shown in Figure 4, requires only "rotating" electrons (the generation of a solenoidal magnetic field) in a magnetized dielectric to effect polarization rotation. Therefore, FRMs reduce the rotating mass that provides modulation by 30 orders of magnitude! Furthermore, these devices are capable of rotating polarized millimeter wave radiation at rates up to 10 kHz—faster than any conceivable time-varying temperature- or optical-gain fluctuation. A superconducting NbTi solenoid wound around the waveguide provides the magnetic field that drives the ferrite into saturation, alternately parallel and anti-parallel to the propagation direction of the incoming radiation. The FRM rotates the CMB polarization vectors by $\pm 45°$ at 1 Hz, well above 1/f-fluctuation timescales (caused by, for example, temperature variations). The Polarization Sensitive Bolometer (PSB) signals are detected using phase-sensitive amplification phase-referenced to the solenoidal field drive waveform.

Faraday Rotation Modulators, like waveplates, have the promise to mitigate optical systematic effects occurring "downstream" from the modulator. Though the FRMs described in this work operate in waveguide, following the primary aperture of the telescope and additional coupling optics, nevertheless the Faraday effect could also be able to be implemented in free-space, perhaps at a cold optical stop. In such an implementation the polarization modulation would be identical to a birefringent waveplate, except that *no* mechanical rotation mechanism would be required.

In free space, or in waveguide, Faraday modulation of the plane of polarization can be effected in a continuous modulation mode or in a step-and-integrate mode. In either mode the effective polarization angle can be commanded with extremely high precision and repeatability which also obviates the need for a servo system as used in a conventional half waveplate modulation.

The Faraday effect in waveguide has also been used to rotate millimeter-wave polarization between two fixed states $-90°$ and $+90°$. Such a mode is equivalent to a 180° phase switch. This implementation has been successfully deployed in the MBI experiment and is further described in this volume [28].

In November 2005 a millimeter-wave polarimeter with 49 spatial pixels was deployed to the United States' Amundsen-Scott South Pole Station. This CMB polarimeter called BICEP[27, 21] required low-systematic, high-reliability, polarization modulation. This led to development and observations with the FRMs as a secondary form of polarization modulation (the primary modulation mechanism being rotation of the BICEP telescope about its boresight axis).

The FRM provides rapid modulation of the plane of linear polarization (the dominant type of CMB polarization) and was operated in a continuously switching mode (as opposed to a continuously rotating mode or a "point-and-integrate mode"). In the continuously switching mode the Stokes parameter measured by a paid of linearly polarized detectors is switched between +Q and -Q. Only the polarized component of the incident flux is modulated and subsequently analyzed with phase-sensitive detection (lock-in amplification). In the BICEP experiment six FRMs; three operating at 100 GHz and three at 150 GHz, were deployed as

[1] US Patent Pending: "Wide Bandwidth Polarization Modulator, Switch, and Variable Attenuator," US Patent and Trademark Office, Serial Number: 60/689,740 (2005)

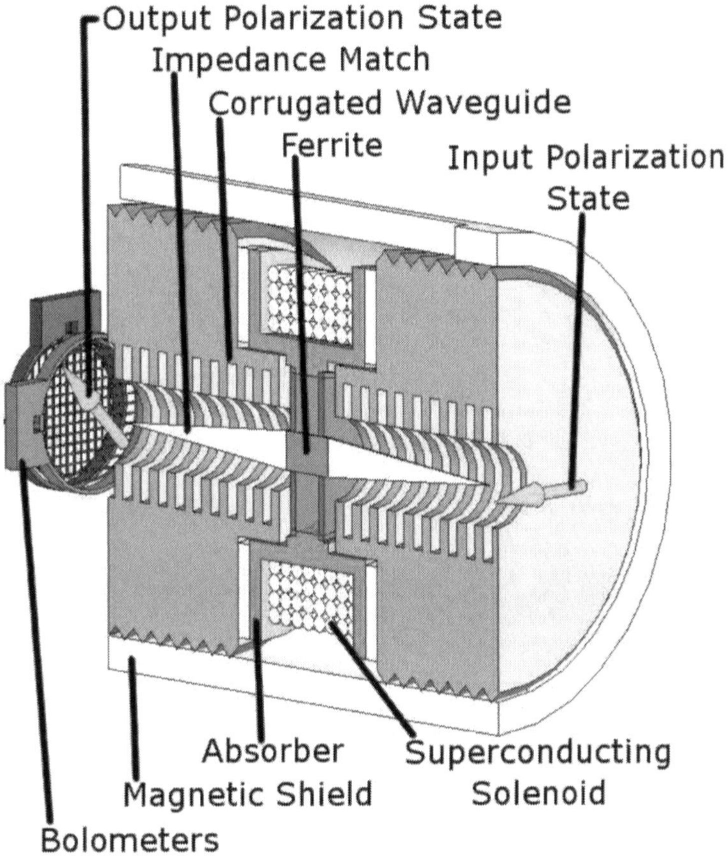

Figure 4. A cross-sectional view of a Faraday Rotation Modulator (FRM) pixel. Polarized light enters from the right, is rotated by $\pm 45°$ and then analyzed (decomposed into orthogonal polarization components, which are detected individually by the polarization-sensitive bolometers, or PSBs). For each FRM pixel, the PSBs are contained within a corrugated feedhorn, cooled to 0.25 K, located approximately 20 cm from the FRM, which is placed in a corrugated waveguide at the interface between two corrugated feedhorns, placed back-to-back and cooled to 4 K – see Figure 5. The (schematic) location of the PSBs in this figure serves to illustrate the coordinate system used as the polarization basis. Figure credit: Thomas Renbarger.

a proof-of-concept test of the FRM technique and to explore operational modes for this new device. Preliminary results from BICEP's FRMs are discussed below.

In the 1970's a polarimeter which used a Faraday modulator and an orthomode transducer as a polarization analyzer was employed by Nanos [29] to conduct early studies observations of the CMB. The modulator described in this white paper (Fig. 4 has a bandwidth $\simeq 30$ times greater than Nanos' or those currently available from commercial vendors such as MPI/QuinStar Technology Inc. Our devices are also capable of operation at cryogenic temperatures, which is an operational advantage not enjoyed by Nanos' devices nor commercially available devices.

3.2. Design

The magneto-optical design of the modulator is based on a room temperature isolator employing a ferrite dielectric waveguide [30] and a cryogenic isolator [31]. These isolators operate in rectangular waveguide allowing precise mode filtration and low spurious mode generation. Operation as a polarization modulator requires matching modal propagation coefficients of the propagating modes in the entrance waveguide (TE_{10}, TE°_{11}, or HE°_{11}) and the propagating HE^{diel}_{11} mode in the dielectric guide. For the tests described in this paper a corrugated circular waveguide with a 2.35 mm diameter was used. The Faraday modulator design is displayed in Fig. 4.

To achieve magnetic saturation with low Joule heating, a solenoid constructed of superconducting copper niobium wire is used. According to the manufacturer of the ferrite this material has a saturation magnetization of $4\pi M_z = 5000$ gauss which is obtained with an applied magnetic field of $H \simeq 30$ Oe. This field is produced by the superconducting solenoid with $N \simeq 200$ amp-turns. The calculated self inductance of the solenoid is $L = \mu_o N^2 A/\ell = 10$ mH where A and ℓ are the area and length of the solenoid. This value agrees well with the inductance measured with an LCR meter and directly by observing the L/R time constant of the solenoid at 4 K. All performance attributes of the modulator are improved with careful attention to assembly details and tolerances. In particular, the alumina cone-ferrite cylinder "toothpick" assembly must be placed along the propagation direction with less than 1° tilt, otherwise resonant mode conversion produces several dB of insertion loss.

3.2.1. Polarimetry with Faraday Rotation Modulators Figure 5 is a schematic illustration of a FRM as implemented in a bolometric polarimeter. Bolometric polarimeters, such as BICEP[21]

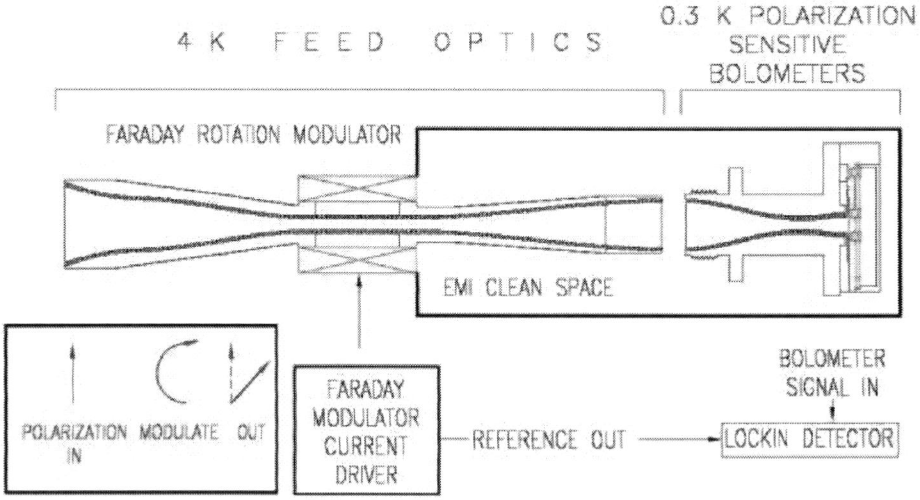

Figure 5. Conceptual schematic of a Faraday polarization modulator coupled to a polarization sensitive bolometer (PSB) used as a polarimeter. The modulator is placed at the waveguide choke and rotates the plane of linear polarization of the propagating HE^{11} spatial mode.

often utilize a back-to-back corrugated feedhorn geometry to provide (electromagnetic interference) EMI shielding between the beam-defining optics (typically cooled to $\simeq 4.2$ K)and the detector coupling optics (typically < 0.3 K). The EMI shield is implemented as a high-pass filter (waveguide choke) between the two back-to-back feedhorns in the beam-defining optics section, preventing radio frequency radiation from being detected by the high-impedance bolometers. The waveguide choke allows a single waveguide mode to propagate, and it is here that the FRM is installed.

Detection after modulation can be accomplished with a bolometric or HEMT amplifier radiometer. For the latter application, the rapid modulation frequencies available with the Faraday rotator allows the radiometer to be Dicke-switched in total power mode, *i.e.*without other forms of phase modulation. The FRM polarimeter, whether used with a HEMT amplifier or bolometer detector system, completely characterizes the linear polarization state of the incident radiation from the same spatial pixel in the focal plane of a telescope. This allows the incident radiation in each linear polarization state to propagate through the same optical train (windows, lenses, filters, detectors), under identical optical loading conditions. Unlike polarization modulation by physical rotation of the polarimeter, polarimeters employing FRMs allow the polarization to be modulated without moving the telescope's antenna beam on the sky. By keeping the spatial beam pattern fixed on the sky we isolate the polarization properties and can ignore variations of the beam response or excess variable pickup from the ground.

We used a pair of polarization sensitive bolometers (PSBs) to simultaneously analyze (separate the incident beam into orthogonal polarization states) and detect the power in each state [32]. For an incident electric field described by

$$\begin{aligned} \vec{E}(x,y,z,t) &= E_x(t)\sin\left[2\pi\left(\frac{z}{\lambda}-\nu t\right)+\phi_x(t)\right]\hat{\mathbf{x}} \\ &\quad +E_y(t)\sin\left[2\pi\left(\frac{z}{\lambda}-\nu t\right)+\phi_y(t)\right]\hat{\mathbf{y}}, \end{aligned} \tag{1}$$

time-averaging over the electric field oscillations, the Stokes parameters of this field are defined as:

$$I \equiv \langle E_x^2\rangle + \langle E_y^2\rangle, \tag{2}$$
$$Q \equiv \langle E_x^2\rangle - \langle E_y^2\rangle, \tag{3}$$
$$U \equiv \langle 2E_xE_y\cos(\phi_x-\phi_y)\rangle, \tag{4}$$
$$V \equiv \langle 2E_xE_y\sin(\phi_x-\phi_y)\rangle, \tag{5}$$

where I is the intensity, Q and U describe the linear polarization, and V quantifies the circular polarization of the electric field. The CMB is linearly polarized so $V = 0$. If the source is a black body radiator in the Rayleigh-Jeans portion of the spectrum, then the observed Stokes parameters are $I \propto (T_x + T_y), Q \propto (T_x - T_y)$ with the proportionality constants fixed by calibrating the receiver system.

The signal produced by differencing the two linear polarization-sensitive bolometers when the microwave signal is Faraday modulated is

$$V_{diff} = S \times \eta_{opt}\eta_{pol}[Q\cos 2\theta(t) + U\sin 2\theta(t)] \tag{6}$$

where the (time dependent) modulation angle is

$$\theta(t) = B_z(t)\gamma l\sqrt{\epsilon}/2c, \tag{7}$$

and l, ϵ, γ, and S are the ferrite's length, dielectric constant ($\simeq 12$), gyromagnetic ratio of the electron, and the bolometric sensitivity in V/W, respectively. $B_z(t) = H_z^{app}(t) + 4\pi M_z(t)$ is the (time-dependent) magnetic induction.

3.3. Performance
In order to be useful as a linear polarization modulator the FRM must rotate mm-wave radiation with high modulation efficiency, low loss, and little degradation of polarimetric fidelity. We divide the performance criteria into four classes (electromagnetic, thermal, modulation, polarization systematics). When multiple devices are used, as with BICEP, we must ensure that the devices are matched in performance and have reasonable operating requirements. To determine if the FRMs could achieve these goals, extensive testing on individual devices (as in Fig. 5) at room temperature, and on arrays of FRMs at cryogenic temperatures (4.2 K) were performed.

3.3.1. High-Frequency Electromagnetic Performance Faraday rotation modulator was constructed for use with polarization sensitive bolometers over the frequency range 80 to 110 GHz. Figure 6) shows the insertion and reflection loss of the rotator versus frequency measured at room temperature on a vector network analyzer (VNA). Rectangular-to-circular transitions were used to couple power from the VNA to the waveguide ports of the FRM. Excellent agreement was found between its measured and performance simulated using Ansoft's High Frequency Structure Simulator. The average microwave transmission efficiency of the rotator is 80% and the band-averaged reflection loss is less than 1% across a 27 GHz band (28% fractional bandwidth). Similar devices operating from 130-215 GHz have been produced, and 49 devices operating in the 100 and 150 GHz bands have been cooled to 4.2 K and used as polarization modulators.

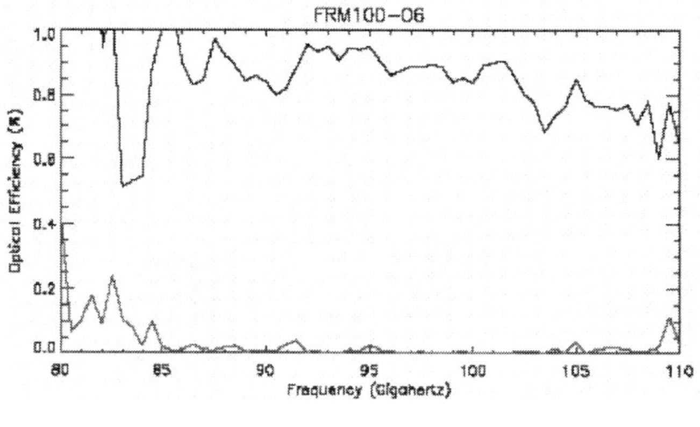

Figure 6. Optical efficiency (transmission) measurements of a 100 GHz Faraday rotation modulator at 300 K measured on a vector network analyzer. The band-averaged transmission for the modulator is 80% warm, and improves to approximately 85% when cooled below 10 K. The reflection (black) and absorption (green line) features at 83 GHz are arbitrary and determined by the desired high-pass cutoff frequency of the corrugated waveguide.

3.3.2. Polarization Modulation Performance To determine the Faraday rotator's suitability for use as a polarization modulator, a 80-100 GHz bolometric test receiver was constructed. The receiver consists of a corrugated feedhorn, Faraday modulator and polarization sensitive bolometers operating at 0.3 K, as shown in Fig. 8. The bolometers were calibrated using unpolarized thermal loads and the receiver was configured to view a 100% polarized source produced by a wire grid polarizer and a 300 K load reflected from the grid and 77 K transmitted through the grid. This source presents a 100% polarized signal with an antenna temperature of 223 K into the receiver which was subsequently modulated by the Faraday rotator.

- **Rotation Angle**
  The Faraday rotator's polarization rotation angle was determined using a polarizing grid. The FRMs were confirmed to continuously rotate polarized millimeter wave radiation by as much as 160°.
  To determine the rotation angle, the maximum and minimum detector output voltages were determined. Since the PSBs are absorbing polarization analyzers, a maximum voltage is produced when the wire-grid axis is oriented perpendicular to the bolometer's sensitivity axis. The bolometer's voltage is minimized when the grid azimuthal angle is rotated by 90° from the maximum. If the grid is positioned midway between the two bolometers, then the polarization rotation angle is given in terms of the bolometer output voltage by:

$$\Theta = \frac{1}{2}\arcsin\Big(\frac{V - (V_{max} + V_{min})/2}{(V_{max} - V_{min})/2}\Big), \tag{8}$$

  where V_{max}, V_{min} are the maximum and minimum signal voltages recorded during complete polarization modulation.
- **Modulation Modes: Continuous Polarization Angle Rotation vs. Switched/Stepped Rotation**
  From equations 3 and 4 we see that when the electric field produced by the grid is at 45° to each bolometer, $Q = 0$ and U is maximized. If the Faraday modulator is then driven into saturation to rotate the incident polarization angle by +45°, the Stokes parameters (in antenna temperature units) will be $Q = 223K$ and $U = 0K$.
  If the current is sinusoidally varied up to the current required for magnetic saturation: $I = I_{sat}\sin\omega t$, the rotation angle will be $\theta(t) = \Theta_o \sin\omega t$ where $\Theta_o = M_{sat}\gamma l\sqrt{\epsilon}/2c$ is in radians, and the difference between the (orthogonally polarized) bolometer signals $V_{x,y}$ is:

$$\begin{aligned}\Delta V(t) &= V_x - V_y \\ &\propto \cos^2\big(\frac{\pi}{4} - \Theta_o \sin\omega t\big) - \cos^2\big(\frac{\pi}{4} + \Theta_o \sin\omega t\big) \qquad (9)\\ &= \sin(2\Theta_o \sin\omega t). \qquad (10)\end{aligned}$$

  Figure 7 shows sample waveforms for three different magnetic bias values (polarization rotation angles).
  Phase sensitive demodulation of $\Delta V(t)$ at the current bias frequency, ω, produces a signal:

$$S = \int_0^{\pi} \Delta V(t)\sin\omega t dt = \int_0^{\pi} -\sin(2\Theta_o \sin\omega t)\sin\omega t dt = gJ_1(2\Theta_o) \tag{11}$$

  where $J_1(x)$ is the first-order Bessel function and g accounts for the gain of the lock-in (and numerical factors). As shown by Nanos [33], S is maximized when $\Theta_o = 52.7°$. Since $\Delta V(t)$ is not a pure sinusoidal signal, higher (odd) harmonics are also present, decreasing as $J_n(2\Theta_o)$. For BICEP however, we chose to modulate in a switched mode where the polarization was rotated by ±45°, having the effect of switching between the Stokes parameter $\pm Q$ in the bolometer coordinate system.
  Taking the difference of the bolometer signals is a measure of the value of the Stokes Q parameter (in the coordinate system defined by the two bolometers). As the grid rotates the difference signal varies exactly as in equation 6, with $\theta(t)$ equal to the azimuthal angle of the grid about the optical axis. This measurement describes the detection of linearly polarized radiation with 100% polarization modulation efficiency (in the sense that no other modulator can produce a larger differential signal).

3.3.3. Polarimetric Fidelity Three primary systematic effects corrupt polarimetry: instrumental polarization, cross-polarization, and depolarization. Instrumental polarization (IP) results from spurious coupling of the polarimeter (modulator, analyzer, and detectors) to unpolarized radiation. The IP for the bolometric receiver shown in Fig. 8 (with the polarizing grid removed) was determined by observing unpolarized thermal loads (either 300 K or 77 K) and measuring the bolometer difference signals produced as the Faraday modulator was square-wave biased at $\sim$ 1Hz. If there is non-zero instrumental polarization produced by either the optical system and/or the FRM , the two orthogonal bolometer signals will be 180° out of phase, and the amplitude of the bolometer signals will scale with the load temperature. The sum of optical and electrical (cross-talk) IP contributions of the polarimeter has been limited to $< 2\%$ for FRMs used in BICEP.

Instrumental polarization can arise from asymmetry in the toothpick structure. Asymmetries arise both in the construction of the toothpick and its placement in the waveguide. Both the tilt of the toothpick and its concentricity in the waveguide must be controlled to minimize spurious mode conversion from the desired HE^{11} mode to higher order modes. Mode conversion produces differential transmission of the two (degenerate) HE^{11} polarization states, resulting in IP. The alumina cones are manufactured with 0.5° tilt and the ferrite cylinder faces are parallel to better than 0.1°. The assembled toothpick is then placed concentrically in the corrugated waveguide.

When viewing unpolarized thermal loads at 300 K, 77 K, and $\sim$ 12K the IP signals can be seen in real time when the FRM is AC biased. The observed $\sim 1\%$ instrumental polarization is consistent with HFSS simulations of the FRM with a $\sim 0.5°$ tilt of the ferrite/alumina toothpick with respect to the corrugated waveguide axis. Since the level IP is constant in time and spatially fixed with respect to the feed and detectors, unlike a polarized celestial source, it can be removed with high-precision. Cross-polarization measurements are also in agreement with HFSS simulations and are consistent with non-orthogonality of the PSBs at the $< 1°$ level.

Depolarization arises due to differences in loss between the two helicity states which propagate in the ferrite. The two helicity states can be written in terms of the linear polarization states

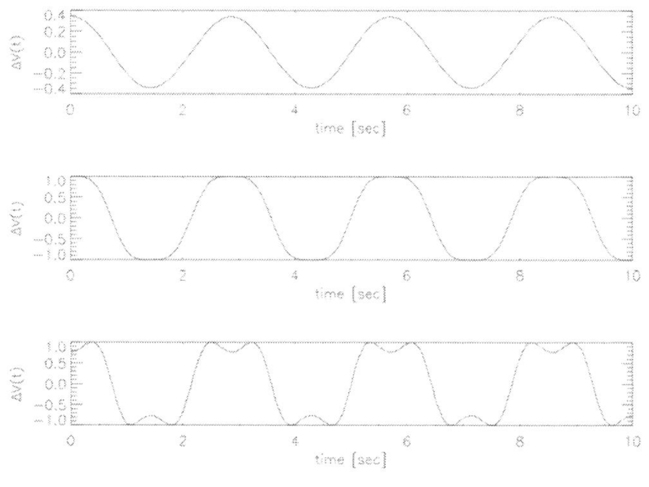

Figure 7. Simulated difference signals (Eq. 10) for orthogonally polarized bolometers viewing a polarized source whose polarization plane is rotated using a Faraday modulator. The bias modulation waveform for all curves is a 2.2 Hz sinusoid with current amplitude set to rotate linear polarization by 10° (top), 45° (middle), and 65° (bottom).

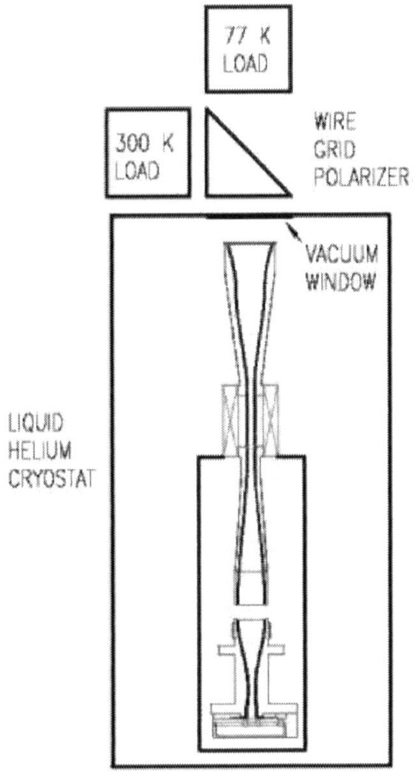

Figure 8. Test set-up for polarization modulation angle, modulation efficiency, and instrumental polarization measurements. Two polarization-sensitive bolometers view a 100% polarized thermal load produced by a wire-grid polarizer tilted at 45° with respect to the optical axis. For measurements of instrumental polarization, the grid is removed and different temperature loads are viewed.

as: RCP $= (\mathrm{E_x}\hat{\mathbf{x}} + i\mathrm{E_y}\hat{\mathbf{y}})/\sqrt{2}$ and LCP $= (\mathrm{E_x}\hat{\mathbf{x}} - i\mathrm{E_y}\hat{\mathbf{y}})/\sqrt{2}$.

At the operating frequencies of interest ($\sim$ 100 GHz) – well-above the Larmor resonance frequency of saturated TT2-111 ferrite ($\simeq$ 11 GHz) – the difference in attenuation coefficients between the two helicity states is $\alpha_+ - \alpha_- < 0.001$ Np/m. This implies that RCP electric fields are attenuated by $e^{\alpha+} = 0.9968$ and LCP fields are attenuated by $e^{\alpha-} = 0.9964$. This differential attenuation therefore results in *depolarization* (as opposed to *instrumental polarization*). The polarimeter's signal is the difference between the power in each polarization state, which is proportional to the difference in intensity between the orthogonal polarizations (which is also proportional to the Stokes Q parameter).
So

$$Q = |E_x^2| - |E_y^2| = 2\mathrm{LCP} \times \mathrm{RCP}$$

implying $Q_{out} = 0.993\,Q_{in}$ and $U_{out} = 0.993\,U_{in}$. This shows that the polarization angle

$$\tan^{-1} Q_{out}/U_{out} = \tan^{-1} Q_{in}/U_{in},$$

Journal of Physics: Conference Series **155** (2009) 012006 doi:10.1088/1742-6596/155/1/012006

Table 1. Measured properties of 100 and 150 GHz Faraday Rotation Modulators.

Property	Value
Absolute RF Transmission (Warm)	80%
RF Reflection (Warm)	$< 3\%$
Current Required for $\pm 45°$ Rotation	± 130 mA
Power Dissipation (4 K)	$\simeq$ 1mW
Polarization Modulation Efficiency (4.2 K)	$\simeq 99\%$
Intrinsic Instrumental Polarization (4.2 K)	$< 2\%$
Cross-polarization (4.2 K)	$< 1\%$
Depolarization (4.2 K)	$< 1\%$

which implies that the output polarized intensity is 99.6% of the input and there is no cross polarization (spurious rotation between $E_x\hat{\mathbf{x}}$ and $E_y\hat{\mathbf{y}}$ or, therefore, between Q and U).

All three polarimetric systematics have been found to be stable over, at least, week-long time periods. This is not surprising as the three systematic effects are intrinsic to the materials used and/or caused by the construction. The stability of these offsets allows them to be modelled and subtracted from the data.

3.3.4. Magnetothermal Performance The Faraday modulator uses a superconducting solenoid that requires a magnetic bias $H_{applied} \simeq 30$ Oersted. Large variations in the magnetic field can produce eddy current heating. Careful thermal and magnetic design is therefore required to minimize eddy currents induced in the modulator. In general, eddy current heating effects are proportional to the sample area transverse to the applied magnetic field. For a solenoid made with copper magnet wire, Joule heating of the coil dominates the electromagnetic loss. Eddy current dissipation in the metallic waveguides is ten times smaller than the coil dissipation ($\sim 200\mu$W) for a solenoid that produces the ~ 30 Oersted field required to saturate a 5 Hz sinusoidally-biased ferrite. For a perfect superconducting coil, the coil loss is zero and the total modulator eddy current heating scales as the modulating frequency squared. Simulations using Ansoft's Maxwell SV are in excellent agreement with measurements of the eddy current dissipation of the Faraday modulator from DC to 20 Hz. Magnetic shielding of the bulk experimental volume/bolometers is accomplished by enclosing the solenoid in a high magnetic permeability jacket fabricated from Cryoperm 10.

3.3.5. Observations of galactic polarization using Faraday Rotation Modulators The faintness of the CMB polarization signal demands exquisite control of instrumental offsets. There are two ways to mitigate offsets: (1) minimize the offset and (2) modulate the signal before detection (Dicke switch) faster than the offset fluctuates. The BICEP experiment [21] does both. A bridge-circuit differences the two PSBs within a single feed, producing a (first) difference signal that is null for an unpolarized input. This minimizes the offset. For six of BICEP's 49 spatial pixels, the polarized signal input is rapidly modulated (second difference) by Faraday Rotation Modulators (FRM) that rotate the plane of linear polarization of the incoming radiation. The FRMs make use of the Faraday Effect in a magnetized dielectric. Polarization modulation allows the polarized component of the CMB to be varied *independently* of the temperature signal, allowing the response of the telescope to remain fixed with respect to the (cold) sky and (warm) ground. This two-level differencing scheme allows for two levels of phase-sensitive detection, allowing optical systematic effects, associated with the telescope's antenna response pattern (which could leak the much-larger CMB temperature signal to spurious CMB polarization), to be distinguished from true CMB polarization.

BICEP conducted observations of the galactic plane using FRMs during the austral winter of 2006. Several hundred hours of data were taken with the FRMs biased with a 1 Hz square wave. This modulation waveform effected $\pm 45°$ of polarization angle rotation. Other modulation waveforms can provide more or less rotation, as desired. To validate the FRM technology we targeted several bright regions of the galactic plane. Results from some of these observations are displayed in Figure 9, where we also show the same region as imaged by WMAP [34].

3.4. Conclusions

The Faraday Rotation Modulator described in this white paper represents a new form of polarization modulation. Due to its wide-bandwidth performance, high modulation rate, and ability to operate at cryogenic temperatures it is uniquely capable as a polarization modulator for CMB applications. The solid-state nature of the device combined with the ability to mass-produce devices also implies it is suitable for array applications – sure to be an essential component of all future probes of the polarization of the Cosmic Microwave Background radiation.

3.5. Acknowledgments

This project has been supported by the Caltech President's Fund and by the University of California, San Diego and NSF PECASE Award # AST-0548262. Evan Bierman assisted in the design, development, and fabrication of the FRMs used in BICEP. Thomas Renbarger and Evan Bierman provided figures used in this paper. Ed Wollack and Neal Erickson provided many useful insights as well as assisting with design considerations. Helpful comments from Jamie Bock, Darren Dowell, William Jones, Chao-lin Kuo, Andrew Lange, and Ki Won Yoon are gratefully acknowledged.

Journal of Physics: Conference Series **155** (2009) 012006 doi:10.1088/1742-6596/155/1/012006

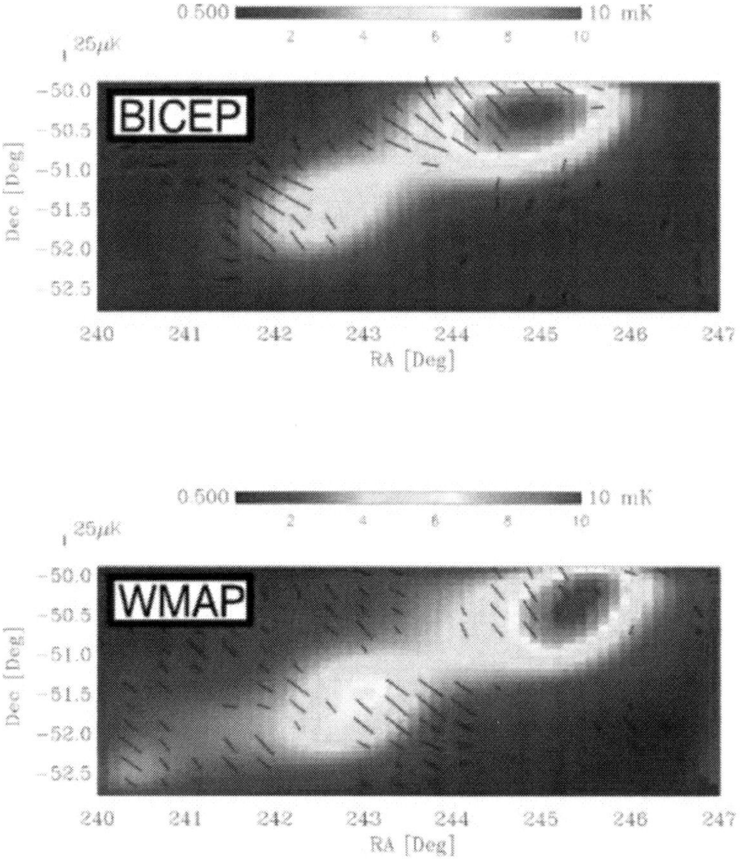

Figure 9. Map of a portion of the galactic plane made using one of BICEP's six Faraday Rotation Modulator (FRM) pixels (top) operating at 100 GHz, compared to WMAP's observations of the same region (bottom). Microwave radiation is polarized by dust grains preferentially aligned by the galaxy's magnetic field. The FRM pixels modulate only the polarized component of the emission and can fully characterize linear polarization *without* rotation of the telescope. The short lines indicate the magnitude and orientation of the plane of polarization, and the color scale indicates the temperature range. For comparison, a scale bar representing 25 μK linear polarization is shown. The galactic plane extends approximately from the lower left to the upper right of each map. Both maps show significant linear polarization orthogonal to the galactic plane[34]. Similar maps were produced for BICEP's 150 GHz FRM pixels. We note that the BICEP FRM data were acquired over the course of only a *week* of observations. Figure credit: Evan Bierman.

Journal of Physics: Conference Series **155** (2009) 012006 doi:10.1088/1742-6596/155/1/012006

4. Quasioptical Reflective Polarization Modulation for the Beyond Einstein Inflation Probe

David T Chuss

NASA Goddard Space Flight Center, Code 665, Greenbelt, MD, 20771 USA

E-mail: David.T.Chuss@nasa.gov

Abstract. Polarization modulators that are based on the introduction of a quasioptical phase delay between two orthogonal linear polarizations are reviewed. The general principle behind the device and the application to the Beyond Einstein Inflation Probe are addressed.

4.1. Introduction

Polarization modulation allows for clean separation of the polarized light from a source and the unpolarized background. This is likely to be an important element in systems designed to search for the B-mode polarization as the unpolarized contribution from the Cosmic Microwave Background (CMB) is much larger than the faint signature of inflation.

It is desirable that a polarization modulator not change the total coherence of the signal that it is modulating. That is, for polarization modulation having low systematics, it is advantageous, and perhaps required, that the total polarization ($P^2 = Q^2 + U^2 + V^2$) remain the same throughout the modulation process. This corresponds to a path on the Poincaré sphere. Such an operation can be physically produced by introducing a phase delay between two orthogonal polarization states. group SU(2) or equivalently by Mueller matrices that are members of

Though there are various ways to do this in practice, this white paper will explore techniques that do this quasioptically and in reflection by adding a phase delay between orthogonal *linear* polarizations. This is most commonly done using a system of polarizers and mirrors. Specifically, it will focus on the Variable-delay Polarization Modulator (VPM). This device modulates the phase between two orthogonal linear polarizations by changing the separation between a grid and mirror. A major advantage of this device is that it can be constructed large enough to be used as the first element in the optical system thereby avoiding most of the instrumental polarization of the system. It can also modulate polarization on time scales (~few Hz) that will allow 1/f mitigation to be done in polarization rather than in a spatial scanning mode. This combination of front-end placement and modulation in polarization is a potential advantage of the VPM since it avoids systematics that mix unpolarized anisotropy into a false polarization signal. Calibration techniques associated with such an architecture may be especially suited to polarimetry at large angular scales where scanning calibration schemes are challenging. In addition, the symmetry of the polarization separation in the VPM can be exploited to provide additional systematic controls.

4.2. Heritage

This concept has heritage in the Martin Puplett interferometer (hereafter MPI) [35] which utilizes a variable phase introduction between two orthogonal linear polarizations to generate an interferogram that, when Fourier transformed, gives spectra of Stokes Q and V. The most common use of this device is as a spectrometer. In such an application, Stokes I of the input signal is mapped to Stokes Q using a polarizing grid (and Stokes V is set to zero) such as in the case of COBE/FIRAS [36]. Even so, the utility of an MPI as a polarization modulator was recognized in early works on the subject [37]. An MPI has been used in the MITOpol CMB spectropolarimeter [38, 39] in combination with a rotating Fresnel double rhomb for the purpose of characterizing the polarization spectrum of the CMB.

A more compact way of introducing a phase delay between two orthogonal polarizations is by placing a polarizing grid in front of and parallel to a mirror. The phase delay is then a function of the mirror-grid separation. Various implementations of this architecture exist in the literature. Howard [40] implements such a system as a polarization transforming reflector (PTR) operating at 1.05 mm. Such setups are common in microwave and radio applications for transforming polarization states. Houde [41] models a similar device (called a "reflecting polarizer" for an OVRO receiver that is used to transform the linear polarization from dust emission into circular polarization that is cross-corellated in the receiver [42]. A similar quarter-wave transformer was also used earlier by Erickson [43] for work at 0.9 mm.

A similar grid-mirror system has been employed by Manabe[44]. In this work, two such systems (here dubbed a "frequency-selective polarizer" or FSP) are used as an alternative to an MPI for interferometry. These authors find that such a system has more alignment tolerance than the MPI, but also introduces an added complexity of treating resonances in the grid-mirror cavity.

Erickson[45] has used a variant called a "Reflecting Polarizing Interferometer" (RPI) in which the grid-mirror cavity is filled with a dielectric. In this work, the RPI is tested as a filter element as an alternative to a Fabry-Perot interferometer.

Polarization modulation with grid-mirror devices has also been done. For such applications, it is possible to modulate polarization in one of two ways. First, one can vary the grid-mirror separation such as is done in the interferometric applications. Second, it is possible to use the device as a half- or quarter- wave plate by fixing the grid-mirror separation (usually to $\lambda/2$ or $\lambda/4$) and rotating the device. The former has been demonstrated in the submillimeter by Krejny[46] in which two VPMs are used to navigate the Poincaré sphere [47]. The Cosmic Foreground Explorer (COFE) used this type of modulation in an early configuration [48], but later moved to a spinning reflective-half wave plate (HWP) modulation [49]. Shinnaga[50] and Siringo[51] have also used the spinning reflective half-wave plate in the submillimeter.

The remainder of this white paper will focus on polarization modulation devices using the parallel grid-mirror system.

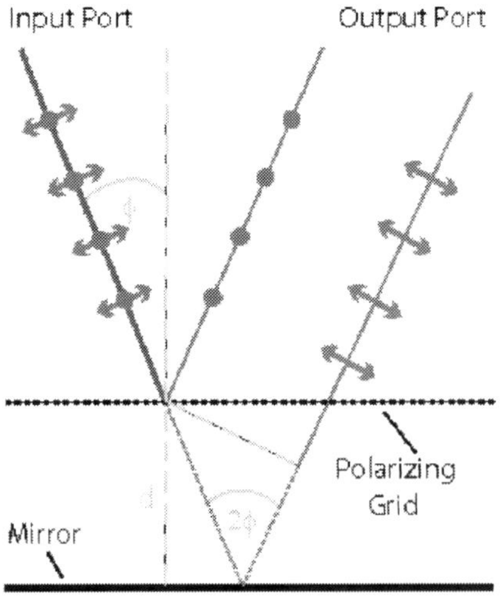

Figure 10. A phase delay between two orthogonal linear polarizations is introduced by an architecture in which a polarizing grid and a mirror are placed parallel to one another. The dotted red line indicates the physical path difference between the two polarizations.

Journal of Physics: Conference Series **155** (2009) 012006 doi:10.1088/1742-6596/155/1/012006

4.3. Operational Principle

The basic architecture that all of the grid-mirror systems discussed in the Introduction is shown in Figure 10. Radiation of an arbitrary polarization state is incident on the grid-mirror structure. The component of the light that is linearly polarized parallel to the grid wires reflects off of the grid. The orthogonal component gets transmitted by the grid, is reflected by the mirror and then passes through the grid again. The two beams recombine with a phase that is related to the path length. The path length is in turn a function of the incident angle and the grid-mirror separation.

$$l = 2d\cos\phi \tag{12}$$

This path difference gives a geometric phase delay of

$$\Delta = \frac{2\pi l}{\lambda} = \frac{4\pi d\cos\phi}{\lambda} \tag{13}$$

Assuming that the wires are perfect polarizers and the grid wires are oriented at an angle θ with respect to the coordinate system of interest, the Mueller matrix for this system is

$$\begin{pmatrix} 1 & 0 & 0 & 0 \\ 0 & \cos^2 2\theta + \cos\Delta\sin^2 2\theta & -\sin 2\theta\cos 2\theta(1-\cos\Delta) & \sin 2\theta\sin\Delta \\ 0 & \sin 2\theta\cos 2\theta(1-\cos\Delta) & -\sin^2 2\theta - \cos\Delta\cos^2 2\theta & -\cos 2\theta\sin\Delta \\ 0 & \sin 2\theta\sin\Delta & \cos 2\theta\sin\Delta & -\cos\Delta \end{pmatrix} \tag{14}$$

For the case of the VPM, the angle is held fixed and the phase is modulated. Setting $\theta = 45°$ gives

$$\begin{pmatrix} 1 & 0 & 0 & 0 \\ 0 & \cos\Delta & 0 & \sin\Delta \\ 0 & 0 & -1 & 0 \\ 0 & \sin\Delta & 0 & -\cos\Delta \end{pmatrix}. \tag{15}$$

This operation is a combination of an inversion operation and rotation about the U-axis on the Poincaré sphere. Because of this, Stokes U is unmodulated, so full linear polarization measurement requires an additional step. Since Δ is dependent on wavelength, the modulation will become less coherent at higher orders, forming an interferogram that contains information about the polarization spectrum of the source.

Conversely, for the reflective HWP, $\Delta = \pi$ and is fixed. The general matrix reduces to

$$\begin{pmatrix} 1 & 0 & 0 & 0 \\ 0 & \cos 4\theta & -\sin 4\theta & 0 \\ 0 & \sin 4\theta & \cos 4\theta & 0 \\ 0 & 0 & 0 & 1 \end{pmatrix} \tag{16}$$

which is recognizable as the Mueller matrix representing a HWP. Note that the path on the Poincaré sphere in this case is a circle in the Q-U plane. Here, linear polarization is completely modulated. Bandwidth effects dilute the signal in a direction perpendicular to the path in the Q-U plane, so the efficiency of the ideal reflective HWP is constant over the modulation, but falls off with increasing bandwidth. This architecture is tunable to multiple bands by changing the grid-mirror separation.

4.4. Current State-of-the-Art

Krejny [46] has reported astronomical use of a dual VPM system at 350 μm. For these observations, the VPM settings were not optimized, resulting in poor modulation efficiency. Nonetheless, an instrumental polarization of less than 1% was measured. Laboratory tests of a

Journal of Physics: Conference Series **155** (2009) 012006 doi:10.1088/1742-6596/155/1/012006

single VPM in this system are shown in Figure 11. In this figure, the data points represent the modulated signal as the grid-mirror separation was varied. The curve represents the polarization transfer function anticipated given the geometrical phase lag (Equation 13.) This has been adjusted in phase and amplitude to match the first peak for illustrative purposes. The difference in the observed phase as compared with that anticipated from the geometrical relationship is most likely due to the fact that the grid wires in this system are a measurable fraction of the wavelength (25 μm vs. 350 μm). This results in a phase shift contribution that is due the reactance of the system. This issue can be resolved by working in a regime where the wires are much smaller than the wavelength. The same VPM has been measured in a 3.15 mm system [52] and this effect was not observed (see the right side of Figure 11). In this plot, the solid line is the polarization transfer function based on the geometric phase delay between the two polarizations.

The polarization modulation efficiency is ~80% for the 350 μm case, but is significantly higher at longer wavelengths. Part of this difference is due to the non-ideal efficiency of the VPM grid at 350 μm. The remainder is most likely due to differences in the optical systems in the two measurements.

This phase-separation relationship needs to be further quantified, especially in the application of VPMs to spectropolarimetry. The wire size, or perhaps more generally the set of polarizer parameters, is a key design consideration in constructing and implementing VPMs.

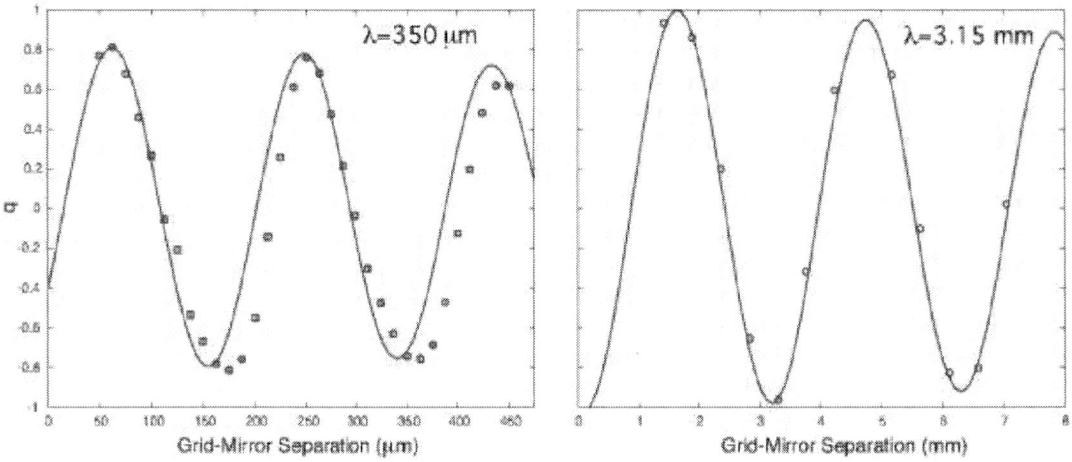

Figure 11. (left) The Hertz/VPM (350 μm) laboratory data is shown [46]. The solid line represents the model for a geometric phase delay that is phase and amplitude matched to the first peak of the data. (right) The same VPM was also tested at 3.15 mm. In this case, no phase or amplitude adjustment was done; the phase more closely matches that of the geometrical case. In each case, the bandwidth is ~10%

4.5. VPMs and CMB Polarization

The systematics of a VPM systematics as they apply to CMB polarimetry are highly-coupled to the instrument architecture in which the VPM is employed, so such a concept is briefly described here. We note as an aside that this differs from the Hertz/VPM implementation mentioned in the preceding section.

Figure 12 shows an optics concept for which the VPM is suited. In this case, the symmetry of the VPM is utilized. The VPM is the first element in the optical path in order to minimize instrumental polarization. This concept requires an array of polarized detectors with the direction of polarization sensitivity oriented at a 45° angle with respect to the wires of the VPM. For the ideal case, this system is described by expression 15. To first order, this system is insensitive to Stokes U, and thus requires an additional parameter (such as spacecraft or instrument rotation) to fully map the linear polarization. However, to counter this disadvantage, this system allows many of the systematic effects to be hidden in Stokes U without affecting the measurements of Stokes Q.

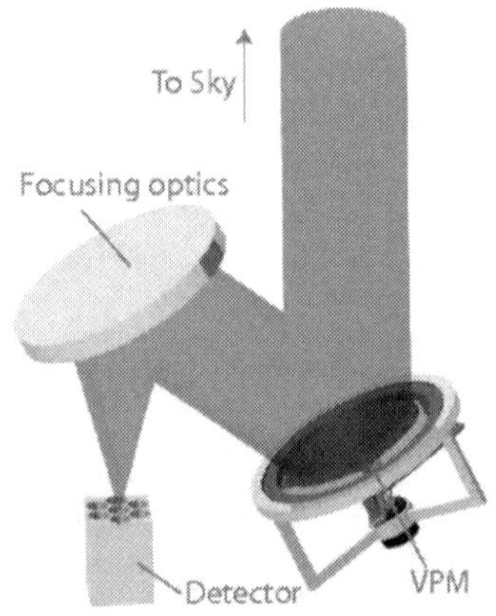

Figure 12. The optical path includes a VPM as the first component in the system.

4.6. Systematic Effects
In many ways, the advantages and disadvantages of reflective HWP are similar to dielectric HWPs, so the following discussion of systematics will focus on the VPM configuration. However, where appropriate below, work on reflective HWPs will be included since it gives valuable insight into the electromagnetics of the grid-mirror system.

4.6.1. Beam Walkoff For the mirror-grid system, the two polarizations are displaced relative to one another a distance $x = 2d \sin \phi$. This quantity, x, is the "walkoff" and for the VPM varies with the modulation variable, d. This leads to a potential synchronous pickup; however this can be mitigated in two ways. First, this effect can be reduced by choosing small incidence angles. Second, if the VPM is located at a pupil in the optical system, then a parallel shifting of the output beams will not change the relative angle between the beams in the two orthogonal polarizations.

4.6.2. Variable Beam Truncation At non-normal incidence, the back optical surface of the VPM will be vignetted by the frame of the front optical surface. The amount of vignetting will depend on grid-mirror separation and so will vary with the modulation. An adequate edge taper will ameliorate this problem. In addition, the symmetry of the concept can help as well. The beam truncation is such that it will affect the Stokes U polarization state. Variations in the light reflected off of the back surface of the VPM will show up in each of the (polarized) detectors as common-mode variations and can be differentiated from the antisymmetric variation of the modulated Q polarization state. Of course, this is only strictly true for the central ray of the system and the effect of beam truncation will vary across the array.

Journal of Physics: Conference Series **155** (2009) 012006 doi:10.1088/1742-6596/155/1/012006

4.6.3. Grid Emissivity and Resonances from Trapped Modes. The emissivity of the grid wires will combine with polarization leakage to cause emission and absorption that will be a function of the grid-mirror separation. Cooling the VPM to 2.73 K will mitigate this risk, but this will impact spacecraft design and will certainly require development effort for cryogenic VPMs. Again, the symmetry of the system helps here as well. Emission and absorption will affect the Stokes U value, but will appear as common-mode in the two orthogonal detectors. The full effect of this will need to be studied in more detail.

An additional effect of imperfect polarization isolation is that radiation can become trapped between the grid and the mirror. This effect has been observed in such grid-mirror systems (for an example, see Houde[41]). This effect is also thought to be responsible for the complex interferometer behavior in Manabe[44]. For broadband operation, this effect is diluted significantly. In addition, optimizing the wire size for the wavelength range will also help mitigate this problem.

4.6.4. Dependence on Grid Properties Extensive literature has been devoted to the electromagnetic modeling of polarizing grids (see Houde [41] and references therein). In the limit where the wavelength is much larger than the wire diameter, the polarizing grid approaches ideal behavior with high isolation between the two orthogonal polarizations. In this case, the phase delay introduced by a VPM is proportional to the path difference between the two polarizations as stated above. As this assumption is violated, the phase delay acquires a contribution due to a reactive component that has a nontrivial dependence on grid-mirror separation. This has been noted by Krejny[46]; however the effect requires more study.

4.7. Technology Development

4.7.1. Parallel Transport Mechanism In order to ensure that the two beams recombine at the same angle, and that the phase difference is well-defined, it is essential that the transport mechanism used to control this mechanism maintain parallelism. This problem is similar to that encountered in Fabry-Perot systems, and similar tolerances apply. One way of constraining parallelism is to employ a system of 3 motors with feedback to control the parallelism [53]. Another is to mechanically constrain all but one degree of freedom and control the position with a single feedback loop [54]. The latter has been implemented in the Hertz/VPM submillimeter polarimeter [55].

The Hertz/VPM devices used piezoelectric motors for motion. The COFE modulator used a linear coil-based motor. In the latter case, the transport stage was driven at resonance to ease control for fast modulation. The feedback can be controlled with a variety of sensors including capacitive devices as well as optical encoders.

4.7.2. Grid Construction In order to implement modulation at the primary aperture for Inflation Probe, VPMs will have to exceed 50 cm in diameter. Grids are typically made by wrapping wire over the grid frame [56]. An alternative technique is based on work by Novak[57] and involves wrapping wires onto a cylindrical mandrel and then transferring them to a grid frame. The latter technique has been used to construct a 50 cm diameter polarizing grid with 63.5 μm wires on a 200 μm pitch (See Fig. 13). The resonant frequency of the wires was greater than 128 Hz, much larger than the $\sim$few Hz modulation frequency. Using the grid flattener, a planarity of the grid surface that was below 50 μm was achieved.

Because the phase definition is of utmost importance with a VPM, the wire surface must be flat to a small fraction of a wavelength. Grid flatteners have been used in both the Hertz/VPM VPMs [55] and in the prototype CMB wire grid [58].

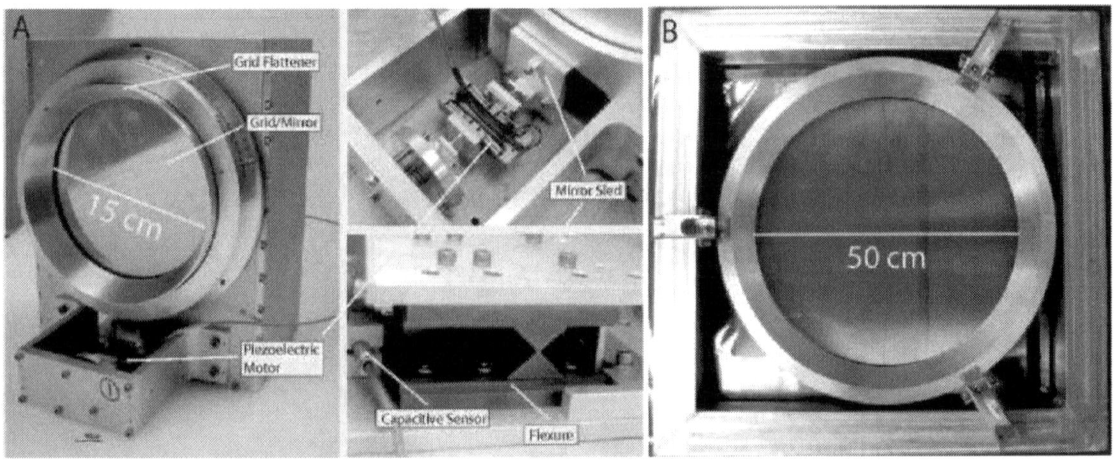

Figure 13. The VPM used in the Hertz/VPM experiment is shown in (A). A 50 cm diameter grid with grid flattener is shown.

4.8. Summary of VPM Advantages/Disadvantages

Advantages

- VPMs are a practical solution for modulation at the front of the optical system because they can be made larger than dielectric HWPs and operate in translation rather than rotation.
- The interferometric modulation scheme provides potential for multi-band, broad-band, or spectropolarimetric use.
- Primary modulation can be done in polarization rather than photometric scanning.
- The translational flexure is frictionless and can be engineered for long, reliable operation in space.
- Systematic control has the potential to be high, due to the capability of "hiding" systematics (e.g. those due to differential loss) in the unmeasured linear Stokes parameter.
- The sensitivity to Stokes V can provide an additional calibration tool.
- "Leakage" term is V rather than U, and therefore VPM may provide good control of (E, B) mixing systematic.
- No transmissive dielectrics are required, thereby mitigating dielectric loss, cavity formation, and large cosmic ray cross section.

Disadvantages

- The VPM is not as tested as the HWP (VPM is at TRL 3).
- Measurement of only one linear Stokes parameter at a time requires spacecraft rotation or cross-calibration with another optical system for complete polarization determination.
- The use of VPMs for front-end modulation would be problematic for high angular resolution experiments. It would be practically limited to lower multipole, large scale B-mode searches.
- Since only one linear Stokes parameter is measured at a time, the modulation efficiency for linear polarization is dependent upon the form of the modulation function. This introduces a potential integration time penalty.
- The VPM is most likely limited to systems with low enough 1/f such that the practical modulation frequencies of ~1-10 Hz would be useful. (This would include bolometer-based systems, but perhaps not many coherent detectors).

Journal of Physics: Conference Series **155** (2009) 012006 doi:10.1088/1742-6596/155/1/012006

4.9. Path to TRL 5

- The flexure concept needs to be adapted for the large aperture VPM. (1.5 Person Years + 50k)
- The continuous ~few Hz modulation/demodulation functionality needs to be tested (1.5 Person Years + 50k)
- The flexure and grid need to be adapted for cryogenic use. (1.5 Person Years + 150k)
- Cryogenic microwave testing must be done. (1.0 Person Years + 100k)
- A system demonstration leading to astrophysical data product on ground-based or suborbital platform must be done.

Journal of Physics: Conference Series **155** (2009) 012006 doi:10.1088/1742-6596/155/1/012006

5. A Metal Mesh Achromatic Half Wave Plate

G Pisano[1], G Savini[2], P A R Ade[2], and V Haynes[1]

[1] The University of Manchester, School of Physics and Astronomy - Alan Turing Building, Upper Brooke street, Manchester, M13 4PL, UK

[2] Cardiff University, School of Physics and Astronomy, Queens Buildings, The Parade, Cardiff, CF24 3AA, UK

E-mail: giampaolo.pisano@manchester.ac.uk

Abstract.
Achromatic Half-Wave Plates can be realised using photolithographic techniques. Prototypes working at 150GHz have been designed, manufactured and tested and preliminary performance results are presented [59]. This type of device could replace the usual birefringent HWP used in mm/submm polarimetry experiments.

5.1. *Introduction*

Birefringent material-based HWPs are used in many present experiments dedicated to the detection of the polarization of the Cosmic Microwave Background radiation. Accurate models [60, 61] have been developed which enable accurate prediction of the HWP performance and enable through design modifications a minimization of systematic effects such as instrumental and cross polarization. However, the demanding requirements for the next generation of CMB instruments, which aim to detect the very low level B-Mode signature, necessitates a new HWP concept which can be utilized with the very large arrays needed to achieve a meaningful sensitivity limit. The costs and the availability of birefringent materials of increasing size is a known problem (crystalline quartz available in ~110 mm diameter boules and crystalline sapphire available in ~280 mm diameter boules) so alternative solutions must be sought. The type of HWP proposed here has a very good chance of replacing the crystalline technology with one constructed from metal mesh components. The manufacture of metal mesh components is a well established technology and will not be discussed in detail here (see filter review of these proceedings).

5.2. *Concept*

A Half Wave Plate introduces a phase shift of 180 degrees between two monochromatic electromagnetic waves that have orthogonal polarizations and that are traveling through it. In an Achromatic HWP, this phase shift is maintained over a broader electromagnetic frequency band. When rotating, a HWP rotates any polarized incident vector at twice the angular velocity of the plate. Thus any linearly polarized incident radiation can be modulated and detected by using detectors sensitive to the polarization (analyzers, OMTs, etc.) after the HWP. The phase shift between the orthogonal axes is normally achieved using the difference between the refractive indices of the ordinary and the extraordinary axes of a birefringent material. Hence , the plate thickness determines the frequency (or multiples of it) at which the crystalline plate acts as a HWP and inherently it is only effective over a narrow frequency band. Since bolometric systems need broad photometric bands to achieve good sensitivity a single crystalline plate would prove very inefficient. A known solution to this was given by Pancharatnam [62], who demonstrated that a broader operational band can be achieved by stacking three or five plates together with appropriate rotation between their optical axes.

Here we propose an alternative solution that consists in using well know metal mesh technology to design and build submillimeter/millimeter HWP. Metal mesh grids with sub-

Figure 14. Pictures of the two stacks of inductive and capacitive grids of a 150 GHz Mesh-HWP. The prototype had an internal diameter of ~35 mm.

wavelength geometries can be designed to manufacture filters for the FIR to submillimeter region. Here, the idea is to extend these studies and design grids which behave differently for orthogonal polarised input and thus to artificially reproduce the behaviour of a birefringent material. Devices of this kind have been realized in the past using other technologies to make microwave retarders at low frequencies [63]. Here we have adopted a particular geometry used in these earlier designs to construct a device able to provide a phase-shift of 180 degrees over a 40% bandwidth around the central frequency of 150 GHz. The concept of a HWP is readily visualized in metal mesh structures as the capacitive and inductive geometries [64] invoke opposite frequency dependent phase shifts. The mesh equivalent of the achromatic HWP thus requires special geometry meshes with delay lines to create a relatively flat phase of 180 degrees over the photometric bands with good in-band transmission of orthogonal polarized vectors. We have used finite-element electromagnetic analysis (HFSS) together with our own transmission line model to optimise the mesh geometries and the design of our prototype achromatic HWP. Here we describe the manufacture of a prototype device for use at frequencies near 150 GHz and compare its measured performance against the model expectations.

5.3. The 150 GHz prototype

The mesh HWP prototype is constructed from twelve meshes whose geometry was determined from the modelling. The device consists of two stacks of grids: one capacitive and the other inductive. The two stacks are built separately and mounted orthogonal to each other so that the axis of one stack is rotated with respect to that of the other stack. Each stack acts on one of the two polarization directions leaving the transmission of the other one almost unperturbed. In addition to the phase effect, the capacitive stack behaves like a low-pass filter and the inductive one as a high-pass filter. There is a spectral region where both the filters have high transmission and their individual phase shifts have opposite signs. The combined difference between phase shifts gives the required 180 degrees delay. Pictures of the capacitive and inductive stacks are shown in Fig.1.

As for many other devices used for the detection of the CMB polarization, one of the most important requirements of the HWP is to have a very low cross polarization. The expected

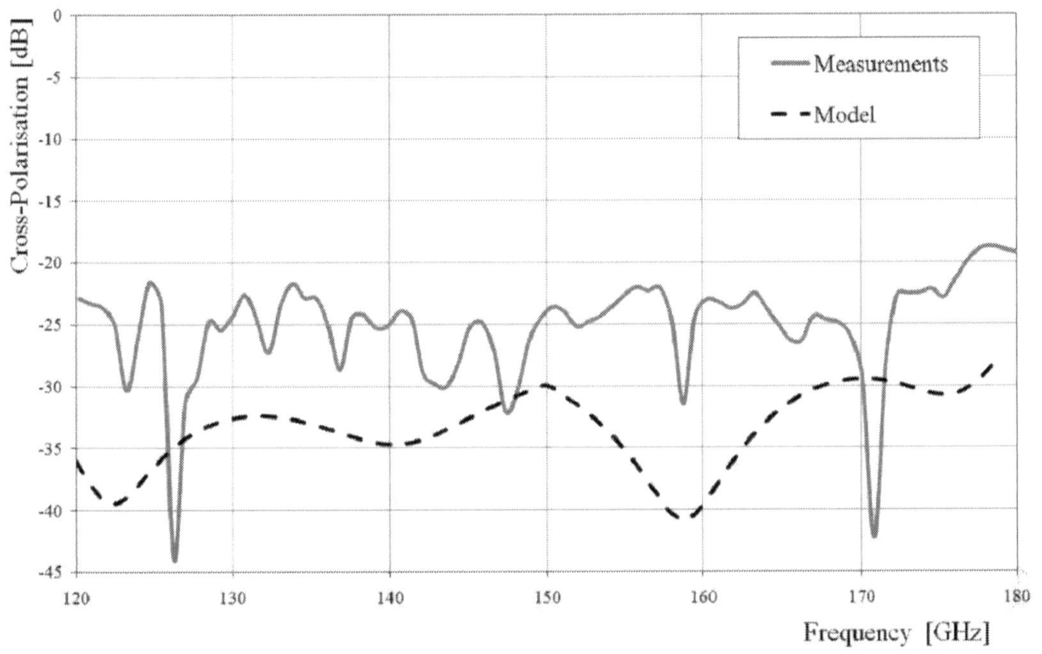

Figure 15. Predicted and measured cross-polarization performance of a 150 GHz Mesh-HWP prototype.

cross polarization for the current device is at the -35 dB level as predicted from the modelling. Measurements of the first prototype are shown in Fig.2 where the cross-polar response is observed to be about -25 dB. The measurements were made using the same experimental setup adopted in [60]. The mesh HWP was held in a rotary stage between two aligned polarizers. Initially, one of the two axes of the HWP was aligned with the polarizers using a laser. The HWP was then rotated by 45 degrees to reach a minimum corresponding to the cross-polarization signal. The discrepancy between model and data can be explained if the alignment between the grids during the assembly of the device was not perfect. However, the measured cross polarization is already comparable to that measured for crystalline devices (for multiplate HWPs we mean the mimimum achievable signal integrated over the band obtained using the same experimental setup). The modulation efficiency of this prototype is greater than 99%.

5.4. Discussion

Mesh filter modelling and its manufacture technology are well understood and filter components made in this way are considered to be mature in terms of flight readiness from usage in a number of missions (ISOLWS, Spitzer, Mars missions, Herschel, Planck). The extension to 'anisotropic' grids to manufacture a HWP has been proved to work well. The designs are scalable to any frequency in the FIR to millimeter region. Conversely, there is not a wide range of birefringent materials available. Moreover, the dimensions of commercially available crystals are limited to ; $\leq$ 28 cm diameter (sapphire only) and even then a complicated manufacture and anti-reflection coating for the three or five bonded plates need to be achieved for acceptable broadband performance. On the other hand the mesh HWP design presented here is an air-gap filter type device. It relies on very thin (1.5 μm thick) copper coated mylar substrates on which the metal mesh geometries have been produced photolithographically. The grids are kept at specific distances by means of annular metallic spacers as used in the air-gap filters used in the cited space borne instruments. However, mesh filters can be also designed without air gaps being completely embedded into dielectric materials. These devices know as 'hot pressed' filters are very robust and like their air-gap counterparts can be cooled cryogenically. We are currently designing a modified 'hot pressed' version of the HWP presented here. Another crucial difference between the technologies is the requirement to cool the devices. For the crystalline devices cooling is essential to remove absorption from phonon bands within the material and hence eliminate emission which could prove a serious sensitivity limitation in terms of the photon noise seen by the detectors. Metal mesh will have minimal absorption throughout the FIR to millimeter region as indeed the filter technology has demonstrated and thus offers the possibility of using the devices warm facilitating simpler rotation mechanism requirements. From the systematics point of view one important aspect to be investigated is the effect of the mesh HWP on the optical beam. The effects due to normal mesh filters are known to be very low. Measurement of the mesh HWP cross polar beams are already planned.

5.5. TRL analysis

The device presented here was built using the air-gap mesh filters technology. The TRL of the present tested device is therefore rated as that of the air-gap filters (see Filter section in these proceedings). Having proved the concept, the next prototypes of mesh HWPs will be made using 'hot pressed' recipes that we have in hand. The hot pressed mesh filter technology is also space qualified. Interestingly, the next devices will therefore only differ in the mesh geometries adopted and should otherwise be considered as a mature space-qualified technology.

Journal of Physics: Conference Series **155** (2009) 012006 doi:10.1088/1742-6596/155/1/012006

6. Polarization Modulation with Photolithographic Modulators

A Kogut

Code 665 Goddard Space Flight Center, Greenbelt, MD 20771

E-mail: `Alan.J.Kogut@nasa.gov`

Abstract. I briefly review the current status for MEMS and SIS switches for CMB polarization modulation. Photolithographic modulators allow development of a "polarimeter on a chip" capable of measuring the Stokes I, Q, and U parameters within a single beam spot on the sky but with no macroscopic moving parts. The modulators place no constraints on the optical design, while the fast switching speed (in excess of 1 kHz) relaxes constraints on beam motion across the sky. However, while some prototype switches have been demonstrated, the devices remain at low TRL.

6.1. Introduction

Reliable characterization of faint polarization superposed on a bright unpolarized foreground requires modulating the polarized signal on time scales short compared to instrumental drifts or $1/f$ noise. Of the various technologies studied in this workshop, microcircuit modulators offer the fastest modulation.

Figure 16 illustrates the concept for phase-sensitive microcircuit modulation. A polarizing microstrip element (e.g., orthogonal probes in a resonant structure) launches voltages into two arms of the polarimeter, proportional to the electric field amplitudes $E_x \cos(kz - \omega t - \phi_x)$ and $E_y \cos(kz - \omega t - \phi_y)$. A half-wave phase switch alternately injects a phase delay 0 or π in one arm. The voltages from the two arms are then combined before square-law detection. When the phase switch is off, the voltages in the two arms are

$$\begin{aligned} V_x &= E \cos\alpha \cos(\omega t - \phi_x) \\ V_y &= E \sin\alpha \cos(\omega t - \phi_y) \end{aligned} \tag{17}$$

where $E = \sqrt{(E_x^2 + E_y^2)}$ and α is the angle between the linearly polarized incident field and the xy coordinate system of the polarimeter. The detector power is

$$\begin{aligned} P_{\rm off} &= \langle (V_x + V_y)^2 \rangle \\ &= E^2(1 + 2\cos\alpha \sin\alpha) \end{aligned} \tag{18}$$

up to an uninteresting constant phase. When the phase switch is on, it introduces an additional half-wave path length so that $V_y = E \sin\alpha \cos(\omega t - \phi_y + \pi) = -E \sin\alpha \cos(\omega t - \phi_y)$. The detector power is then

$$P_{\rm on} = E^2(1 - 2\cos\alpha \sin\alpha) \tag{19}$$

As the switch chops, the detector produces a slowly-varying offset

$$\begin{aligned} P_{dc} &= (P_{\rm on} + P_{\rm off})/2 \\ &= E^2 \quad (i.e.\ \text{Stokes I}) \end{aligned} \tag{20}$$

proportional to the unpolarized intensity, plus a rapidly modulated term

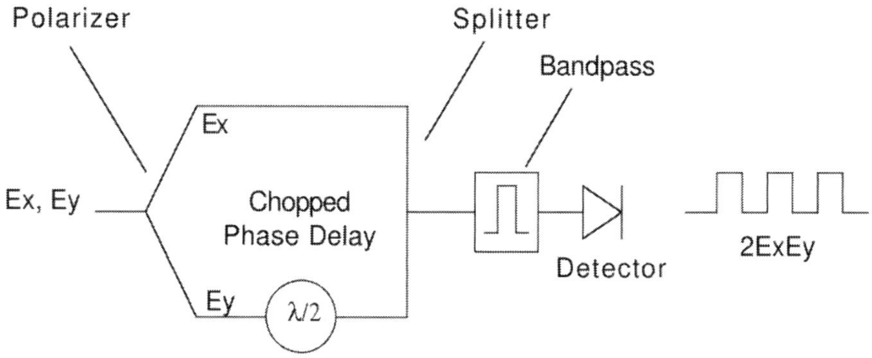

Figure 16. Schematic microcircuit polarimeter. A phase switch introduces a half-wave delay in one arm before the signals are combined and detected. The detector output has a dc term proportional to the unpolarized intensity (Stokes I) plus an ac term at the switch frequency proportional to the linear polarization. In practice, the circuit would be symmetrized with a 90° switch in each arm.

$$
\begin{aligned}
P_{ac} &= (P_{\rm on} - P_{\rm off})/2 \\
&= 2E^2 \cos\alpha \sin\alpha \\
&= 2E_x E_y \;\; (i.e.\ \text{Stokes U})
\end{aligned} \tag{21}
$$

proportional to the linear polarization (Stokes U). We thus unambiguously separate the polarized and unpolarized components in a single measurement with a single detector.

The concept can readily be extended to provide simultaneous measurements of the Stokes I, Q, and U parameters. Figure 17 shows one such design. A polarizing element again launches voltages onto microstrip components proportional to the x and y components of the incident electric field. Each component is then split to separate detection chains. Detector A (top) is

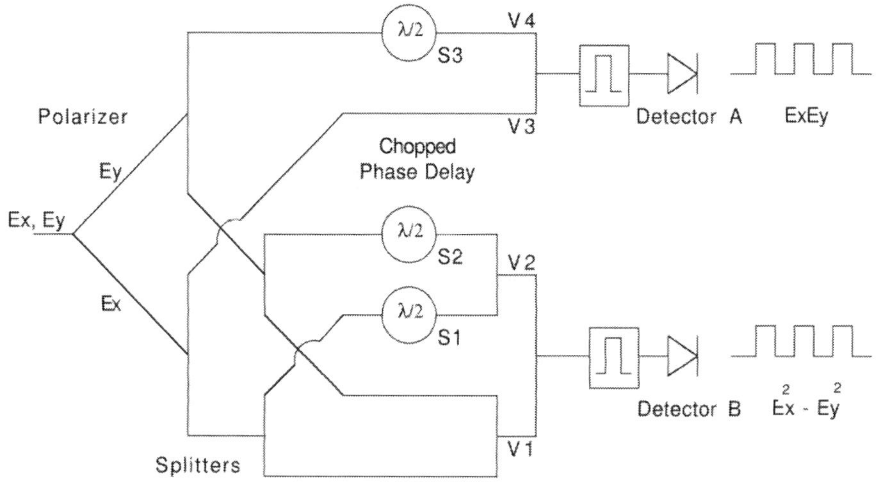

Figure 17. Schematic planar polarimeter to measure Stokes IQU components. Detector A measures the Stokes U parameter as in Figure 16. Switches S1 and S2 inject a half-wave delay out of phase with each other to allow Detector B to measures Stokes Q.

identical to the IU polarimeter in Figure 16. Detector B has an additional pair of phase switches run out of phase with each other (e.g., S1 has phase delay 0 when S2 has delay $\lambda/2$, or else S1 has phase delay $\lambda/2$ when S2 has delay 0). As switches S1 and S2 chop, detector B has a dc term proportional to Stokes I, and a modulated term $P_{ac} = E_x^2 - E_y^2$ proportional to the Stokes Q parameter. Other designs are possible. By simultaneously measuring I, Q, and U, the phase-switched polarimeter fully characterizes the incident linear polarization. This decoupling of the polarization measurement from the instrument orientation or beam scan motion greatly reduces the problem of aliasing unpolarized anisotropy into a spurious polarization signal.

To reduce systematic effects, the modulator circuit should be symmetrized so that all paths from the input to the detector encounter similar switching circuits. Figure 18 shows a symmetrized version of an IQU polarimeter. The upper circuit is configured to detect Stokes U, while the lower circuit detects Stokes Q. Each transmission line has one switch to minimize differential loss between the orthomode transducer and the detector.

6.2. Modulator Circuit

The microcircuit polarimeters described above require phase modulation in individual arms of the circuit. Delay lines of different path length provide a simple, convenient way to produce the phase shift. The simplest such circuit would synchronously switch the transmission path between two legs differing by a fixed length, which in turn would create a phase difference in the transmitted signal. The phase difference, however, would depend on wavelength so that the resulting device would operate as intended only over a narrow wavelength band. Overcoming photon noise statistics requires broad-band operation and thus more complicated circuits are required.

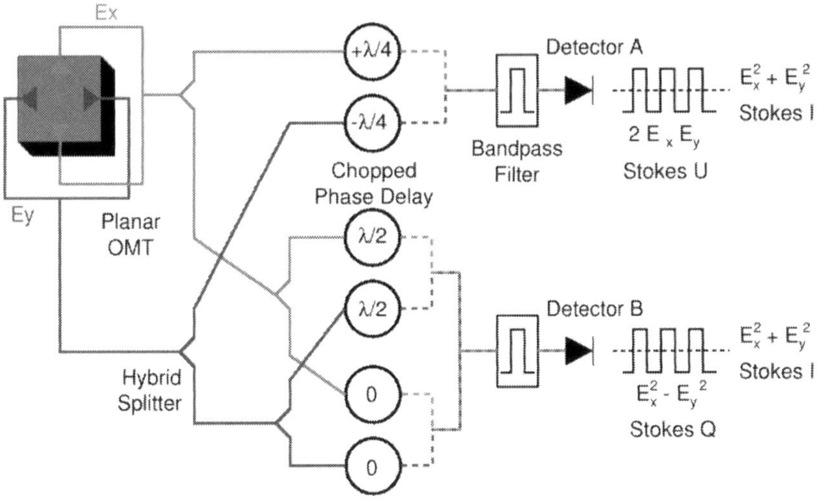

Figure 18. Symmetric version of planar IQU polarimeter. Each leg has one switch to minimize differential loss.

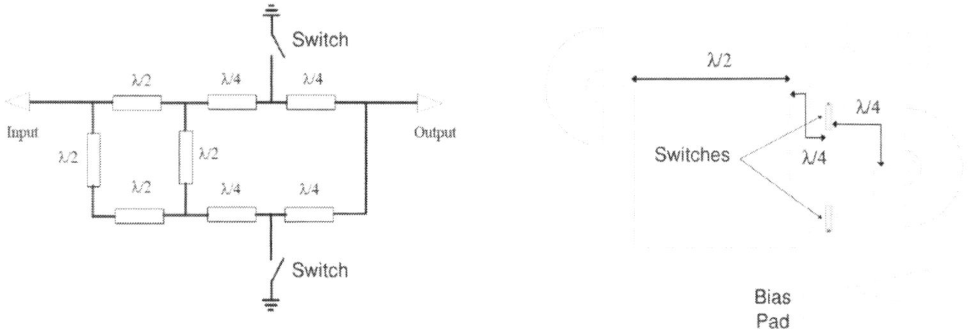

Figure 19. Broad-band phase circuit for CMB polarimetry. (left) Schematic representation. (right) Implementation at 3 mm wavelength.

Figure 19 shows a 90° phase shift circuit for CMB applications, along with a specific implementation of the circuit at 3mm wavelength. Figure 20 shows the predicted performance. Phase stability of a few degrees across a 30% bandpass is achievable. The insertion loss is low (typically 0.3 dB or better).

6.2.1. MEMS switches A planar polarimeter requires active switching. Micro Electro Mechanical Systems (MEMS) switches are one technology capable of operation at cryogenic temperatures and millimeter wavelengths. MEMS switches are miniature surface micromachined components providing controlled motion over a short distance to create either an open circuit or a short across a transmission line. Figure 21 shows a schematic switch for the PAPPA

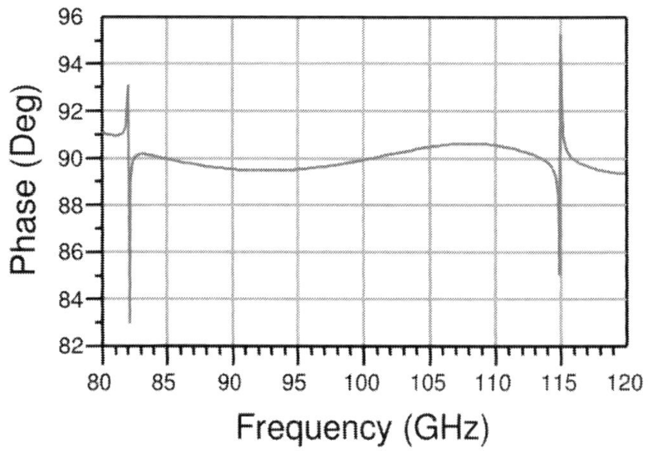

Figure 20. Calculated performance for the 90° phase circuit. Phase stability of a few degrees across a 30% bandpass is achievable.

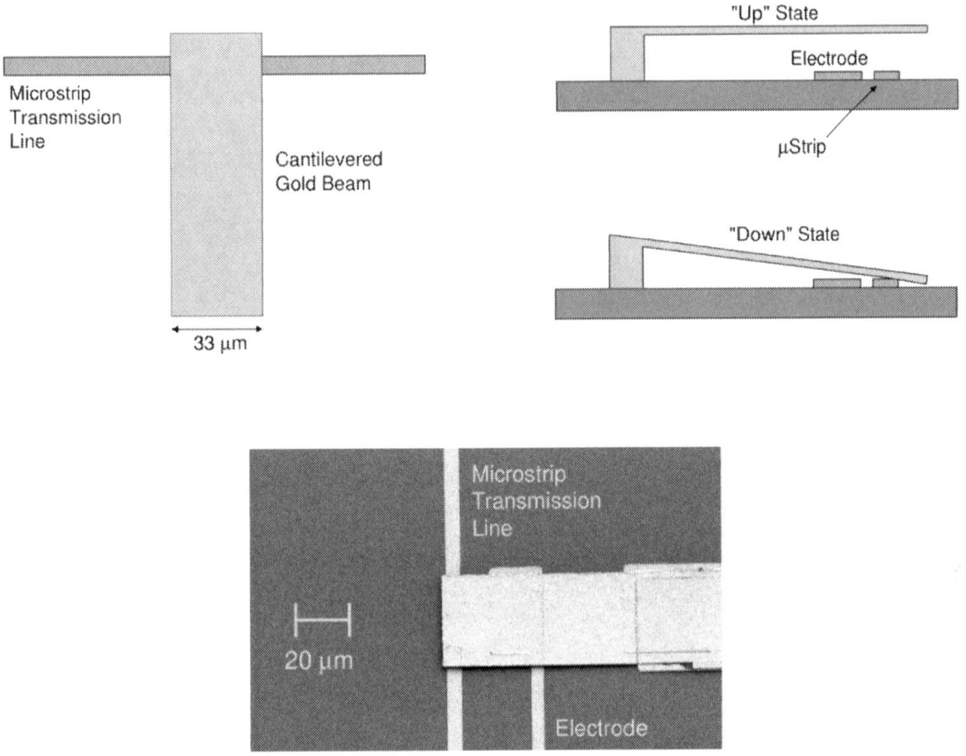

Figure 21. Schematic showing MEMS switch. (Top Left) Top view showing metal beam cantilevered above transmission line. (Top Right) Side view showing the beam in the "up" and "down" positions. (Bottom) Phase switches have reliably been fabricated using photolithographic techniques for operation at 3 mm wavelength.

polarimeter. It consists of a metal beam 33 μm $\times$ 100 μm $\times$ 1 μm thick cantilevered 1 μm above the transmission line. An electrostatic voltage applied to a pull-down electrode pulls the beam down to short out the transmission line. Other designs are possible, including non-contacting designs which modulate the capacitance between the cantilever and the transmission line.

MEMS switches are an attractive technology for CMB polarimetry. The switching transient time is small (a few μs) so that phase switching at kHz frequencies is possible. Commercial versions of RF-MEMS switches are available for frequencies up to approximately 20 GHz. Commercial MEMS switches operating at 1.5 K flew on the ARCADE suborbital mission. Prototype devices for operation at the higher frequencies typical for CMB polarization have been built at GSFC and the University of Virginia.

6.2.2. SIS switches Superconductor-Insulator-Superconductor (SIS) junctions can also be used as the switching element for a microcircuit modulator. Figure 22 shows performance of a prototype SIS switch developed at JPL. Efficiencies of order 80% have been demonstrated across a 15% fractional bandwidth at 3 mm wavelength.

6.3. Advantages and Disadvantages

Both MEMS and SIS switches share a number of advantages and disadvantages for use in CMB polarimetry. Advantages include the following:

- **Scalable Technology** The switches are produced using the same photolithographic techniques on the same wafer as the detectors, and are scalable to kilo-pixel focal plane arrays. No hand assembly is required to produce multiple copies of a "polarimeter-on-a-chip."
- **Fast Modulation** Switching speeds are inherently rapid (100 Hz or faster). The actual speed is likely to be limited by the power dissipation per switching cycle, since the total dissipation from all switches must not exceed the available cooling capacity.
- **Repeatability** Once the initial switch design is tested and verified, performance variability from one device to another should be small. This will reduce systematic effects associated with the overall circuit symmetry.
- **Cryogenic Reliability** There are no macroscopic moving parts (or for SIS switches, no moving parts at all), eliminating the risk associated with cryogenic mechanisms.

Disadvantages are dominated by the issue of complexity. Switch fabrication alone easily adds 10–20 steps (masking, deposition, etching, etc) to the photolithographic circuit production,

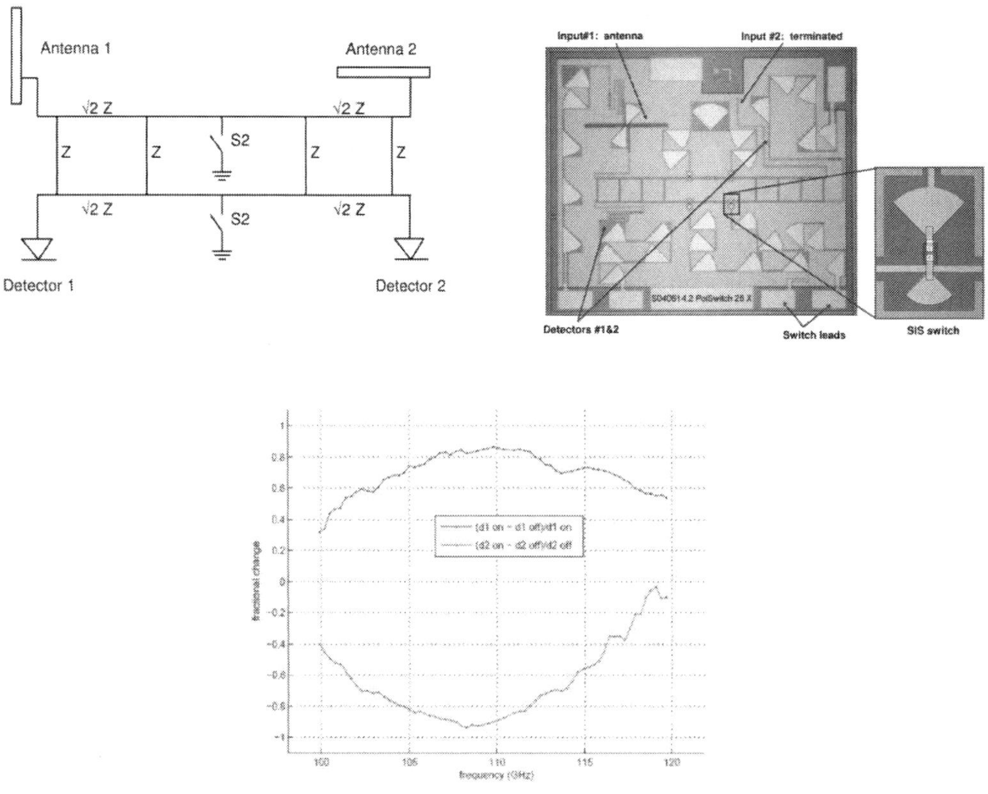

Figure 22. Demonstration of SIS switch. (Top Left) Circuit diagram for SIS test circuit. (Top Right) Photograph of fabricated circuit. (Bottom) Fractional signal modulation as the switches couple and decouple the detectors.

with resulting higher costs and lower yields. In addition, the circuit design necessary to achieve broad-band ($\Delta\nu/\nu \geq 0.15$) signal switching is complicated both for design and verification.

Largely due to the complications involved, these devices remain at low TRL. Phase-switched MEMS circuits have been demonstrated at GSFC and the University of Virginia, which retains a small program devoted to MEMS switch development. A prototype SIS switch has been demonstrated at JPL, but there is no longer an active program developing SIS switches. Both MEMS and SIS switches are currently at TRL ~3.

Raising the modulators to TRL 5 would require 4 or more years development at funding levels roughly $400K per year per project. Both the time and cost are dominated by the fixed costs associated with photolithographic fabrication for multi-mask processes. Testing costs are sub-dominant compared to fabrication.

Journal of Physics: Conference Series **155** (2009) 012006 doi:10.1088/1742-6596/155/1/012006

7. Millimeter Wave Orthomode Transducers

E J Wollack

Observational Cosmology Laboratory, NASA/GSFC, Greenbelt, MD 20771

E-mail: edward.j.wollack@nasa.gov

Abstract. An orthomode transducer (OMT) is a polarization selective splitter or diplexer. The function of this device in an imaging system is to separate the incident radiation into its orthogonal polarization states. Symmetric waveguide OMT devices have found widespread usage in radioastronomy, communications, and metrology at microwave through millimeter wavelengths due to their demonstrated performance. In this paper, the current status of OMT architectures and their application in the measurement of the Cosmic Microwave Background polarization is summarized.

7.1. Introduction

The term orthomode transducer (OMT) is traditionally reserved in the microwave literature to describe a guided wave device which operates on the linear polarization basis. Radiation collected by an antenna structure enters the OMT structure via the dual-mode or main-arm waveguide port and is separated into its horizontal and vertical polarization components. Related waveguide variants which separate the incident radiation field into left and right circular polarization can also be realized. In waveguide this can be achieved though the use of a septum polarizer, however, a quarterwave-plate followed by an OMT can have larger bandwidth. Ideally, these polarization discrimination devices should not adversely impact the systems isolation, optical efficiency, or operational bandwidth. In addition, when used in an imaging array, the OMT structure should be compatible with device integration and facilitate realization of the required environment for the sensors end use.

The role of waveguide polarization diplexers can be illuminated by considering their function in the more general technological context. The OMT is used in a system to separate linearly polarized light into its horizontal and vertical components. In a quasi-optical system this polarization splitting function is traditionally performed by a wire grid. A plane wave incident on the polarizer is separated by reflection and transmission of its linearly polarized components. An orthogonal pair of linearly polarized aerials in free space plays a similar function; however, strictly speaking as a dual-mode waveguide component, the OMT, does not play a role in defining the systems angular response. Spurious modal generation by an antennas feed structure can adversely influence the antenna response and must be limited by design. The OMT structure may also serve as a transition between transmission line types and can be used as a means of conveniently integrating functionality. The fundamental difference between these examples is the underlying mode space or basis on which the ideal device operates. One approach to make a polarization sensitive sensor is by defining the angular resolution with an antenna, then separating the desired polarization states with an OMT, and finally detecting these signals. An OMT allows a clean separation of optical beam definition and mode coupling functions required to define the sensor response.

7.2. Symmetric Orthomode Junctions

Receiver systems for radioastronomy and other precision coherent detector applications have historically used waveguide based polarization discrimination components due to their performance, volume, and mass. In practice, it is possible to keep the polarization isolation small compared to that arising from the optical elements, realize a good coupling, and achieve low

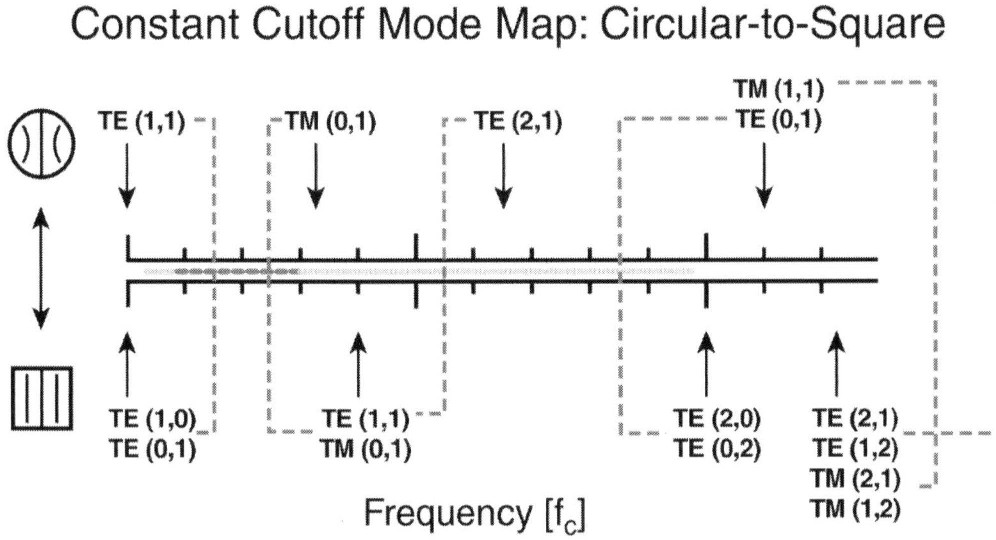

Figure 23. Circular and Square Waveguide Mode Map. The red line indicates the normalized frequency range available when the over-moded common-arm section is asymmetric with respect to rotation. The green line indicates the range available when the common-arm junction is symmetric. The blue dashed lines indicate self similar modal mappings between circular and square guide upon continuous deformation of their boundaries [67].

loss. Full waveguide bandwidth designs for cryogenic operation are known and are in relatively wide spread use. A key to achieving these properties is the use and understanding of symmetric feed and OMT structures.

The performance of all waveguide components are closely related to the underlying junction and electromagnetic symmetry in the structure [65],[66]. Asymmetric OMTs are commonly found in commercial microwave applications. The OMT main or common-arm port is coupled to dual-mode, typically circular or square, waveguide with the modes allowed by the structures symmetry. The bandwidth of asymmetric structures is limited in practice to a ~10-20% fractional bandwidth by the excitation of higher-order modes. The underlying issue can be seen in the waveguide mode map for square and circular guide depicted in Figure 23. In an asymmetric structure the geometry has overlap and can excite modes in the desired pass band. To respond to this bandwidth reduction, a designer is driven to reduce the common-arm guide diameter which indirectly increases guide dispersion and loss. The use of symmetric OMTs enable the full >40% fractional bandwidth of the single-mode (horizontal and vertical polarization) ports to be used. This is achieved by limiting the excitation of higher-order modes in the dual-mode (common-arm) port structure.

In a symmetric OMT the polarization isolation is high and typically limited by residual misalignments. The co- and cross-polar responses are set by the design and realization of the beam-forming optics. The polarization isolation is defined as the leakage between the horizontal and vertical polarization ports. As such it is a scalar quantity. The cross-polarization on the other hand, is a measure of polarization response of the beam as a function of angle [68]. These terms can be found used interchangeably in the open literature; however, in general they describe fundamentally different properties. This distinction can provide valuable insight into the physical origin and can influence the control over systematic errors in polarimetric imaging systems.

The development of four-fold symmetric structures can be traced to the early work on turnstile junctions by R. Dicke while at the MIT radiation laboratory. The waveguide turnstile junction is a natural candidate for wideband OMT performance as a result of its high symmetry; however, the devices inherent three-dimensional topology presents manufacturing challenges, complexity, and cost. Structures of this type are presently an area of active study for millimeter wavelength receiver applications [69]. The Robinson two-fold symmetric OMT used an orthogonal finline transition to pick off each polarization sequentially [70],[71]. An example of a four-fold symmetric adaptation of this approach is the quad-ridge OMT. By folding a part of the turnstile arms, Bifot [72],[73] was able to realize an OMT in a split-block configuration. A recent interesting variation on this design has been presented by Dunning [74],[75] which eliminates the need for the pins and septum used in the Bifot configuration. A more complete review of the historical development of symmetric OMT waveguide structures can be found in [66],[76],[77]. The historical development of these devices reveals a general trend toward reducing the number of components and assembly complexity while retaining the desired electromagnetic properties. These advances have enable realization and use of these structures to $\sim$1 mm.

We note that symmetric waveguide polarization diplexers have polarized outputs which topologically lie in orthogonal planes. For packaging most coherent receivers this does not pose a fundamental issue due to the existing chip fabrication, test and integration procedures, as well as number of receivers. As such, the structures described have been demonstrated and are suitable for use in CMB polarimetry in this context. However, large arrays of incoherent detectors require a higher level of integration than is practical with these inherently three-dimension waveguide structures. This has led to the investigation of planar topologies which allow on chip integration of the waveguide coupling structures and the sensors detectors. Several notable efforts to realize planar OMT junctions can provide useful insight to potential challenges. In avoiding the use of vias or crossovers, an OMT structure violating the desired four-fold symmetry was fabricated and realized an extremely narrow band response [78]. The symmetry of the structure is the key to controlling modal excitation and must be respected. A planar L-band OMT scale model prototype encountered relatively high power loss [79],[80],[81]. This issue was traced to coupling to dielectric substrate and loss in the power combiners. Recent innovations have enabled control over these design concerns and have greatly reduced the potential implementation risks with this approach [82],[83].

7.3. Technology Description

For incoherent polarization sensors having a single planar interface between the dual-mode waveguide port and the sensor chip, which contains a pair of detectors, is highly desirable. This allows a natural separation of items requiring tight photolithographic and loose mechanical tolerances in the integrated package. One approach to achieving this configuration is the use of a planar OMT with four-fold symmetry. The development efforts toward realizing OMT structures for incoherent detectors for use in characterization of the CMB polarization are briefly summarized.

The Clover instruments 97 GHz channel realizes its transition between the feed horns circular waveguide port and rectangular waveguide by using an electroformed turnstile junction OMT [84]. This design accepts a dual-mode circular waveguide inputs and then outputs the vertical and horizontal polarizations into single mode rectangular waveguide. The measured return loss is -20 dB and a polarization isolation of better than -40 dB. A finline transition is used between the waveguide and the 225 μm thick silicon substrate on which the detector resides. The silicon detector chip is mounted in E-plane split block rectangular waveguide. The edge of the chip is covered with metal serrations in the region where the chip is held in a groove in the block to prevent the formation of parasitic resonances in the final device response. The silicon chip has a tapered point to form an adiabatic transition from unloaded waveguide to waveguide loaded with silicon. Next, this mode is converted to antipodal finline and then to microstrip [85],[86] before being terminated in the bolometers absorber.

For Clovers higher frequency channels (150 and 225 GHz) the OMT configuration will be a cylindrical waveguide with four-fold symmetric probes [87]. Signals from opposite probes will be combined in planar circuitry to form the vertical and horizontal polarizations that are connected directly to the TES detectors. This choice allows the OMT probes and TES detectors to reside on the same chip. Simulations have predicted a return loss of -20 dB and isolation of -60 dB. These predictions have been validated for scale models of the structure [88].

Goddard Space Flight Center is currently investigating polarimeter chips with four-fold symmetric probe antennas which transfer signals from the square waveguide to the microstrip lines. See Figure 24. These four-probe antennas are suspended on a thin silicon membrane. This approach uses superconducting niobium microstrip circuits to appropriately combine the desired modes. Broadband crossovers and terminations are used to symmetrize the microstrip lines. After combining, the signals are low-pass filtered and detected with a TES bolometer. The waveguide cut-off defines the lower edge of the operating band. Currently, the filter structure is under test in a custom cryogenic package using commercial microwave probes. A disordered metal alloy is used for the broadband terminations and loads. In this configuration, control over the chip-to-waveguide interface is an important consideration. In this design a photonic waveguide joint [82] with four-fold symmetry is used to prevent leakage of power from the waveguide into the volume of space in which the planar circuitry for the detectors reside. Appropriate attention to the details of the substrate and choke parameters allows control over coupling between the waveguide and the planar chip with the detectors and supporting circuitry. Full wave simulations indicate negligible insertion loss, high isolation and -20 dB return loss over greater than a 35% fractional bandwidth is achievable with this approach.

7.4. Planar OMT Benefits and Disadvantages

7.4.1. Benefits

- In a symmetric OMT the polarization isolation (leakage between the horizontal and vertical modes) is small and independent of the cross-polar response defined by the feed assembly. In operation, this consideration can be used to help in defining a system which limits the influence of spatial structure on the retrieved polarization signal.
- Elements requiring high tolerance are realized photolithographically and the remainder are realized by traditional machining techniques. This allows integration of planar band-pass filters, terminations, absorbers, and bias circuits required to tailor and define the detector response.
- The technology allows a natural partition between the beam-forming and detection circuitry which can be used for optimization of the system performance, characterization, and risk mitigation. The approach leverages existing fabrication approaches for fabrication of corrugated platelet feed horn arrays. Additional waveguide components (e.g., broadband waveplates [89],[90], active phase shifters or Faraday rotators [91], and filtering [92],[93]) can be accommodated in the sensor, if desired.
- The metal structures used in the feed array assembly can be used to support and shield the dielectrics used in the detectors. This design choice provides a natural path to achieving an environment for the sensor which is compatible with both the electromagnetic and ionizing radiation environments as well as providing thermal inertia to limit the impact of environmental changes on the instrument response.
- The detector chip resides in proximity to the antenna feed array interface. The use of a symmetric choke structure can be used to limit propagation of the signal and its sensitivity to the details of the gap geometry. This interface also can provide a convenient interface for a thermal break for the detectors.

Journal of Physics: Conference Series **155** (2009) 012006 doi:10.1088/1742-6596/155/1/012006

- The planar OMT architecture is potentially compatible with, but does not require, modulators on the chip or in the dual-mode waveguide region of the assembly. As with all modulation approaches, the lowest level of complexity compatible with the required system performance and margin is advocated. In making such system level choices the risks associated with common and differing systematic effects need to be carefully considered in the context of how non-ideal responses, if present, will be mitigated. An answer to this the issue of how to appropriately partition the system such that systematic influences on the data are not degenerate presently remains open.

7.4.2. Disadvantages

- As with other approaches, throughput with illumination matched to optics ultimately links the number of sensors, resolution, and cost/risk. This places high importance on sensor performance, yield, and reliability to achieve science goals with acceptable margin.
- Mechanical alignment between mechanical interfaces must be achieved upon cooling.
- Focal plane area trade for beam quality leads to modest optical throughput.
- Planar OMT structures are presently at a low demonstrated level of technical readiness.

7.5. Technological Readiness Level

Conventional symmetric waveguide based orthomode junctions are presently at a relatively high level of technical readiness. They have been widely deployed in coherent millimeter wave systems for radioastronomy, however, a practical path toward the use of these three-dimensional structures in large scale arrays of bolometric detectors is lacking. Planar OMT structures compatible with usage in such arrays are presently under development. Scale models have been demonstrated and millimeter wave designs have been completed which are presently undergoing fabrication and test. As such, these devices are at relatively modest levels of technological readiness, somewhere in the range of ~2-3, and will require additional investment to implement and deploy in systems.

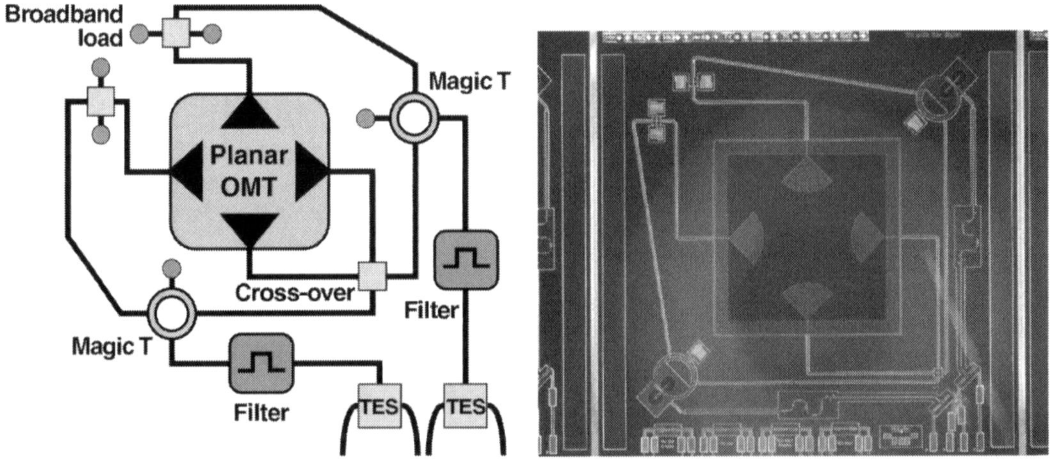

Figure 24. Symmetric OMT Circuit Diagram (Left) and Prototype Device (Right). In the implementation depicted the signals from a pair of waveguide probes are combined in the magic-T difference port. This enforces the desired symmetry and prevents excitation of the modes which can limit the desired response. An additional pair of crossovers has been added to symmetrise the phase between the vertical and horizontal OMT ports prior to filtering. An alternative implementation where the detector is differentially driven can be used if appropriate phase and amplitude matched filters can be achieved.

References

[1] Bock J, Church S, Devlin M, Hinshaw G, Lange A, Lee A, Page L, Partridge B, Ruhl J, Tegmark M, Timbie P, Weiss R, Winstein B and Zaldarriaga M 2006 *eprint arXiv:astro-ph/0604101*
[2] Montroy T E, Ade P A R, Bock J J, Bond J R, Borrill J, Boscaleri A, Cabella P, Contaldi C R, Crill B P, de Bernardis P, De Gasperis G, de Oliveira-Costa A, De Troia G, di Stefano G, Hivon E, Jaffe A H, Kisner T S, Jones W C, Lange A E, Masi S, Mauskopf P D, MacTavish C J, Melchiorri A, Natoli P, Netterfield C B, Pascale E, Piacentini F, Pogosyan D, Polenta G, Prunet S, Ricciardi S, Romeo G, Ruhl J E, Santini P, Tegmark M, Veneziani M and Vittorio N 2006 **647** 813–822 (*Preprint* `arXiv:astro-ph/0507514`)
[3] Hinderks J, Ade P, Bock J, Bowden M, Brown M L, Cahill G, Carlstrom J E, Castro P G, Church S, Culverhouse T, Friedman R, Ganga K, Gear W K, Gupta S, Harris J, Haynes V, Kovac J, Kirby E, Lange A E, Leitch E, Mallie O E, Melhuish S, Murphy A, Orlando A, Schwarz R, O' Sullivan C, Piccirillo L, Pryke C, Rajguru N, Rusholme B, Taylor A N, Thompson K L, Tucker C, Wu E Y S and Zemcov M 2008 *arXiv*
[4] Johnson B R, Collins J, Abroe M E, Ade P A R, Bock J, Borrill J, Boscaleri A, de Bernardis P, Hanany S, Jaffe A H, Jones T, Lee A T, Levinson L, Matsumura T, Rabii B, Renbarger T, Richards P L, Smoot G F, Stompor R, Tran H T, Winant C D, Wu J H P and Zuntz J 2007 **665** 42–54 (*Preprint* `arXiv:astro-ph/0611394`)
[5] Johnson B R, Collins J, Abroe M E, Ade P A R, Bock J, Borrill J, Boscaleri A, de Bernardis P, Hanany S, Jaffe A H, Jones T, Lee A T, Levinson L, Matsumura T, Rabii B, Renbarger T, Richards P L, Smoot G F, Stompor R, Tran H T, Winant C D, Wu J H P and Zuntz J 2007 **665** 42–54 (*Preprint* `arXiv:astro-ph/0611394`)
[6] Wu J H P, Zuntz J, Abroe M E, Ade P A R, Bock J, Borrill J, Collins J, Hanany S, Jaffe A H, Johnson B R, Jones T, Lee A T, Matsumura T, Rabii B, Renbarger T, Richards P L, Smoot G F, Stompor R, Tran H T and Winant C D 2007 **665** 55–66 (*Preprint* `arXiv:astro-ph/0611392`)
[7] Lau J M 2007 *CCAM: A Novel Millimeter-wave Instrument Using a Close-Packed TES Bolometer Array* Ph.D. thesis Princeton University
[8] Hecht E and Zajac A 1997 *Optics* (Addison Wesley)
[9] Savini G, Pisano G and Ade P A R 2006 *Applied Optics* **45(35)** 8907 – 8915
[10] Fowler J 2008 Personal communication
[11] Lee A 2008 Personal communication
[12] Pancharatnam S 1955 Achromatic combinations of birefringent plates Tech. Rep. 71 Raman Research

Institute, Bangalore 137-144
[13] Hanany S, Hubmayr J, Johnson B R, Matsumura T, Oxley P and Thibodeau M 2005 *Applied Optics* **44**(22) 4666–4670
[14] Pisano G, Savini G, Ade P A R, Haynes V and Gear W K 2006 *Applied Optics* **45**(27) 6982–6989
[15] Johnson B R, Abroe M E, Ade P, Bock J, Borrill J, Collins J S, Ferreira P, Hanany S, Jaffe A H, Jones T, Lee A T, Levinson L, Matsumura T, Rabii B, Renbarger T, Richards P L, Smoot G F, Stompor R, Tran H T and Winant C D 2003 *New Astronomy Review* **47** 1067–1075 astro-ph/0308259
[16] Rennick T S, Vaillancourt J E, Hildebrand R H and Heimsath S J 2002 *Proceedings of the 36th Aerospace Mechanisms Symposium*
[17] Hanany S, Matsumura T, Johnson B, Jones T, Hull R and Ma K B 2003 *IEEE Trans. Appl. Supercond.* **13** 2128–2133
[18] Oxley P, Ade P A, Baccigalupi C, deBernardis P, Cho H M, Devlin M J, Hanany S, Johnson B R, Jones T, Lee A T, Matsumura T, Miller A D, Milligan M, Renbarger T, Spieler H G, Stompor R, Tucker G S and Zaldarriaga M 2004 *Infrared Spaceborne Remote Sensing XII* (*Presented at the Society of Photo-Optical Instrumentation Engineers (SPIE) Conference* vol 5543) ed Strojnik M pp 320–331
[19] Hanany S 2008 Personal communication
[20] Keating B G, O'Dell C W, Gundersen J O, Piccirillo L, Stebor N C and Timbie P T 2003 *Astrophysical Journal Supplement Series* **144** 1–20
[21] Yoon K W, Ade P A R, Barkats D, Battle J O, Bierman E M, Bock J J, Brevik J A, Chiang H C, Crites A, Dowell C D, Duband L, Griffin G S, Hivon E F, Holzapfel W L, Hristov V V, Keating B G, Kovac J M, Kuo C L, Lange A E, Leitch E M, Mason P V, Nguyen H T, Ponthieu N, Takahashi Y D, Renbarger T, Weintraub L C and Woolsey D 2006 *Millimeter and Submillimeter Detectors and Instrumentation for Astronomy III* (*Presented at the Society of Photo-Optical Instrumentation Engineers (SPIE) Conference* vol 6275) ed Zmuidzinas J, Holland W S, Withington S and Duncan W
[22] Leitch E M, Kovac J M, Pryke C, Carlstrom J E, Halverson N W, Holzapfel W L, Dragovan M, Reddall B and Sandberg E S 2002 *Nature* **420** 763–771
[23] Caderni N, Fabbri R, Melchiorri B, Melchiorri F and Natale V 1978 **17** 1908–1918 URL `http://adsabs.harvard.edu/cgi-bin/nph-bib_query?bibcode=1978PhRvD..17.1908Cdb_key=AST`
[24] Murray A G, Flett A M, Murray G and Ade P A R 1992 *Infrared Physics* **33** 113–125
[25] Hildebrand R H, Davidson J A, Dotson J L, Dowell C D, Novak G and Vaillancourt J E 2000 **112** 1215–1235 URL `http://adsabs.harvard.edu/cgi-bin/nph-bib_query?bibcode=2000PASP..112.1215Hdb_key=AST`
[26] Lubin P M and Smoot G F 1979 *Physical Review Letters* **42** 129–132 URL `http://adsabs.harvard.edu/cgi-bin/nph-bib_query?bibcode=1979PhRvL..42..129Ldb_key=AST`
[27] Keating B G, Ade P A R, Bock J J, Hivon E, Holzapfel W L, Lange A E, Nguyen H and Yoon K 2003 *Polarimetry in Astronomy. Edited by Silvano Fineschi. Proceedings of the SPIE, Volume 4843.* pp 284–295
[28] Timbie P and Tucker G 2008 *this volume*
[29] Nanos G P 1979 **232** 341–347 URL `http://adsabs.harvard.edu/cgi-bin/nph-bib_query?bibcode=1979ApJ...232..341Ndb_key=AST`
[30] Barnes C E 1961 *IEEE Trans. Microwave Theory Tech.* **9** 519
[31] Erickson N R 2001 *IEEE Trans. Microwave Theory Tech.* **1** 1
[32] Jones W C, Bhatia R, Bock J J and Lange A E 2003 *Millimeter and Submillimeter Detectors for Astronomy* vol 4855 ed Phillips T G and Zmuidzinas J pp 227–238
[33] Nanos G P J 1974 *Ph.D. Thesis*
[34] Page L, Hinshaw G, Komatsu E, Nolta M R, Spergel D N, Bennett C L, Barnes C, Bean R, Doré O, Dunkley J, Halpern M, Hill R S, Jarosik N, Kogut A, Limon M, Meyer S S, Odegard N, Peiris H V, Tucker G S, Verde L, Weiland J L, Wollack E and Wright E L 2007 *Ap. J. Supp. Ser.* **170** 335–376 (*Preprint* `arXiv:astro-ph/0603450`)
[35] Martin D and Puplett E 1970 *Infrared Physics* **10** 105–109
[36] Mather J C, Fixsen D J and Shafer R A 1993 *Proc. SPIE Vol. 2019, p. 168-179, Infrared Spaceborne Remote Sensing, Marija S. Scholl; Ed.* (*Presented at the Society of Photo-Optical Instrumentation Engineers (SPIE) Conference* vol 2019) ed Scholl M S pp 168–179
[37] Martin D 1974 *Infrared and Millimeter Waves* vol 6 ed Button J (Academic Press)
[38] Battistelli E, DePetris M, Lamagna L, Maoli R, Melchiorri F, Palladino E and Savini G 2002 *astro-ph/0209180v1*
[39] Catalano A, Conversi L, de Gregori S, de Petris M, Lamagna L, Maoli R, Savini G, Battistelli E S and Orlando A 2004 *New Astronomy* **10** 79–89 (*Preprint* `arXiv:astro-ph/0405399`)
[40] Howard J, Peebles W and NC Luhmann J 1986 *IJIMW* **7** 1591–1603
[41] Houde M, Akeson R L, Carlstrom J E, Lamb J W, Schleuning D A and Woody D P 2001 *PASP* **113** 622–638

(*Preprint* arXiv:astro-ph/0311414)
[42] Akeson R L, Carlstrom J E, Phillips J A and Woody D P 1996 *ApJL* **456** L45+
[43] Erickson N 1978 *IEEE MTT-S Intl. Microwave Symposium Digest* 438–439
[44] Manabe T, Inatani J, Murk A, Wylde R J, Seta M and Martin D H 2003 *IEEE Transactions on Microwave Theory and Techniques* **51** 1696–1704
[45] Erickson N 1987 *IJIMW* **8** 1015–1025
[46] Krejny M, Chuss D, Drouet d'Aubigny C, Golish D, Houde M, Hui H, Kulesa C, Loewenstein R F, Moseley S H, Novak G, Voellmer G, Walker C and Wollack E 2008 *Applied Optics* **in press** (*Preprint* 0803.3759)
[47] Chuss D T, Wollack E J, Moseley S H and Novak G 2006 *Applied Optics* **45** 5107 5117
[48] Lubin P 2005 *Private Communication*
[49] Leonardi R, Williams B, Bersanelli M, Ferreira I, Lubin P M, Meinhold P R, O'Neill H, Stebor N C, Villa F, Villela T and Wuensche C A 2006 *New Astronomy Review* **50** 977–983
[50] Shinnaga H, Tsuboi M and Kasuga T 1999 *PASJ* **51** 175–184
[51] Siringo G, Kreysa E, Reichertz L A and Menten K M 2004 *Astronomy and Astrophysics* **422** 751–760
[52] Chuss D and Wollack E 2008 *In preparation*
[53] Staguhn J G, Benford D J, Pajot F, Ames T, Allen C A, Chervenak J A, Lefranc S, Maher S, Moseley Jr S H, Phillips T, Rioux C G, Shafer R A and Voellmer G M *Millimeter and Submillimeter Detectors for Astronomy II*
[54] Bradford C M, Stacey G J, Swain M R, Nikola T, Bolatto A D, Jackson J M, Savage M L, Davidson J A and Ade P A R 2002 *Applied Optics* **41** 2561–2574 (*Preprint* arXiv:astro-ph/0205159)
[55] Voellmer G M, Chuss D T, Jackson M, Krejny M, Moseley S H, Novak G and Wollack E J 2006 *Optomechanical Technologies for Astronomy. Edited by Atad-Ettedgui, Eli; Antebi, Joseph; Lemke, Dietrich. Proceedings of the SPIE, Volume 6273, pp. 62733P (2006).* (*Presented at the Society of Photo-Optical Instrumentation Engineers (SPIE) Conference* vol 6273)
[56] Lahtinen J and Halilikainen M 1999 *IJIMW* **20** 3
[57] Novak G, Sundwall J L and Pernic R J 1989 *Applied Optics* **28** 3425–3427
[58] Voellmer G M, Bennett C L, Chuss D T, Eimer J, Hui H, Moseley S H, Novak G, Wollack E J and Zeng L 2008 *in preparation* Presented at the Society of Photo-Optical Instrumentation Engineers (SPIE) Conference
[59] Pisano G, Savini G and Ade P A R 2008 *Applied Optics* 6251–6256
[60] Pisano G, Savini G, Ade P A R, Haynes V and Gear W K 2006 *Applied Optics* **45** 6982–89
[61] Savini G, Pisano G and Ade P A R 2006 *Applied Optics* **45** 8907–15
[62] Pancharatnam S 1955 *Proceeding of Indian Academy of Sciences* **41** 137–44
[63] Shatrow A, Chuprin A and Sivov A 1995 *IEEE Trans. Ant. Prop.* **43** 109–13
[64] Ulrich R 1967 *Infrared Physics* **7** 37–55
[65] Montgomery C G, Dicke R H and Purcell E M 1948 *MIT Radiation Laboratory Series* vol 8 (New York: McGraw-Hill) chap 12
[66] Uher J, Bornemann J and Rosenberg U 1993 *Waveguide components for Antenna Feed Systems: Theory and CAD* (Norwood, MA: Artech House)
[67] Thompson D W 1917 *On Growth and Form* (Cambridge: Cambridge University Press)
[68] Ludwig A C 1973 *IEEE Transactions on Antennas and Propagation* **21** 116–119
[69] Navarrini A and Plambeck R 2006 *IEEE Transactions on Microwave Theory and Techniques* **54** 272–277
[70] Robertson S D 1956 *IRE Transactions on Microwave Theory Techniques* **4** 263–267
[71] Chattopadhyay G and Carlstrom J 1999 *IEEE Microwave and Guided Wave Letters* **9** 339 – 341
[72] Bifot A M, Lier E and Schaug-Pettersen T 1990 *Proceedings of the IEEE* **137** 396400
[73] Bifot A M 1991 *European Transactions on Telecommunications and Related Technologies* **2** 503 – 510
[74] Dunning A 2002 *Workshop on the Applications of Radio Science*
[75] Moorey G e a 2006 *Radioastronomy Proceedings Applications of Radio Science Conference*
[76] Wollack E 1996 *NRAO EDIR Memo Series*
[77] Wollack E and Grammer W 2003 *Proceedings of the 14th International Symposium on Space TeraHertz Technology in Tucson, AZ* 169–176
[78] Jackson R W 2001 *IEEE Microwave and Wireless Components* **11**
[79] Bock D 1999 *BIMA Memo #74 Letters* **11** 483–485
[80] Engargiola G and Plambeck R L 2002 *BIMA Memo #87*
[81] Engargiola G and Plambeck R L 2003 *Review of Scientific Instruments* **74** 1380 – 1382
[82] Hesler J 2001 *IEEE-MTT-S International Microwave Symposium Digest* **2** 783–786
[83] U-yen K, Wollack E J, Papapolymerou J and Laskar J 2008 *IEEE Transactions on Microwave Theory and Techniques* **56** 172–177
[84] Pisano G, Pietranera L, Isaak K, Piccirillo L, Johnson B R, Maffei B and Melhuish S 2007 *IEEE Microwave and Wireless Components Letters* **17** 286288

[85] North C E, Yassin G and Grimes P K 2006 *Proceedings of the 17th International Symposium. on Space Terahertz Technology*
[86] Yassin G, Withington S, Buffey M, Jacobs K and Wulff S 2000 *IEEE Transactions on Microwave Theory and Techniques* **48** 662
[87] Grimes P K, King O G, Yassin G and Jones M E 2007 *Electronics Letters* **43**
[88] North C E *et al.* 2007 *18th Int. Symp. on Space Terahertz Technology*
[89] Kovac J M 2003 *Detection of Polarization in the Cosmic Microwave Background using DASI* Ph.D. thesis University of Chicago, Chicago, IL
[90] Kovac J M and Carlstrom J E 2005 Broadband high precision circular polarizers and retarders in waveguides US patent number 6963253
[91] Keating B G 2008 Faraday rotation modulators published in these proceedings.
[92] Amari S and Bornemann J 2001 *Microwave and Optical Technology Letters* **31** 334336
[93] Vanin F M *et al.* 2006 *IEEE MTT-S International Microwave Symposium Digest* 127130

Journal of Physics: Conference Series **155** (2009) 012007 doi:10.1088/1742-6596/155/1/012007

Optical Elements for a CMBPol Mission

Editors: H. Tran[1] & L. Page[2]

[1] Dept. of Physics, University of California, Berkeley
[2] Dept. of Physics, Princeton University

E-mail: huantran@berkeley.edu

Abstract. We present an overview of selected optical technologies that are candidates for use in the CMBPOL satellite mission, indicate how the technologies are being tested, and estimate their Technology Readiness Level (TRL).

1. Introduction

There are at present multiple proposed configurations for a satellite to measure the B-mode polarization of the CMB. Schemes range from bare corrugated horns aimed directly at the sky[1] to 2 m reflector telescopes[2], and include interferometers[3] and smaller reflector/refractor systems. Each optical configuration is optimized for a specific scientific or technological purpose. We can draw on experience from COBE and WMAP (and soon PLANCK) to evaluate some technologies such as corrugated feedhorns and large composite reflectors, but there are other technological components that have not yet flown, most notably millimeter AR coatings, which may or may not be needed for a space mission.

In the next few years a number of new CMB polarization experiments will come online as shown in Table 1. The QUAD and BICEP experiments have already taken enough data to investigate the performance of on-axis reflectors and small AR coated refractors coupled to multiple corrugated feedhorns. In the next few years the launch of the EBEX and SPIDER balloon payloads is anticipated. PLANCK is planned for a 2009 launch. With the array of techniques that will be tested soon, in combination with the lessons already learned, one expects that near 2012 multiple detailed instrument designs for CMBPOL that are grounded in experience will be possible.

This paper containers a collection of studies on potential optical elements for a satellite to measure the B-mode polarization of the CMB. The contributions are organized as follows.

§2 "Offset Dual Reflector Systems for CMBPol," by Huan Tran, University of California Berkeley.

§3 "Suitability of On Axis Reflector Designs for a Future CMB Polarization Space Mission," by Clem Pryke, University of Chicago.

§4 "Layered Anti-Reflection Coatings," by Chao-Lin Kuo, Stanford University.

§5 "Simulated Dielectric Anti-Reflection Coatings," by Jeff McMahon, University of Chicago.

Journal of Physics: Conference Series **155** (2009) 012007 doi:10.1088/1742-6596/155/1/012007

In each entry, the current status of optical technologies relevant to CMBPol is reviewed. Based on the author's and our read of the state of the technology we summarize here the Technological Readiness Level (TRL) of the various technologies. To be ready for a satellite proposal, the technology must have a TRL of 5.

Corrugated feeds without lenses. TRL 5. These have flown in space and there are accurate codes to model the polarization response.

Corrugated feeds with lenses. TRL 4-5. These have been used on the ground in numerous experiments though there has not been an analytic study of the polarization properties.

Corrugated feeds with dual offset reflector systems and codes to assess them. TRL 7. These have flown in space and there are accurate codes to model the polarization response.

Antenna feeds with reflectors or lenses TRL 3. Antenna based focal planes will soon be used on the ground in both lens and dual reflector systems. The tools used to simulate polarized feed patterns are still maturing.

On-axis Reflector Systems TRL 3 to 5. On-axis design using conventional rigid feedleg secondary support: TRL 5 — but does not meet requirements. On-axis design using "zero blockage" secondary support: TRL 3 — research required to prove feasibility, and may not meet requirements anyway.

Single-layer AR coatings TRL 4-5. Single layer AR coatings for large plastic and silicon lenses estimated to be at TRL 5, with the possible exception of vibration testing and radiation testing. They have been produced and are in use. They may be modeled fairly well.

Multi-layer broadband AR coatings TRL 2-3. Multi-layer broadband AR coatings for large plastic and silicon lenses are being developed and will soon be tested in the field.

Simulated dielectrics on HDPE lenses for single bands TRL 5.

Simulated dielectrics for wide bands TRL 4. Wide band tapered designs (grooves in LDPE, pyramids in flat rexolite) have been prototyped, but have not yet been fielded in CMB polarization experiments.

Simulated dielectrics for silicon TRL 2.

Table 1. CMB Polarization Experiments Post 2007[1]

Name	Status	Frequency	Optical Configuration	AR coatings
Ground:				
QUAD	Done	100&150 GHz	Feed-fed refractive receiver on an on-axis 2.6 m Cass[2]	Yes
BICEP	Observing	100&150 GHz	Feed-fed refractive receiver	Yes
QUIET	Launching	40&90 GHz	Feed fed CD[3]	No
ABS	Building	150 GHz	Feed-fed cryo 1 m CD	No
BICEP2	Building	100&150 GHz	Antenna-fed refractive receiver	Yes
CLOVER	Building	97,150,&225 GHz	Feed-fed 1.8 m CD	No
MBI	Building	90 GHz	Bolometric Interferometer	No
POINCARE	Building	40 & 150 GHz	CD	No
POLARBEAR	Building	150& 220 GHz	Greg[4]/ Refractor	Yes
ACTPOL	Planning	TBD	TBD	TBD
SPTPOL	Planning	TBD	TBD	TBD
Balloon:				
EBEX	Building	150, 250,& 410 GHz	Greg/ Refractor	Yes
SPIDER	Building	40, 90, 145, & 220 GHz	Refractor	Yes
PAPPA	Building	89, 212, & 302 GHz	Single offset reflector	No
Current Satellite:				
WMAP	Observing	22-95 GHz	Feed-fed 1.4 m Greg	No
PLANCK	Building	30-680 GHz	Feed-fed 1.5 m Greg	No

([1]) The table focuses on just the optical components discussed in this section. There are many other distinguishing features between experiments. ([2]) Cass is for a Cassegrain reflector system. ([3]) CD stands for Crossed Dragone reflector system. ([4]) Greg is for an offset Gregorian reflector system.

2. Offset Dual reflector systems for CMBPOL by Huan Tran

2.1. Background on Offset Dual Reflector Systems

Reflector based systems are the only practical method of achieving apertures larger than ~60 cm. Dual reflector systems have long been used in astronomy to produce either a more compact design or to produce a larger field of view (FOV). Offset dual reflector designs have the added benefit of an unobstructed aperture, leading to lower sidelobe levels. Furthermore, offset dual reflectors will have a larger FOV when compared to similar centered systems with small obstructions. Offset systems do, however, violate circular symmetry, leading to more polarization effects. Detailed simulations show that polarization systematics induced by the reflectors is at or below the level induced by the feeds, although more investigation is necessary.

Two broad configurations of offset dual reflectors have been considered in the community, the Gregorian and the Crossed Dragone. The most common configuration is the Gregorian, shown in figure 1(b), which generally has a small secondary and a chief ray that traces a zig-zag pattern. An alternative configuration, the Crossed or Side-Fed Dragone, has received some recent attention [4, 5], particularly for apertures of ≈1.5 m. The Crossed Dragone uses a large secondary in a folded design, shown in figure 1(a), with a chief ray that crosses itself.

We discuss the relative merits of the Crossed and Gregorian Dragone configurations and indicate the possible future work needed to bring a dual reflector system to TRL 5 is considered.

2.2. Gregorian

The Gregorian is the dominant reflector configuration used in CMB experiments. The following sub-orbital experiments made use of Gregorian style optics: ACME, ACT, ARCHEOPS,

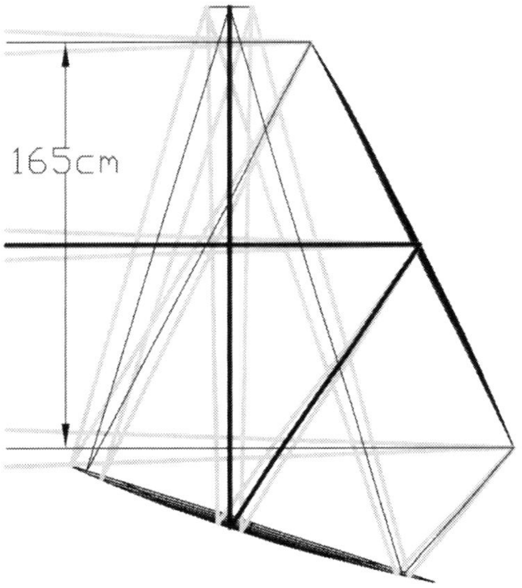

(a) Crossed

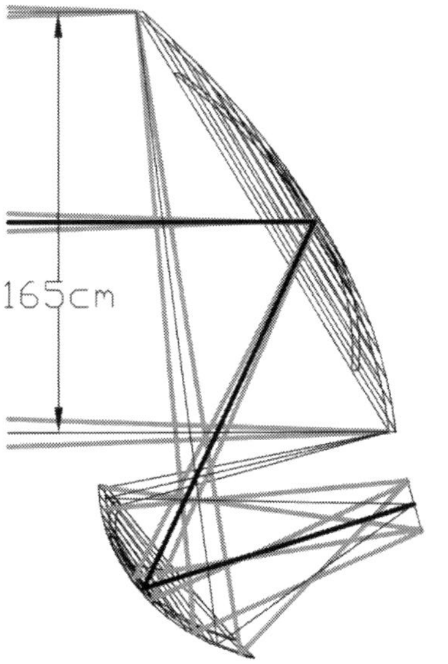

(b) Gregorian

Figure 1. Diagrams of two Mizuguchi-Dragone configurations. For each, the chief ray is shown as the thick black line. The ±2° off-axis beams are shown in grey. Both have the same aperture diameter and Effective Focal Length (EFL), and thus, the same F/#. Figure taken from [6].

BOOMERANG, CAPMAP MAX, MAXIMA, SPT and VIPER. WMAP has now flown with two back to back Gregorian telescopes, and have extensively characterized the in-flight performance. The PLANCK mission will also feature a single Gregorian telescope, and they have performed a significant amount of optical analysis to ensure that the beams will perform as needed.

There are many sub-classes of Gregorian telescopes. The simplest offset Gregorian has a parabolic primary and an elliptical secondary, where the axes of symmetry of the parent conics are co-linear. A simple improvement on the design is the aplanatic, where the conic section of the primary is allowed to deviate slightly from parabolic. An alternative improvement is to allow the axes of symmetry of the primary and secondary to tilt relative to each other, satisfying the Mizugutchi-Dragone condition. Most modern CMB systems are based on these conditions, or are slightly modified from these conditions.

2.2.1. WMAP *Reflectors.* The WMAP optical systems is composed of two back-to-back 1.4 m Gregorian systems, fed by scalar feed horns. The specific configuration of the reflectors starts with the Mizuguchi-Dragone condition, and is then shaped to make the beams smaller and more symmetric [7].

The optical performance has been extensively analyzed, including figure imperfections, by comparing in-flight beams to models computed using DADRA [8]. A model for the sidelobe performance of the full satellite was also developed [9], however the theoretical simulation alone was not sufficient to model the full performance. The reason is that structural elements that affect the beam at low levels were not included in the model.

At the level that WMAP detected polarization, the optical effects were negligible [10]. The largest effect is from a difference in the beam size and shape between the pairs of telescopes as a result of differences in telescope distortion and to passband mismatch.

The main limitation of the WMAP optics, the surface deformations, is directly traceable to the manufacturing processes for the composite optics. Other manufacturing processes have been demonstrated to result in better performance upon cooling.

2.2.2. PLANCK *Reflectors.*

The PLANCK design is similar to WMAP's but instead of starting with a Mizuguchi-Dragone the base system was an aplanatic with two ellipsoidal reflectors. As with the Dragone, the axes of the two reflectors are then allowed to tilt relative to one another [11]. The resulting design provides a larger FOV than a Dragone, but has more cross polarization for the central beam.

PLANCK team members have published simulations of the telescopes. Methods have been developed to analyze the beams [12, 13] with particular attention paid to polarization performance. The key conclusion is that beam maps from orthogonal linear polarizations from a single polarization sensitive bolometer (PSB) pair within a given feedhorn should not differ by more than 0.9% in amplitude, including the effects of finite conductivity and alignment error. Beams from separate horns could differ significantly due to astigmatism induced by elliptical reflectors [14]. Thus a single non-modulated PSB horn is required to measure either Q or U. Pairs of horns are required to construct E and B. A mismatch within a horn leads to leakage from T to E and B, whereas mismatch between horns leads to leakage between E and B. The effect is largest at high ℓ, but is not negligible at low-ℓ near the gravitational wave bump.

2.2.3. Gregorian Systems with Reimaging Optics.

Both WMAP and PLANCK operate without any reimaging optics. The FOV of the Gregorian telescopes is big enough to feed 10's of widely spaced detectors, and the corrugated feed horns provide enough edge taper and at the same time are individually oriented to properly illuminate the primaries.

It is unlikely that a Gregorian reflector and feedhorn system alone will have enough throughput to achieve the CMBPOL sensitivity goals. To achieve higher focal plane area utilization, the individual pixels must be smaller, leading to higher edge taper. This problem has been traditionally solved with a Lyot stop, created with reimaging optics. Some new generations of focal plane arrays also require a flat, telecentric field, also created with reimaging optics.

Reflectors have been used for reimaging optics in both BOOMERANG and MAXIMA. As throughput grows and F/# shrinks, however, reimaging with reflectors becomes more difficult because of the limited amount of space for folded beams to clear obstructions. Cold lenses are the natural solution, but they are not without their own problems.

The millimeter-wave Gregorian telescopes with cold reimaging lenses include ACT, SPT and GREEN BANK with POLARBEAR and EBEX to soon follow. For the EPIC mission concept study, a large 3 m system with large silicon reimaging optics was designed. While all designs produce large throughput at acceptable levels of optical performance, all systems are still in the initial operational phase, and the full polarization performance has yet to be determined.

2.3. Crossed Dragone.

A promising offset dual reflector configuration is the Crossed Dragone, also known as the Compact Range Antenna, or side fed antenna. This configuration has been chosen for QUIET , CLOVER and ABS , although many other experiments have seriously considered it. The biggest advantage of this configuration is the 2× larger FOV, as shown in figure 2. The Crossed Dragone has the added advantage of having a distant exit pupil, meaning that the bare focal plane is already near telecentric (see figure 1(a)). QUIET , CLOVER and ABS plan on using the Crossed Dragone without reimaging optics, relying instead on scalar feedhorns to taper the illumination at the edge

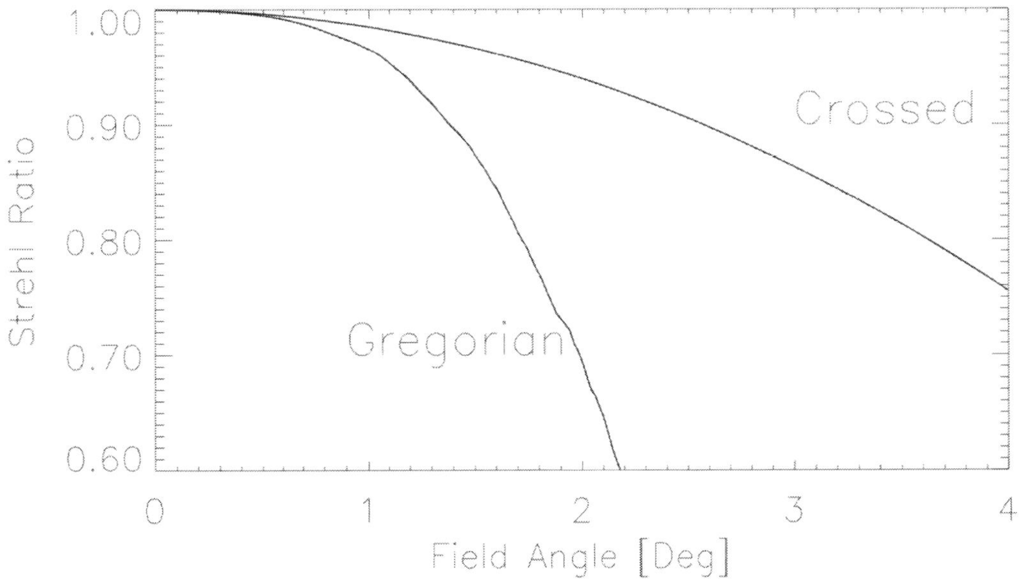

Figure 2. Strehl ratios across the focal plane, calculated at 150 GHz. The Crossed Dragone shows a clear advantage in terms of DLFOV. A design is considered diffraction limited if the Strehl ratio is above 0.8. Figure taken from [6].

of the primary. The absence of reimaging optics is a major simplification over equivalent Gregorian systems, allowing an accurate simulation of optical performance using tools such as `DADRA` and `GRASP9`.

The Crossed Dragone does suffer from a few disadvantages. Gregorian systems are more mature. A thorough design study of a Crossed Dragone configuration for a satellite mission has not yet been performed, but will be a major topic of the next phase of the CMBPol mission study.

The reflectors on a Crossed Dragone are more shallow and possibly easier to fabricate than those of the Gregorian design. The Crossed Dragone, however, does have a large secondary reflector, which makes large telescopes (~2 m and above) impractical. The large secondary also places some difficult constraints for mounting within a rocket shroud. Packing constraints will likely set the largest diameter possible for the Crossed Dragone configuration.

Another issue with the Crossed Dragone system is the tradeoff between F/# and beam clearance. There is a minimum practical F/# that is allowed before the beam either interferes with the primary or the secondary. This has implications for the quality of baffling the optics from stray light, and the level of expected sidelobes.

A third challenge for the Crossed Dragone system is the need for a large, cold aperture. To optimally use the available FOV, it is likely that closely packed pixels that maximize the focal plane illumination will be necessary. This leads directly to larger beams emanating from each pixel. In the Gregorian case with reimaging optics, the pixel sidelobes are truncated with a Lyot stop. Without a Lyot stop, the simplest solution to this problem is to use a large, cold, absorbing ring surrounding the entrance pupil.

Journal of Physics: Conference Series **155** (2009) 012007 doi:10.1088/1742-6596/155/1/012007

2.4. Work required to reach TRL 5

2.4.1. Mechanical. The Gregorian configuration has already been flown in WMAP with composite reflectors. The PLANCK mission will also fly a composite Gregorian reflector system. The mechanical aspects of the reflectors themselves are already at TRL 5.

The Crossed Dragone configuration is expected to require a large cold aperture. Cooling this ring down to levels that will keep bolometer loading low will pose some technical challenges, we expect to reach TRL 3 soon.

The Gregorian configuration requires dielectric lenses. While there have been many space-qualified refractive telescopes, a technology for broadband millimeter AR coatings that can survive cryogenic cycling and launch vibration has not yet been demonstrated.

2.4.2. Simulation. Understanding the full polarization systematic properties of full dual reflector systems is still, however, immature, especially when considering the tiny signals and extreme contrast of the B-mode signature. The behavior of the reflectors alone can be precisely simulated with available physical optics tools. Simulation tools for feeds, however, are at varying levels of maturity based on the technology used. Systems requiring "Detailed analysis" are categorized by NASA as TRL 2, although detailed simulation is not strictly required before bringing the system to a higher TRL.

The success of the WMAP beam modeling depended on the highly accurate physical optics simulation packages available that can analyze corrugated feed horns and highly conductive reflector surfaces. No such code is widely available to analyze the *dielectric* reimaging systems. In order to understand the role of reimaging lenses with the Gregorian configuration, physical optics simulations of dielectric lenses with account of their actual physical properties (absorption, internal reflection, polarization effects at reflection, any kinds of antireflection coatings, actual complicated shape etc) must be developed. A quasi-Gaussian beam analysis or similar "quasi-optical" extensions are inadequate to the real dielectric lens problem. Real lens properties are all significant and are all properly accounted for by the "lens physical optics" approach being developed [15].

Simulation tools for waveguide coupled corrugated feedhorns are mature. Tools for PSB's and antennas are not yet at the same level of accuracy. Tools for three dimensional structures with cold thin films or small dielectric lenses are still under development.

There is also possible work involved with simulations of beam forming elements in the focal plane. Discussion these simulations is outside the scope of this whitepaper, but possible areas of future development include advanced modeling of thin film polarized absorbtion in PSB's and advanced simulation of beam effects in lenslet-antenna coupled bolometers. The detailed polarization behavior of feed elements is relevant at the level of the B-mode signature. However, such systems are straightforward to characterize in the lab and in the field.

3. Suitability of On Axis Reflector Designs for a future CMB Polarization Space Mission by Clem Pryke.

This document will attempt to answer the question "Should on-axis reflector based designs be considered for use in a future CMB Polarization Space Mission?". This question can be re-phrased as follows: "Which is better — on-axis or off-axis?". The community consensus answer to this latter question is clear — all the up-coming generation of reflector based polarization experiments use off-axis designs, including EBEX , CLOVER , POLARBEAR , and QUIET .

On-axis designs have the following disadvantages versus off-axis:

- The blockage of the primary reflector by the secondary produces near sidelobes and a loss of aperture efficiency. Large field of view requires large secondaries making these problems worse.
- The secondary support structure scatters light to large angles producing far sidelobes (see below for experience from QUAD). In addition blockage by the secondary supports results in additional loss of aperture efficiency.
- The secondary reflector illuminates the annulus between the edge of the receiver entrance aperture and the start of the primary reflector potentially producing far sidelobes. (These effects can be reduced by putting a hole in the middle of the secondary, and/or baffling around the receiver entrance aperture.)
- Diffraction from the edge of the secondary reflector produces far sidelobes.

One therefore needs very compelling reasons to consider on-axis designs further. Dragone [16] has shown that if the angles in an offset antenna are chosen correctly, then aberrations and cross-polarization effects are the same as those in a conventional on-axis antenna with the same diameter and focal length. Excellent field of view is possible using off-axis designs [17].

The non-ideality of the polarization properties of telescope systems is normally discussed in terms of instrumental polarization and cross polarization. The former is the degree to which the optical system causes an un-polarized source to appear polarized. The latter is the rotation of the apparent polarization angle of a polarized source.

The net amount of instrumental polarization induced is related to the degree of symmetry in the optics, and hence off-axis designs generally give poorer performance. However the relevance of this statistic needs to be carefully considered — in an experiment which measures polarization through the differencing of two separate orthogonal detectors instrumental polarization is completely degenerate with the relative detector gains, which need to be calibrated to excellent precision anyway. For systems which employ polarization modulation or switching downstream of the optical system the optically induced instrumental polarization is potentially relevant. However by orienting the OMTs (or similar) symmetrically with respect to the optical system the effect can be nulled (as was done in CAPMAP).

It's not really the case but let's suppose that on-axis designs gave lower cross polarization — would that be sufficient reason to consider them further? To first order cross-pol is simply a rotation of the response angle of a given detector. To reach the required levels of accuracy for next generation experiments these angles will almost certainly need to be empirically measured anyway, rather than relying on design or model values. Therefore it is only if off-axis designs gave cross polarization which was *more unstable* that on-axis designs that the latter would have an advantage. Cross polarization is generated by curvature of the optical elements so it should be extremely stable, especially in a space environment.

The only CMB polarimeters so far to use on-axis reflector systems are COMPASS [18] and QUAD [19]. Rather than use a conventional tripod assembly both experiments supported the secondary reflector on a cone made from a foam material known to have low absorption and scattering at microwave frequencies, the idea being to have the secondary reflector effectively "float" in front of the primary. COMPASS used Styrofoam and reported adequate performance.

QUAD initially attempted to use Styrofoam but encountered severe problems obtaining sufficient mechanical stiffness and switched to a Zotefoam design. (Zotefoam is a polypropylene foam expanded using dry nitrogen.) Since Zotefoam is only available in 6' x 3' flat sheets an elaborate design was devised gluing two offset nine segment layers together using a very thin layer of adhesive. The foam adhesive laminate was measured to have scattering of $\sim 2\%$.

QUAD is the most sensitive CMB polarimeter to date for multipoles $\ell > 200$ and has reported good results [20]. Nevertheless it has some optical problems and is not a good model for future experiment. Ground pickup was encountered which is believed to be associated with scattering around the cryostat entrance window. In addition scattering by the foam cone produces a ring sidelobe at 100° from the optical axis which is highly polarized ($\sim 50\%$). The brightness of a point source located in this ring lobe is $\sim 10^{-4}$ relative to the main beam. This allows the moon to produce serious contamination, and at the sensitivity required for future experiments, the galactic plane would also be a problem.

The experience of QUAD is arguably not relevant to a space mission — likely a one piece foam cone support with much better properties could be developed. However it would only be worth considering this if truly compelling practical advantages for an on-axis design were identified. As outlined above no such in principle reasons are apparent. It doesn't seem that there are any practical or engineering based reasons either — while ground based off-axis telescopes are typically around three times more expensive than their on-axis equivalents this is due to the additional strength required from the mechanical structure to resist Earth's gravity, and hence doesn't apply in space. Off-axis telescopes have a proven record in space (e.g., WMAP), and the designs being considered for next generation space missions are no larger than the up-coming PLANCK mission.

3.1. Technology Readiness Level of On-axis Reflector Systems

On-axis design using conventional rigid feedleg secondary support: TRL 9 — but does not meet requirements. On-axis design using "zero blockage" secondary support: TRL 3 — research required to prove feasibility, and may not meet requirements anyway.

4. Layered Anti-Reflection Coatings by Chao-Lin Kuo

4.1. Overview

Reflections in optical systems cause loss and lead to systematic effects. To reduce these effects, optical elements can be antireflection (AR) coated by direct bonding of dielectric layers with the right thickness and index of refraction. Such an approach has been successfully applied to optical elements such as lenses, metal-mesh or bulk infrared (IR) blocking filters, half-wave plates, and the lenslets or silicon substrates that admit radiation in lithographic arrays of slot antenna-coupled bolometers. The most common lens/substrate materials include HDPE, PTFE (Teflon), silicon, quartz, and sapphire.

The concept of layered AR-coatings is quite simple, but the implementation can be tricky. The details depend on the coating materials, but the procedures usually involve (i) identifying and characterizing candidate coating materials with the right index of refraction; (ii) vacuum/thermal forming, or pre-machining of the AR layers; and (iii) for cryogenic applications, developing a bonding that is robust against multiple thermal cycles between room temperature and cryogenic temperatures. For multi-layer broadband AR coating, these procedures have to be repeated as many times as the number of layers. While the development of these processes can be costly and time consuming, once a recipe is found, layered AR-coatings can be produced significantly faster and less expensively than machining sub-wavelength structures onto the surfaces. Sections §4.2.1, §4.2.2, and §4.2.3 summarize successful recipes for single-layered AR-coatings of HDPE, PTFE, silicon, and progress in multi-layer AR coating silicon substrates.

4.1.1. Design and Modeling. The most important parameter in AR coatings is the index of refraction n of the substrates and the coating materials. Without AR coatings the reflective loss is given by $R = [(1 - n_s)/(1 + n_s)]^2$ per surface at normal incidence. The loss R is 30% for silicon ($n = 3.4$), and 4.5% for HDPE ($n = 1.54$). It is easy to show that an ideal single-layer AR coating at frequency ν is a dielectric layer with $n_c = \sqrt{n_s}$ and thickness $d = c/4n_c\nu$. Even with such ideal single-layer coating, the index of refraction dictates the working bandwidth of the coating. In Table 2 we list the index of refraction for several common dielectrics for substrates and coating.

The most convenient theoretical tool to calculate wave propagation in multiple layers of dielectrics is the characteristic matrix method, given in §1.6 of Born and Wolf [21]. This treatment includes full polarization effects and arbitrary incident angles. It assumes that light consists of plane waves, and the sample is planar and infinite. These are good approximations if the wavelength is much smaller than the radius of curvature of the lenses, the beam waist, and the sample size.

Polarization effects. It is a well known prediction of Fresnel equations that an incident plane wave at the dielectric interface is polarized. At the Brewster angle the reflected light is 100% polarized. The degree of polarization for the transmitted radiation is not as extreme, but can still be significant without AR coating. Figure 3 plots the induced polarization in the transmitted wave for a PTFE substrate at 5° and 20° with and without AR coating. The induced polarization can contribute to the instrumental polarization in some observing modes. Similarly, AR coating of the half-wave plate is often necessary because it significantly reduces artifacts associated with differential reflection in the ordinary and extraordinary axes.

4.1.2. Measurements. A reflectometer measurement is a quick and straightforward way to confirm the reduced reflection, though it is usually limited to only one frequency and room temperature. Transmission measurements with a Fourier Transform Spectrometer coupled to a broadband bolometer are necessary to understand the indices and the absorption losses, and they can be made at a wide range of frequencies. The beam incident on the samples needs to be slow enough such that the wavefront curvature is negligible. Spectra with and without the sample in the beam are usually taken. The "off" spectrum is used for calibration, which can be divided out.

Table 2. Dielectric properties of the materials for AR coating.[1]

Material	Index n	$\tan\delta^2$
Silicon	3.42	2.2×10^{-4}
Sapphire	3.07/3.40	$2.3/1.2 \times 10^{-4}$
TMM 10i	3.13	
TMM 10	3.03	
TMM 6	2.55	
TMM 4	2.12	
TMM 3	1.81	
Quartz (Herasil)	1.87	0.001
Cirlex	1.84	0.008
Stycast 1266	1.68	0.023
HDPE	1.54	3×10^{-4}
PTFE	1.44	3×10^{-4}
Expanded PTFE[3]	1.2	

[1] Data from [22, 23, 24]
[2] At or around 150 GHz.
[3] Zitex, or Porex Mupor.

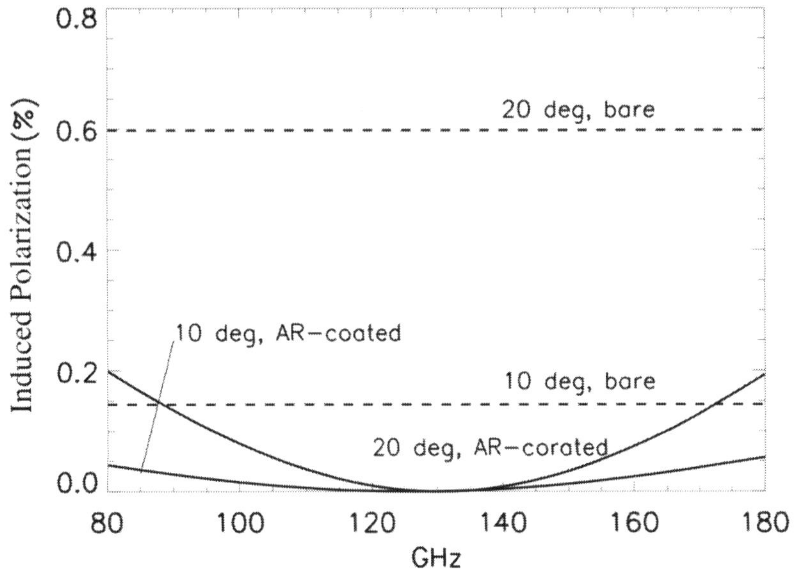

Figure 3. Induced polarization for an unpolarized plane wave transmitted through a dielectric interface, uncoated (dashed) and AR-coated (solid). The dielectric is PTFE ($n = 1.44$), and the AR coating is a single layer of expanded PTFE optimized for normal incidence at 125 GHz.

4.2. Implementation and Results.
In the next 3 sections, we describe the procedures and performance of AR coatings for HDPE, PTFE, and silicon implemented by several groups.

4.2.1. AR-Coating HDPE Lenses and PTFE Filters for BICEP The BICEP optics include 2 HDPE lenses and 2 PTFE thermal filters, all AR coated with single-layer expanded PTFE membranes using thermal lamination techniques. Without AR-coating, the IR blocking filters at LN_2 and the vapor-cooled shield (VCS) stages would have incurred over 10% loss and ghost beams at a similar level [25]. The two HDPE lenses, if left uncoated, would produce higher reflections then the PTFE filters because of the higher index ($n = 1.54$ vs 1.44). In addition, reflection from lenses cause polarized ghosting which must be reduced.

The flat PTFE filters are relatively straightforward to AR coat. Because of the low index of refraction, a single layer, $\lambda/4$ coating provides enough bandwidth for both of BICEP's frequency bands, 100 and 150 GHz. An ideal single layer coating for PTFE should have an index of $\sqrt{1.44} = 1.2$. Micro-porous PTFE (Mupor or Zitex) is an excellent material for this purpose. It has very low loss in the millimeter wave band. The manufacturers supply a variety of void fractions, pore sizes, and thicknesses. A void volume of 40% provides roughly the right index ($n \sim 1.2$) to AR coat PTFE [26]. The target thickness d is optimized for 125 GHz ($\lambda = 2.4$ mm) to cover both frequency bands. We identify the Mupor PM23DR membrane supplied by Porex Inc. to be a good approximation to the ideal $d = 2.4/4n = 0.5$ mm.

Because of their similar coefficients of thermal expansion, it is relatively straightforward to bond the Mupor membranes onto the surfaces of the filters. One slight complication in bonding PTFE is surface treatment. Because of its non-stick surface properties, the PTFE filters need to be chemically etched (Tetra-Etch) before bonding. The etching procedure changes the appearance and the color of the surface. However, the etched layer is only a few molecules thick, and no measurable increase in microwave loss is observed after etching. Two bonding techniques are tested. The first approach uses Eccobond 24 (a clear, low viscosity epoxy similar to Stycast 1266), estimated to be 0.02-0.05 mm thick, to join the substrate and the membrane. Another method uses thermal lamination, with a layer of LDPE between the substrate and membrane as the bonding film. Both methods provide excellent bonding that survives thermal cycling between room temperature and LN_2. We found it easier to apply uniform bonding for large aperture components with the heat lamination approach. (Fig. 4 a) is the measured transmission of a PTFE sample that is AR-coated on both sides with the Mupor membranes. Compared to the blank substrates (Fig. 4 b), the Fabry-Perot fringes are reduced significantly. The measured spectra agree very well with the model spectra.

We apply the same AR coating technique on the HDPE lenses and also obtain good results. An ideal single layer coating for HDPE should have an index of $\sqrt{1.54} = 1.24$, similar to the requirements for PTFE. No etching is required for bonding HDPE with Mupor membranes. However, the lamination temperature needs to be controlled accurately, because of the similarity in the melting temperatures of HDPE and LDPE. To bond Mupor membranes onto the curved lens surfaces, it is helpful to use thermal or vacuum forming techniques to pre-form the membranes into the right shape. The thermal laminations were done with an aluminum press shaped to match the lens surfaces. The coated PTFE filters and HDPE lenses have survived numerous thermal cycles.

In-band Scattering. The pores in the expanded PTFE sheet Rayleigh scatter far-IR, and to some extent, millimeter wave radiation. While scattering in short wavelengths provides additional IR blocking capability, scattering in-band can reduce efficiency and create loading from stray-light coupling. According to the vendor, the pore size of Mupor PM23DR membrane is $\sim 1.5\mu$m, and filling factor $f \sim 0.6$. To estimate the effect of Rayleigh scattering in mm-wave, we model the medium as being made of PTFE spheres with a characteristic radius of $a = 0.75\mu$m. The attenuation factor due to Rayleigh scattering [27] is given by $\alpha_R = 8\pi^3 N\gamma^2/3\lambda^4$, where N is the number density of the "molecules", given by $3f/4\pi a^3$. The polarizability γ of a dielectric sphere is given by $\gamma = 4\pi a^3(\epsilon - 1)/(\epsilon + 2)$. From these relations and parameters, we obtain a completely negligible attenuation factor $\alpha_R \sim 2 \times 10^{-9}$ cm^{-1} at 2 mm wavelength.

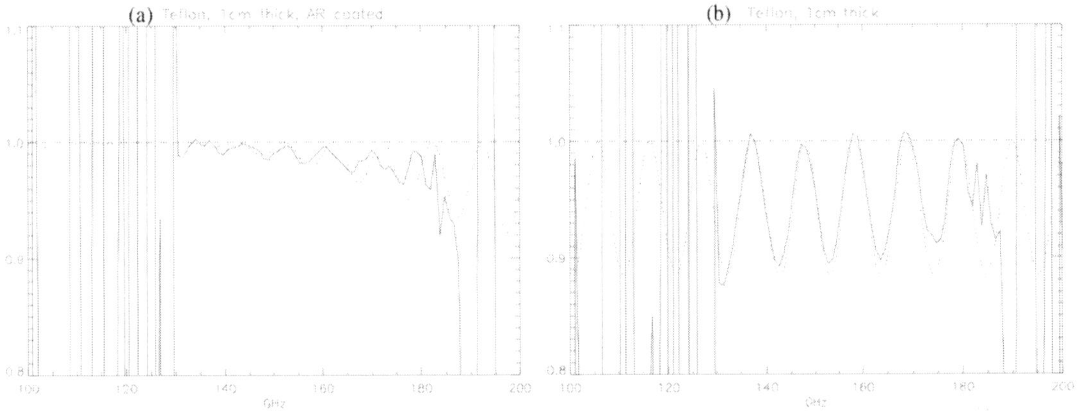

Figure 4. The transmitted spectra for a 1-cm thick PTFE sample, *left:* with Mupor AR coating, and *right:* without AR coating. The spectra are normalized by dividing by the "sample out" measurement. Outside the bandwidth of the bolometric radiometer (130 GHz - 180 GHz) the response drops to zero and the spectra become noisy. The dashed lines are the theoretical calculations assuming $n_s = 1.44$ and $n_c = 1.2$. This coating is optimized for 125 GHz but performs well for both the 100 GHz and 150 GHz bands of BICEP .

4.2.2. AR-coating Silicon Lenses for ACT A cryogenic AR coating for silicon lenses was developed for the Atacama Cosmology Telescope (ACT) [23]. The coating, a machined piece of Cirlex® [28] polyimide glued to silicon, reduces reflection to < 1.5% per lens at the design wavelength of 2 mm while maintaining > 90% transmission at $\nu < 300$ GHz. This section describes the development and performance of the coating.

The ideal AR coating at 2 mm has $d = 270\,\mu$m and an index $n_c = 1.85$. The low-frequency (~ 1 kHz) dielectric constant and loss reported in the Kapton polyimide data sheet suggest that polyimide and silicon could be combined in an AR configuration. Cirlex is a black pressure-formed laminate of Dupont Kapton® polyimide film readily available in sheets up to 60 cm square, and a variety of thicknesses. The mm-wave index and loss of Cirlex are measured both at room temperature and cryogenic temperature. Like many materials, the index of refraction increases slightly from its room temperature value of ~ 1.84 to ~ 1.95 at 5K. These measurements confirm that Cirlex is a good AR coating material for silicon.

The Cirlex layer is glued onto the silicon surface with Stycast 1266, a two-component, low viscosity epoxy made by Emerson Cuming. Silicon and Stycast 1266 thermally contract from 296 K to 4 K with $\Delta L/L = 2.2 \times 10^{-4}$ and 110×10^{-4}, respectively. Cirlex is expected to have a coefficient of thermal expansion comparable to that of a plastic, *i.e.* approximately that of Stycast 1266 and ten to one hundred times that of silicon. Differential thermal contraction often shears apart the substrate upon cooling to cryogenic temperatures. Through experimenting, the Princeton group found that if a thin layer of Lord Ap-134 adhesion promoter [29] is applied before Stycast 1266, the coating can endure multiple cryogenic cycles [23].

The transmission spectrum of a coated silicon sample is measured with an FTS. In Figure 5, the measured spectrum is shown along with a model. The measurement is the ratio of a sample to a reference spectrum, which are averages over two and six spectra, respectively. The model is determined from the measured Cirlex, Stycast, and silicon properties given in Table 2 and by

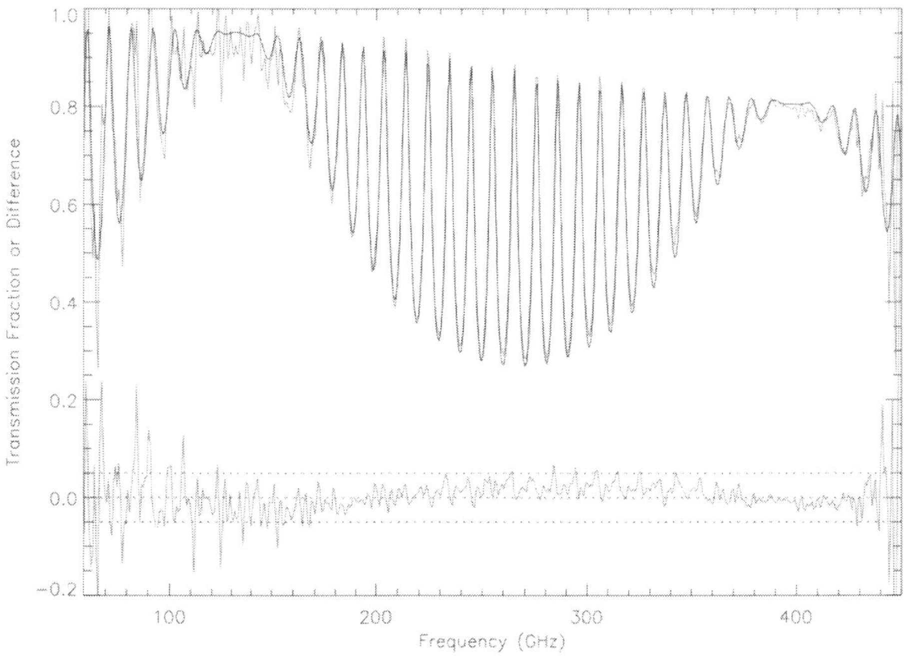

Figure 5. The room temperature transmission of a coated 4 mm-thick silicon flat, both modeled (black) and measured on the FTS (gray). The lower curve shows that the difference (measurement minus model) is within 5% of zero through the well-measured range. The slow reduction in transmittance with increasing frequency is due to increasing loss in the coating and glue [23].

measurements of the component thicknesses. The silicon loss was treated as an unknown and varied to fit the measured transmission.

For the curved lens surface, a piece of Cirlex approximately 1 cm thick is machined to the curved shape and then held in a Teflon gluing jig shaped to match the lens surface while the epoxy cures. Multiple antireflection-coated samples were made to test their robustness, including five flats—four 100 mm in diameter and one over 200 mm in diameter—and three plano-convex lenses. The lenses have been cycled four times to less than 4 K in a dewar, also without damage. The silicon lenses in ACT are all AR-coated using the procedure described here. They have been thermal cycled multiple times to 300 mK and have again shown no sign of damage.

4.2.3. Broadband AR-Coating of Silicon The CMBPol mission may well to require broadband AR coating of silicon, either for the large throughput re-imaging optics or for the focal plane arrays, depending on the details of the design. The UC Berkeley group has been developing a multi-layer AR coating for silicon using TMM, a Rogers Corporation product. TMM is commercially available as rigid sheets in a series of dielectric constants, listed in Table 2. A coating made from flat pieces of TMM and an additional layer of Zitex (expanded Teflon) was theoretically calculated to reduce percentage reflection from 30% to 3% at each silicon-vacuum interface, integrated over the three bands at 90, 150, and 220 GHz.

In a prototype testing, AR coatings were applied to two sides of a flat silicon sample. The transmittance spectrum is measured at 1.2 K in a light-pipe coupled FTS. The AR coating consisted of three layers of TMM and one layer of Zitex. The transmittance spectra of coated and bare silicon are shown in Figure 6. The silicon data exhibits periodic Fabry-Perot interference. The presence of the sample introduces coupling which causes peaks in excess of 100%. The location of these peaks in frequency space is dependent on sample mounting position. This geometrical effect can also be seen in the peak and trough heights of the bare silicon data. It is apparent however that the Fabry-Perot interference is greatly reduced over a wide frequency range for the coated sample. Applications of this technology on curved surface are in development by thermosetting TMM on a curved form.

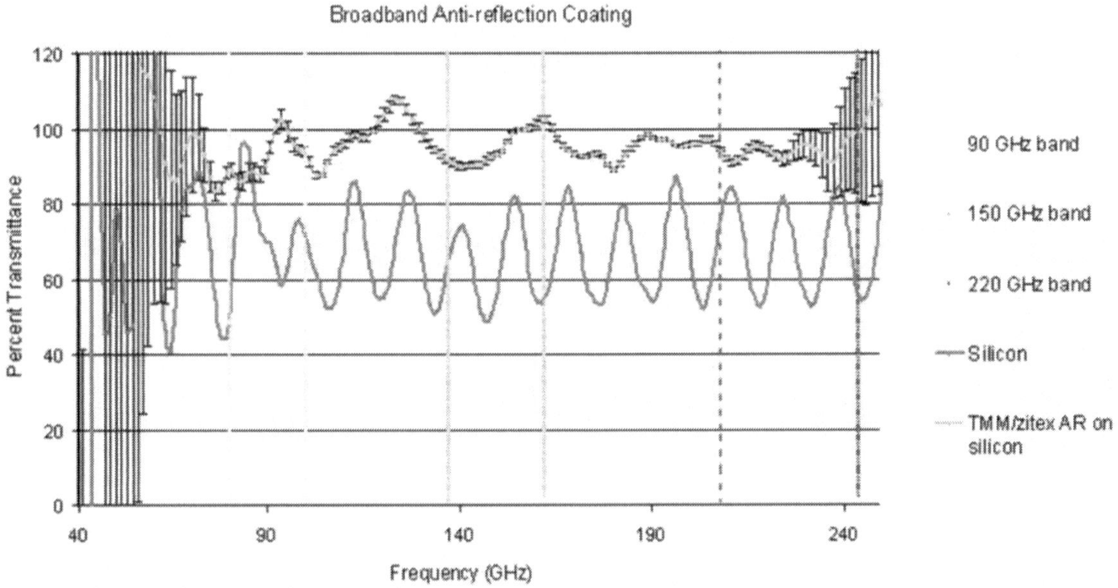

Figure 6. The transmittance spectra for a reference silicon sample and a AR-coated sample. The normalization of the measurement is affected by the optics, but the 4-layer coating significantly reduces the reflection fringes over the entire frequency range (E. Quealy).

5. Simulated Dielectric Anti-Reflection Coatings by Jeff McMahon

It is well known that when light in a vacuum strikes a dielectric substrate with an index of refraction n at normal incidence, a fraction of the power given by $R = \left(\frac{n-1}{n+1}\right)^2$ is reflected. When light is incident at an angle θ to the normal, R depends on both θ and the linear polarization state. This leads to two issues for CMB polarization experiments. First reflected power amounts to a loss of signal which reduces sensitivity. Second, differences in the reflection coefficients for the two linear polarizations mix the Stokes parameters I, Q, and U leading to $T \to E$, $T \to B$, and $E \to B$ conversion. These effects can be reduced with AR (anti-reflection) coatings.

Several practical difficulties arise in the application of AR coatings. First, it is sometimes difficult to identify materials with the necessary index of refraction ($n' \approx \sqrt{n}$). Second, in cryogenic applications, it is necessary to match the coefficient of thermal expansion (CTE) between the matching layer and the substrate. Simulated dielectric (SD) layers offer a solution to these problems. A SD layer is manufactured by machining sub-wavelength structure into the surface of the substrate creating a layer with an effective index of refraction n' which can be tuned between n and 1.

In this note we review the principles of AR layers and then discuss the principles and technological readiness of SD-AR layers.

5.1. AR Coating Basics

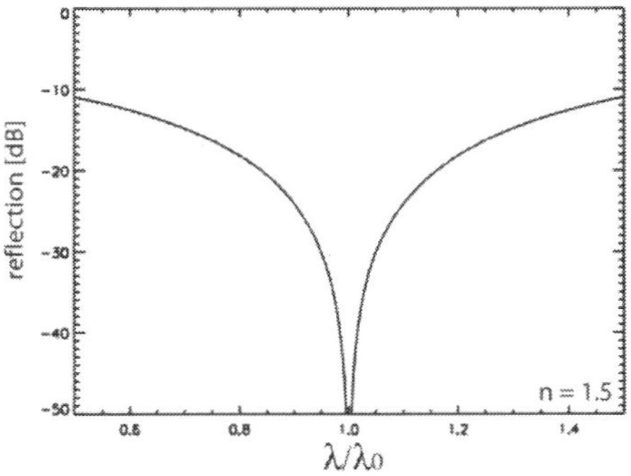

Figure 7. The reflection for a $n = 1.5$ surface AR-coated with a single layer $\lambda/4$ layer as a function of wavelength. The wavelength has been divided by the design wavelength λ_0.

The simplest AR coating consists of a $\frac{1}{4}$-wavelength ($\lambda/4$) thick layer of dielectric with index $n' = \sqrt{n}$ placed at the vacuum-dielectric boundary. The magnitude of the reflections from the front and back side of this layer are equal amplitude, and at the design wavelength, perfectly out of phase leading to a complete cancellation of the reflected power. This design is narrow-band but for $n \approx 1.5$ reflections can be suppressed below -20 dB over $\approx 20\%$ bandwidth (see Figure 7).

In many cases (e.g., lenses) it is necessary to design AR coatings that work at non-normal angles of incidence (see [30]). The angle of incidence θ is defined as the angle between the incoming ray and the local surface normal. The polarization and θ dependence of reflection and refraction from a dielectric lead to the following (different) requirements for the index of refraction of a matching layer for the two linear polarization states

$$n_{||}'^{\,2}(\theta) = \frac{n^2 + \sqrt{n^4 - 4n^2 \cos\theta \sin^2\theta \sqrt{n^2 - \sin^2\theta}}}{2\cos\theta\sqrt{n^2 - \sin^2\theta}}, \tag{1}$$

for light parallel to the plane of incidence, and

$$n_{\perp}'^{\,2}(\theta) = \sin^2\theta + \cos\theta\sqrt{n^2 - \sin^2\theta}, \tag{2}$$

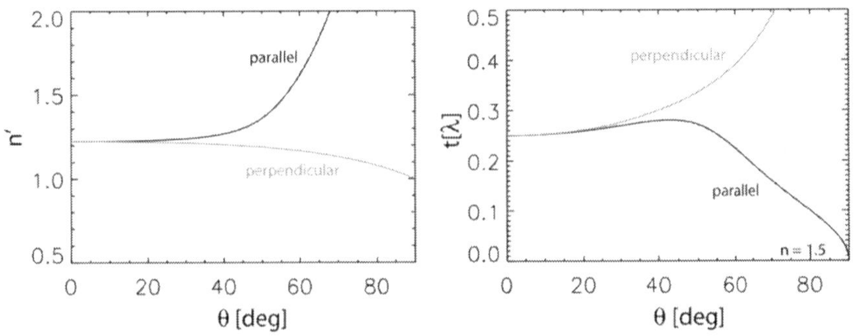

Figure 8. This figure shows the index of refection (left) and design thickness (right) of a $\lambda/4$ AR layer on HDPE as a function of angle of incidence for both linear polarization states (red and black curves). The thickness t is given in units of wavelength within the matching layer.

for light perpendicular to the plane of incidence. The thickness of the matching layer t is given by

$$t(\theta) = \frac{\lambda}{4\sqrt{n'^2 - \sin^2\theta}} \tag{3}$$

where n' is given by either Equation 1 or 2 depending on the choice of linear polarization. The dependance of n' and t on angle and linear polarization is shown in Figure 8 for the case of a substrate with $n = 1.5$. Note that the requirements for n' and t for the two linear polarization states diverge for $\theta > 30°$. Use of $\lambda/4$ AR layers at high angles of incidence suffer from reduced performance and lead to differences in the reflected power between linear polarization states generating cross-polarization. Application of $\lambda/4$ AR layers above about 30° should be avoided.

Wider bandwidth and greater immunity to the polarization dependance of boundary conditions at large angles of incidence can be achieved using graded-index AR coatings. Graded-index AR coatings consist of a dielectric layer over which the index of refraction varies approximately adiabatically between 1 and the index of the substrate n. The performance of these AR coatings improves with increases in the thickness of the layer, but offer a factor-of-two improvement even if the thickness is only $\lambda/4$ [31]. These coatings can suppress reflections below -25dB over $> 100\%$ bandwidth (see *University of Minnesota* in Section 5.3).

5.2. Description of Technology.

Both the $\lambda/4$ and graded-index coatings discussed in the previous section can be made using simulated dielectric techniques. An SD consists of a layer of dielectric which has been perturbed by removal of material (typically by simple machining) to form sub-wavelength structure. The geometry of these structures is chosen to tune the effective index of refraction n' of this layer to any value between the index of refraction of the substrate n and that of vacuum. Typical $\lambda/4$ SD-AR layers are made by cutting grooves, drilling periodic arrays of holes, or making an array of posts on the surface of the optic to be treated. See Figure 9 for examples.

An overview (summarized here) of the theory and performance of $\lambda/4$ SD-ARs can be found in [30]. Approximate analytic expressions for n' can be derived by calculating the polarizability of the SD layer when it is placed in a uniform electric field, treating the layer as infinitely thick, and neglecting the coupling between neighboring dielectric structures. Enforcing the condition $n' = \sqrt{n}$ yields equations for the 'fill factor' f needed for a quarter wave layer. These expressions depend on

θ, but at normal incidence reduce to

$$f_{g,\perp} = \frac{1}{n+1} \tag{4}$$

$$f_{g,||} = \frac{n}{n+1} \tag{5}$$

for grooves (light polarized perpendicular and parallel to grooves respectively), and

$$f_h = \frac{n^2+1}{2n(n+1)} \tag{6}$$

for holes. The required fill factor for posts has been experimentally found to be 0.88 of that of holes. We emphasize that grooves are birefringent ($f_{g,\perp} \neq f_{g,||}$) and therefore not suitable for polarization measurements. Holes and posts do not share this problem when placed on a square grid, though asymmetric grids can also lead to birefringence. These formula are accurate to $\approx 10\%$ provided the scale of the features are less than $\approx \lambda/2$.

Graded index SD-AR layers can be made by tapering the width of the grooves, cones, or posts. This amounts to cutting grooves with a triangular cross-section, drilling conical holes, or making square bottomed pyramids on the surface to be treated. This taper causes the fill factor (and therefore n') to vary with depth. It has also been proposed to make wide-band SD-ARs by drilling a large number of holes with uniformly distributed random locations and random depths chosen from a distribution such that the n' is appropriately tapered with depth in the layer. The bandwidth of such layers depends on both the thickness and taper of the index. Calculations of the performance of these layers are possible using electromagnetic simulators such as HFSS or by approximating the system as a cascade of thin layers of material with varying effective index. The latter calculation shows that the performance improves as the layer is made thicker and is only weakly dependent on the functional form of the taper[31]. The scale of features in these layers should be kept smaller than $\approx \lambda/2$.

5.3. Status of this Technology

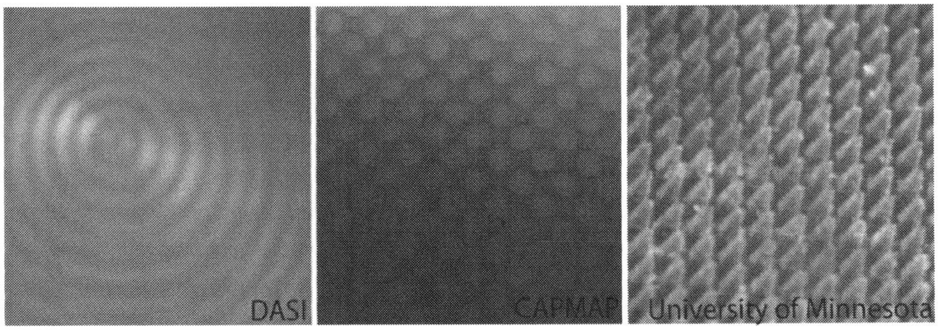

Figure 9. Examples of SD layers from several groups: $\lambda/4$ grooves in HDPE from DASI, $\lambda/4$ holes in HDPE from CAPMAP, and an adiabatic transition consisting of tapered square posts in rexolite from the University of Minnesota.

In this section we note and briefly discuss SD-AR designs which have been used or considered by various groups in the CMB community. For each group we describe the geometry of the SD layer, summarize the performance (when possible), and note if the technology was fielded. The order of this list is roughly chronological.

DASI High density poly ethylene (HDPE) meniscus lenses were AR coated using a $\frac{1}{4}$-λ thick (at 32 GHz) SD layer consisting of concentric grooves. Measured reflections were below 1% across 30% bandwidth. The birefringence of these grooves lead to a peak cross-polarization of -22 dB. Published polarization science data were taken with this system.

CAPMAP Bi-convex lenses made from HDPE were AR coated using a $\frac{1}{4}$-λ thick SD layer consisting of a square grid of holes. Lenses were made for the 40 and 90 GHz band. Measured reflections were below 1% across 20% bandwidth in both cases. The measured cross-polarization of these lenses were below -40 dB. Published polarization science data were taken with this system.

SPT / APEX HDPE lenses were AR coated using a wide-band graded index SD layer consisting of concentric triangular grooves. The ratio of the width to depth of these grooves is 1:1. Measurements are not available, but simulations suggest that the peak transmission is 99% at 150 GHz with greater than 98% transmission between 100 and 300 GHz This design is not optimized for polarization science due to the birefringence of the grooves. This system is currently fielded.

University of Minnesota has prototyped a wideband SD-AR consisting of a square grid of square bottom pyramids (see Figure 9) on a flat Rexolite ($n \approx 1.5$) surface. The hight of the pyramids is 500 μm and the grid spacing is 250 μm. Calculations predict a reflectance of less than 0.2% over a band between 100 and 500 GHz. Noise-limited lab measurements at frequencies between 120 and 300 GHz using a Fourier transform spectrometer imply the reflection is less than $\approx$ 2% over the entire bandwidth. No measurements of polarization purity are available.

UC Berkeley An attempt to mill a square grid of 4 : 1 (depth to width) square based pyramids on small samples of TMM (plastic) lenses was aborted due to difficulties with machining. Laser and EDM machining in Silicon were considered but not attempted.

Miller Group The Miller group at Columbia has considered micro-machining holes into a thin layer of silicon and 'slumping' this SD layer onto a silicon optic. This approach is expected to work on a flat substrate, but is believed to be more difficult than direct machining.

5.4. Benefits and Disadvantages.

The benefits of this technology are 1) its mechanical robustness, 2) ability to achieve arbitrary index of refraction, and 3) automatic matching of the CTE of the AR layer to the CTE of the lens. For SD layers which are highly symmetric, low-cross polarization has been demonstrated. The disadvantage of this technology is that its applicability is limited to materials that can be machined appropriately (currently HDPE, LDPE and Rexolite). The implications of this restriction depend on the design of the full optical system. If high magnification lenses are required this restriction could be a severe disadvantage while for designs using plastic lenses this is not an issue. It is likely that direct machining of silicon could be developed with some amount of experimentation.

5.5. Technical Readiness: Estimates for Time to TRL 5

The TRL level varies for the different SD designs with different materials. Designs using $\lambda/4$ SD-AR layers consisting of grooves and holes on HDPE have been used in CMB polarization experiments and achieved TRL 5. Wide band tapered designs (grooves in LDPE, pyramids in flat rexolite) have been prototyped, but have not yet been fielded in CMB polarization experiments. These designs are at TRL 4. Use of SD layers on silicon are TRL 2.

Several steps required to bring a new application of SD technology to TRL5. Electromagnetic simulations are the natural starting point. These could be completed in one to two man-months. The new application will then need to be fabricated. Changing materials or SD geometry will require extra time for development of appropriate machining techniques. For plastics, fabrication is likely to require several man months. For silicon this could easily require a year or more of development. Once a prototype has been fabricated several measurements are needed to verify its performance. These measurements include 1) transmission and reflection, and 2) cross-polarization. The transmission and reflection measurements can be made using either a Fourier transform spectrometer or a network analyzer. Cross-polarization measurements could be made by measuring the transmission of the SD layer while rotating the sample or using beam measurements using a polarized (or strictly unpolarized) point source. These measurements could be completed in a few man-months at a well equipped facility, but would require more time if the test equipment needs to be acquired and set up.

Acknowledgments
I want to thank the members of the CMB groups listed in Section 5.3 for providing detailed information about the design and status of their SD-AR layers.

6. Bibliography

[1] Colombo L and The SPOrt Collaboration 2004 *Nuclear Physics B Proceedings Supplements* **134** 133–135
[2] Bock J, Cooray A, Hanany S, Keating B, Lee A, Matsumura T, Milligan M, Ponthieu N, Renbarger T and Tran H 2008 *ArXiv e-prints* **805** (*Preprint* 0805.4207)
[3] Timbie et al *EPIC Interferemoter Study*
[4] Newburgh L B, UC B C, Caltech Collaboration, Columbia U Collaboration, GSFC Collaboration, Harvard-Smithsonian CfA Collaboration, JPL Collaboration, Kavli Inst U C C, Kavli Inst S U C, U Miami Collaboration, U Oxford Collaboration and Princeton U Collaboration 2005 *Bulletin of the American Astronomical Society* pp 1429–+
[5] Taylor A C, Challinor A, Goldie D, Grainge K, Jones M E, Lasenby A N, Withington S, Yassin G, Gear W K, Piccirillo L, Ade P, Mauskopf P D, Maffei B and Pisano G 2004 *ArXiv Astrophysics e-prints* (*Preprint* astro-ph/0407148)
[6] Tran H, Hanany S, Lee A, Milligan M and Renbarger T 2008 *Appl. Opt.* 103–109 (*Preprint* ttp://www.opticsinfobase.org/abstract.cfm?URI=ao-47-2-103)
[7] Page L, Jackson C, Barnes C, Bennett C, Halpern M, Hinshaw G, Jarosik N, Kogut A, Limon M, Meyer S S, Spergel D N, Tucker G S, Wilkinson D T, Wollack E and Wright E L 2003 *Ap. J.* **585** 566–586 (*Preprint* arXiv:astro-ph/0301160)
[8] Hill R S, Weiland J L, Odegard N, Wollack E, Hinshaw G, Larson D, Bennett C L, Halpern M, Page L, Dunkley J, Gold B, Jarosik N, Kogut A, Limon M, Nolta M R, Spergel D N, Tucker G S and Wright E L 2008 *ArXiv e-prints* **803** (*Preprint* 0803.0570)
[9] Barnes C, Limon M, Page L, Bennett C, Bradley S, Halpern M, Hinshaw G, Jarosik N, Jones W, Kogut A, Meyer S, Motrunich O, Tucker G, Wilkinson D and Wollack E 2002 *Astrophysical Journal Supp* **143** 567–576 (*Preprint* astro-ph/0301159)
[10] Page L, Hinshaw G, Komatsu E, Nolta M R, Spergel D N, Bennett C L, Barnes C, Bean R, Doré O, Dunkley J, Halpern M, Hill R S, Jarosik N, Kogut A, Limon M, Meyer S S, Odegard N, Peiris H V, Tucker G S, Verde L, Weiland J L, Wollack E and Wright E L 2007 *Ap. J. Suppl.* **170** 335–376 (*Preprint* arXiv:astro-ph/0603450)
[11] Villa F, Sandri M, Bersanelli M, Butler R C, Mandolesi N, Mennella A, Marti-Canales J and Tauber J 2003 *ArXiv Astrophysics e-prints* (*Preprint* astro-ph/0304137)
[12] Yurchenko V B, Murphy J A and Lamarre J M 2004 *Optical, Infrared, and Millimeter Space Telescopes. Edited by Mather, John C. Proceedings of the SPIE, Volume 5487, pp. 542-549 (2004).* (*Presented at the Society of Photo-Optical Instrumentation Engineers (SPIE) Conference* vol 5487) ed Mather J C pp 542–549
[13] Noviello F 2006 *CMB and Physics of the Early Universe*
[14] Rosset C, Yurchenko V B, Delabrouille J, Kaplan J, Giraud-Héraud Y, Lamarre J M and Murphy J A 2007 *Astron. Astrophys.* **464** 405–415
[15] Yurchenko V and Altintas A 2007 *Proc. 6th Intl. Conf. on Antenna Theory and Techniques (ICATT'07)*
[16] Dragone C 1982 *IEEE Transactions on Antennas and Propagation* **30** 331–339
[17] Hanany S and Marrone D P 2002 *Applied Optics* **41** 4666–4670
[18] Farese P C, dall'Oglio G, Gundersen J, Keating B, Klawikowski S, Knox L, Levy A, O'dell C, Peel A, Piccirillo L, Ruhl J and Timbie P 2003 *New Astronomy Review* **47** 1033–1046 (*Preprint* arXiv:astro-ph/0305608)
[19] Hinderks J, Ade P, Bock J, Bowden M, Brown M L, Cahill G, Carlstrom J E, Castro P G, Church S, Culverhouse T, Friedman R, Ganga K, Gear W K, Gupta S, Harris J, Haynes V, Kovac J, Kirby E, Lange A E, Leitch E, Mallie O E, Melhuish S, Murphy A, Orlando A, Schwarz R, O' Sullivan C, Piccirillo L, Pryke C, Rajguru N, Rusholme B, Taylor A N, Thompson K L, Tucker C, Wu E Y S and Zemcov M 2008 *ArXiv e-prints* **805** (*Preprint* 0805.1990)

Journal of Physics: Conference Series **155** (2009) 012007 doi:10.1088/1742-6596/155/1/012007

[20] Pryke C, Ade P, Bock J, Bowden M, Brown M L, Cahill G, Castro P G, Church S, Culverhouse T, Friedman R, Ganga K, Gear W K, Gupta S, Hinderks J, Kovac J, Lange A E, Leitch E, Melhuish S J, Memari Y, Murphy J A, Orlando A, Schwarz R, O'Sullivan C, Piccirillo L, Rajguru N, Rusholme B, Taylor A N, Thompson K L, Turner A H, Wu E Y S and Zemcov M 2008 *ArXiv e-prints* **805** (*Preprint* 0805.1944)

[21] Born M and Wolf E (eds) 1999 *Principles of optics : electromagnetic theory of propagation, interference and diffraction of light*

[22] Lamb J W 1996 *International Journal of Infrared and Millimeter Waves* **17** 1997–2034

[23] Lau J e a 2006 *Applied Optics* **45**

[24] Stycast 1266 *Emerson and Cuming, 869 Washington Street, Canton, MA 02021*

[25] Yoon K W, Ade P A R, Barkats D, Battle J O, Bierman E M, Bock J J, Brevik J A, Chiang H C, Crites A, Dowell C D, Duband L, Griffin G S, Hivon E F, Holzapfel W L, Hristov V V, Keating B G, Kovac J M, Kuo C L, Lange A E, Leitch E M, Mason P V, Nguyen H T, Ponthieu N, Takahashi Y D, Renbarger T, Weintraub L C and Woolsey D 2006 *Millimeter and Submillimeter Detectors and Instrumentation for Astronomy III. Edited by Zmuidzinas, Jonas; Holland, Wayne S.; Withington, Stafford; Duncan, William D.. Proceedings of the SPIE, Volume 6275, pp. 62751K (2006).* (*Presented at the Society of Photo-Optical Instrumentation Engineers (SPIE) Conference* vol 6275)

[26] Benford D J, Gaidis M C and Kooi J W 2003 *Applied Optics* **42** 5118–5122

[27] Jackson J D 1975 *Classical electrodynamics* (New York: Wiley, 1975, 2nd ed.)

[28] Fralock, Division of Lockwood Industries, Inc *21054 Osborne Street, Canoga Park, CA 91304.*

[29] Chemlok AP-134 *Lord Corporation, 111 Lord Drive, P.O. Box 8012, Cary, NC 27512*

[30] Cohn S B 1961 *Lens Type Radiators: Antenna Engineering Handbook* (McGraw-Hill, N.Y.)

[31] Plagge T 2008 *tplagge@berkeley.edu, private comunication*

Journal of Physics: Conference Series 155 (2009) 012008 doi:10.1088/1742-6596/155/1/012008

Cryogenic Technology for CMB-Pol

M DiPirro[(a)], D L Johnson[(b)], P Shirron[(a)]

(a)NASA/Goddard Space Flight Center
Greenbelt, MD 20771

(b)Jet Propulsion Laboratory
California Institute of Technology
Pasadena, CA 91109

E-mail: mike.dipirro@nasa.gov

Abstract. Future space telescopes such as CMBPol, SAFIR, DARWIN, SPICA and XEUS will require cooling to very low temperatures. Staged cooling is the most efficient means of achieving low temperature in an observatory or instrument with the least cost and mass. The first stage is usually passive radiators taking advantage of views to deep space. In the past stored cryogen systems provided the next lower stagesof cooling. Mechanical cryocoolers represent a significant enabling technology, especially at the lower temperatures where the passive coolers' effectiveness is limited. These coolers are in general lighter, have more cooling capability, and more operationally flexible than stored cryogens. Sub Kelvin cooling is required for many of the most sensitive detectors. For fundamental reasons, microcalorimeters and bolometers must be cooled to extremely low temperature to achieve their ultimate resolution and, eventually, background-limited detection. The state of the art for these cryogenic cooling technologies are presented along with plans to advance the technology readiness level to enable these future missions.

1. Passive Cooling

When designing a low temperature system for an instrument or an observatory it is most efficient to use multiple stage of cooling – both for thermodynamic efficiency and to utilize optimal cooling techniques for the various temperature ranges. The first (warmest) stage of cooling is usually passive. Passive cooling systems block parasitic heat inputs from the Sun, Earth, spacecraft and other warm components from reaching the low temperature stages. Passive cooling systems also take advantage of the biggest heat sink in the universe to radiate heat to deep space. V-Groove radiators provide one or more stages of radiation in an optimal direction defined by the "V" while blocking radiated heat from the other directions[1]. Planck is a good example of a multi-stage, fixed V-groove radiator[2]. A larger, very light weight, deployable version of a V groove radiator can be found in the James Webb Space Telescope (JWST) sunshield[3]. (See Figure 2.) The shield is made up of 5 layers of aluminized or Si coated Kapton to provide more than 40 dB of attenuation of the incident sunlight power while radiating to space out the gaps between layers. The size and mass limitations limit the membrane shape and layer separation that can be achieved on the ground. This, in turn, limits the testability of the design.

Because of the proximity, angular coverage and varying direction of thermal radiation and reflected sunlight from the Earth, low Earth orbit (LEO) is thermally inferior to a deep space orbit like the Earth-Sun L2 Lagrange point or a drift-away orbit. L2 was used by the Wilkinson Microwave Anisotropy Probe (WMAP) which radiatively achieved 90 K[4], and will be used by Herschel[5] and Planck, followed by JWST (27 K on the instrument radiators). With a very simple fixed shield and radiator, Spitzer has achieved 34 K in a drift-away solar orbit[6], while the outer shell of the COBE dewar in LEO reached 138 K.

Note that while the cosmic microwave background temperature is 2.73 K, deep space is effectively 7 K due to inner solar system dust[7]. This difference is not a practical limitation, however, due to the T^4 dependence of radiation and the other practical limitation on radiators, such as conductive supports. The difference between the two space temperatures on the radiation from the Spitzer outer shell at 34 K is only 0.2%.

The Spitzer observatory consists of a 0.85 m diameter telescope cooled to 5.5 K by the boil-off gas from a superfluid helium dewar. The telescope is mounted directly to the dewar and supported from a radiative cooled outer shell with a vapor cooled shield in between. See Figure 1. The helium dewar is described in the next section. The outer shell, which is cooled entirely by radiation, is silvered on the side facing the solar array and the sun and is painted with Ball InfraRed Black (BIRB) on the deep space facing side. The radiative performance turned out to be as predicted based on heat load from the solar array, the conduction of the outer shell and the emissivity of the BIRB[6].

Figure 1. The Spitzer Space Telescope outer shell cools radiatively to 34 K.

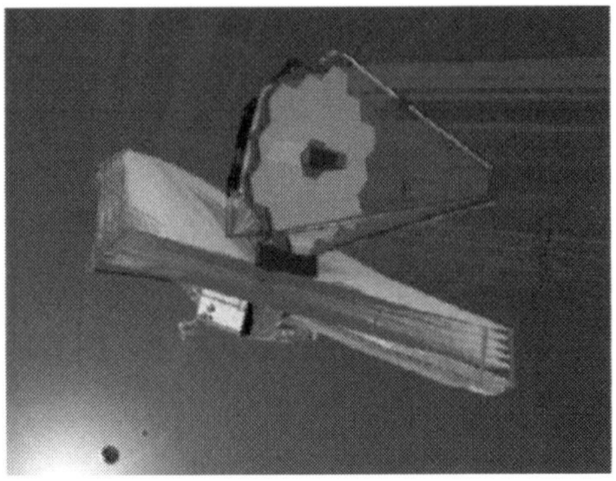

Figure 2. The James Webb Space Telescope will cool its telescope radiatively to 40 K and its instruments to 27 K.

The thermal testing of passive cooling systems is problematic not only due to the size and gravity effects, but also due to the difficulty of achieving and measuring milliwatt heat loads in the presence of kilowatts of thermal radiation from room temperature spacecraft components within the same thermal/vacuum chamber. A very big thermal driver is the difference between real chamber walls and empty space. The warm spacecraft and any solar simulator can be seen by the cold portions of the instrument through reflections off these walls. The chamber also needs to be cold enough so that radiation from the wall itself is not significant. Even a great observatory like Spitzer could not afford a test that would adequately simulate the thermal environment of space. The "inexpensive" compromise test that was run on Spitzer resulted in a heat load to the instrument that was 10 times the actual on-orbit value, so was of little use in predicting on orbit behavior.

To overcome this limitation it is proposed that the problem be attacked in two ways: make and test a subscale hardware model of the thermal system of a realistic large space telescope and perform a thermal model on this subscale model and test conditions; and outfit the test chamber with the appropriate thermal baffling and cold black body surfaces to achieve a more space-like thermal

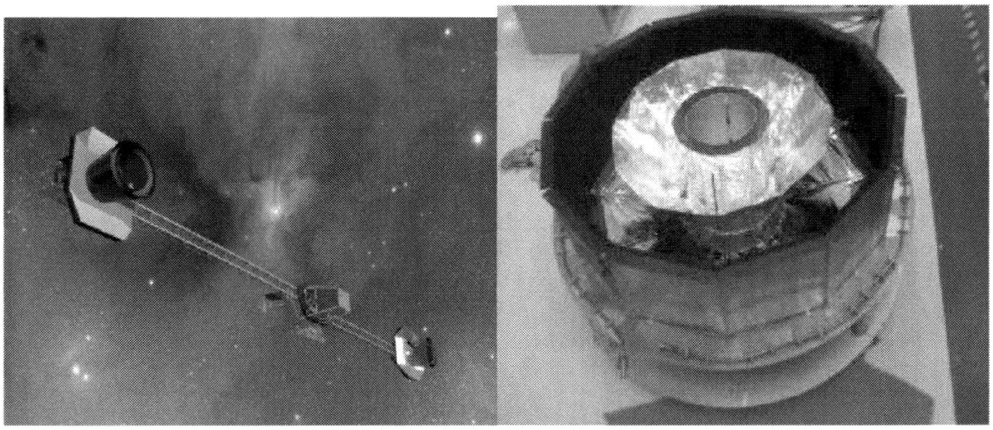

Figure 3. Demonstration of subscale thermal testing of a large sunshield system.

environment. The first enables use of a smaller chamber, which lowers the cost and increases the opportunities for test. It also helps with the one-g problem – in general, the smaller the system, the less effect gravity has on it. The second requires a careful design of a chamber/experiment baffle. This is also aided by making the test article smaller. One can vary the boundary conditions on the resulting test article to simulate conditions like LEO or deep space. JWST will test its sunshield design on a one third scale model. The size was chosen to make use of existing thermal vacuum chambers, and to provide the proper scaling. That is, too small a model will be dominated by conduction and boundary effects while too large a model will be affected by gravity and facility availability.

To demonstrate the feasibility of this scheme, an 18% scale (the largest diameter shield was 1.5 m) version of the SPIRIT Origins Probe telescope shield was tested in a low cost helium shroud within a LN2 shrouded vacuum chamber. See Figure 3. Performance data showed that less than 10 microwatts of radiative heat leaked from the warm to cold sides of the shields during the test implying that the boundary conditions were well understood. Excellent agreement was obtained between the data and the thermal models[8].

2. Stored Cryogen Cooling

Several cryogens, solid and liquid, have been used as consumables to cool instruments and observatories. Volumes have ranged from 36 liters (Astro-H) to over 2500 liters (Herschel). See Figure 4. Most were great successes, but there have been some notable failures. IRAS, COBE, CLAES, IRTS, ISO[9], GP-B[10], and Spitzer[6] are in the latter category. The Hubble Near Infrared Camera and MultiObject Spectrograph (NICMOS) which used solid nitrogen had a much shorter on-orbit lifetime due to a thermal short caused by the solid nitrogen itself. The Wide field InfraRed Explorer (WIRE) which was cooled by two stages of solid hydrogen had a premature release of its aperture cover which enabled sunlight to enter the cold volume and exhausted the hydrogen within one day. The Suzaku XRS2 instrument, cooled by a combination of mechanical cooler at 100 K, solid neon at 16 K and superfluid helium, suffered a complete loss of helium four weeks into the mission when vented gas reentered the dewar vacuum space. It is noted that all of these failures were peculiar to using stored cryogens, and would not have happened to a mechanical cryocooler system. In fact, NICMOS was brought back into service by attaching a cryocooler to the dewar in 2002 and is still successfully operating[11].

The successor to the Japanese Suzaku xray mission is Astro-H. The current thinking on the cryogenic system is to use both a superfluid helium tank and a Joule-Thomson cryocooler to achieve technological redundancy. While this does provide redundancy, it has a distinct mass disadvantage to either of the two cooling systems alone.

The Spitzer cryogenic system approaches the optimum engineered solution for stored helium systems. The combination of excellent passive cooling and careful thermal design produced a cryogenic system whose lifetime is expected to exceed its 5 year goal using only 360 liters of superfluid helium. The Spitzer mirror is cooled to 5.5 K by the boil off vapor from a superfluid helium tank that operates at 1.2 K. A heat load to the superfluid helium of 6 mW coincidently produces about 6 mW of cooling by the vapor at 5.5 K, the enthalpy difference of the gas from 1.2 K to 5.5 K being approximately equal to the latent heat at 1.2 K. The average instrument dissipation is about 3 mW which is directly deposited into the liquid. The outside shell of the dewar is itself cooled by the boil-off, resulting in a negligible parasitic heat load to the superfluid. The remainder of the 6 mW is generated by a heater (2.5 mW average) and by excess superfluid film flow (0.5 mW).

The liquid is phase separated from the gas by a porous plug in the vent line. For low flow rates, the superfluid film flow leaking past the porous plug is a substantial fraction of the total flow of gas vented. In Spitzer this was controlled by instituting a 3.5 mm diameter tube immediately downstream of the porous plug, which limited the film flow to a calculated 0.5 mW equivalent. More drastic measures were used to eliminate the film from the XRS instrument on Astro-E and Suzaku[12].

From the Spitzer example one can see that cooling by vapor alone would not be efficient for a CMB-Pol telescope operating at 2.7 K. Some of the cooling would be generated by a weak thermal link directly from the telescope/instrument into the superfluid helium tank.

Figure 4. The large and the small. On the left is the Herschel observatory with its 2560 liter helium tank in black in the center. On the right is the Suzaku observatory with the 32 liter shiny helium tank near the bottom. The pictures are roughly to scale. Herschel is expected to last over 5 years while Suzaku was designed for a 2-3 year cryogen lifetime. By using very low temperature cryocoolers, the successor to Suzaku, Astro-H, is expected to have a lifetime in excess of 5 years using only 36 liters of helium.

At these low flow rates the vapor cooling is also small. The inner vapor cooled shield just outside the telescope settles at 24 K. This implies a vapor cooling of only 27.5 mW at this shield.

It is useful to estimate the realistic heat loads for a system like CMB-Pol. Assume a 3 m diameter primary mirror operating at 2.7 K. One can get a handle on the expected heat loads by scaling from Spitzer. The Spitzer primary is 0.85 meter in diameter. The cooled mass includes the entire helium dewar including the vacuum shell, and is 240 kg. Of this about 140 kg is the dewar and helium itself. Roughly one third of the cold surface area is due to the dewar. The parasitic heat loads come from conduction in structural supports and electrical leads and radiation from the surrounding 24 K shield which is cooled by the boil-off vapor. The mirror and instrument mass is roughly 100 kg. The total heat load at 5.5 K for Spitzer is 6 mW. The structural support may be scaled linearly with mass and with the square of the shield temperature that is used to intercept conducted heat. The radiation may be scaled linearly with surface area and as the 4th power of the shield temperature, and the electrical leads which feed the instruments and telescope mechanisms will be roughly the same. The mass of the instrument will be assumed to be constant at 75 kg, and the mirror and structure will scale as the cube of the dimension. The outside surface area will scale with the square of the linear dimension. If one arbitrarily assigns 2 mW each to the structure conduction, lead conduction, and the radiation for Spitzer, then the values expected for CMB-Pol are: 17 mW conduction and 25 mW from radiation for a total of 44 mW. For a 5 year mission this would lead to a 2700 liter dewar - on the order of the size of the ISO, Herschel and GP-B dewars. Such a large dewar would probably force the observatory into the heavy lift category of rockets.

Note that it is not necessary for the stored cryogen system (or mechanical cooler for that matter) to reach the telescope/instrument operating temperature of 2.7 K. A closed cycle JT expander and sorption cooler or an Adiabatic Demagnetization Refrigerator (ADR) could provide the final stage of cooling from 10 K or so. In that case a SH_2 dewar could be used with its higher latent heat, and consequently, lower mass.

3. General Comments

Large uncertainties exist in the thermal designs for space flight cryogenic systems. To accommodate these uncertainties the AIAA Spacecraft Thermal Control Handbook recommends design margins for cryogenic systems of 50% at project start, falling to 35% at the critical design review, and 25% at launch[13]. Here margin is defined as the predicted heat load subtracted from the cooling capability divided by the cooling capability. These margins have been derived from flight project experience. It is noted that the last preflight predictions for lifetimes of both IRAS and COBE were 25% above the actual lifetime achieved on orbit.

Mechanical cryocoolers provide the competition for stored cryogen systems over the range from room temperature to a few K. The trade-off between the two usually comes down to cost, mass (which can equate to launch cost), technology readiness level (TRL), reliability and redundancy, servicing, and operational flexibility. A stored cryogen is a large potential cooling capacity: it is equivalent to energy. To compare with continuously produced cooling, from a passive radiator or mechanical cooler, for instance, it is necessary to know the requirement and goal for mission lifetime. In general, for missions expected to last more than one year, a mechanical cryocooler will be the more cost effective solution.

The total mass of a cryocooler system must include the power system (solar arrays, power conditioning electronics, batteries, and cryocooler controllers, the radiators to dissipate the cryocooler power, as well as the cryocooler components itself. Even so, the overall mass trade came down in favor of a cryocooler system for JWST (240 kg vs. 50 kg). In addition, the dewar mass and volume must all be absorbed in the cold section of the instrument or observatory. Mass here is at a premium due to the need to limit parasitic heat loads through the structure.

An assessment of the maturity of a technology is usually made through the use of its Technology Readiness Level (TRL). The TRL of a component or subsystem is defined on a numerical score of 1-9 with 1 being the least mature (the device exists as a concept only) through TRL 9 (the device has

Table 1. Definitions of Technology Readiness Levels (TRLs).

TRL	Definition	Hardware Description	Exit Criteria
1	Basic principles observed and reported.	Scientific knowledge generated underpinning hardware technology concepts/applications.	Peer reviewed publication of research underlying the proposed concept/application
2	Application and/or operating concept formulated	Invention begins, practical application is identified but is speculative, no experimental proof or detailed analysis is available to support the conjecture.	Documented description of the application/concept that addresses feasibility and benefit
3	Analytical and/or experimental proof-of-concept demonstrated	Analytical studies place the technology in an appropriate context and laboratory demonstrations, modeling and simulation validate analytical prediction.	Documented analytical/experimental results validating predictions of key parameters
4	Low-fidelity functionality demonstrated in laboratory	A system/component breadboard is built and operated to demonstrate basic functionality and critical test environments and associated performance predictions are defined relative to the final operating environment.	Documented test performance demonstrating agreement with analytical predictions. Documented definition of relevant environment.
5	Mid-fidelity functionality demonstrated in a relevant environment	A system/component brassboard is built and operated to demonstrate overall performance in a simulated operational environment with realistic support elements that demonstrates overall performance in critical areas.	Documented test performance demonstrating agreement with analytical predictions. Documented definition of scaling requirements
6	Mid-fidelity functionality and scaled form/fit demonstrated in a relevant environment	A system/component prototype that adequately addresses all critical scaling issues is built and operated in a relevant environment to demonstrate under critical environmental conditions.	Documented test performance demonstrating agreement with analytical predictions
7	High-fidelity functionality and scaled form/fit demonstrated in its operational environment	An engineering unit that adequately addresses all critical scaling issues is built and operated in a relevant environment to demonstrate performance in the actual operational environment and platform (ground, airborne or space).	Documented test performance demonstrating agreement with analytical predictions
8	Final product in mission configuration qualified through test and evaluation	The final product in its final configuration is successfully demonstrated through test and analysis for its intended operational environment and platform (ground, airborne or space).	Documented test performance verifying analytical predictions
9	Final product validated through successful mission operations	The final product is successfully operated in an actual mission	Documented mission operational results

flown in its present form). See Table 1. It is generally assumed that stored cryogen systems have a TRL of 7-9 because many dewars have flown before. In fact helium dewar systems are generally TRL 5 at best, because the design is always somewhat different. As a case in point, the COBE dewar was supposed to be a rebuild of the successful IRAS dewar, but in fact 50% of the drawings were changed including the cryogen tank size.

A stored cryogen system in general is not redundant. It is extremely cumbersome for a dewar system to include redundancy without providing a completely separate cryostat. Even components

Journal of Physics: Conference Series 155 (2009) 012008 doi:10.1088/1742-6596/155/1/012008

such as valves have limited redundancy due to increased mass and size. Most helium dewars' boil off rates have been dominated by parasitic heat loads which does not allow a dewar to be held back for use later in the mission. In contrast, most cryocooler systems can be made fully or partly redundant. A spare cryocooler can remain off or in use with limited power with only a small penalty in parasitic heat load on the primary cooler. Joule Thomson expanders are usually isolated by long lengths of low conductivity material and thus have a negligible extra parasitic. One can make the electronics part of the cryocooler system redundant (the path that JWST has taken with its cryocooler) since there are no cold moving parts and the warm moving components in the compressors have very high reliability and have been successfully used on many missions.

One must consider the ground operations up to launch that a cryogenic system must go through. Stored cryogen systems need constant care and are more vulnerable to leaks, vent line plugs, and ground support equipment failures. Before launch the servicing equipment that maintains the dewar close to its operating temperature in space must be removed. This limits the length of time that the dewar may sit on the launch pad before servicing must be reinitiated, thus limited the launch schedule flexibility.

During the mission one must consider the parasitic load on the cryogen when scheduling events. Decontaminating optics or detectors is difficult or impossible given the much larger heat leaks to the cryogen. With a known or estimated end date, on orbit check out and mission planning has an extra level of urgency. One need only look at the WIRE and Astro-E2 (Suzaku) dewar failures to see that one mistake can lead to the end of a mission or instrument.

4. Mechanical Cryocoolers - Introduction

There have been some significant advancements in cryocooler development since Ron Ross presented a paper entitled "Aerospace Coolers: A 50-Year Quest for Long-life Cryogenic Cooling in Space" at the Cryogenic Engineering Conference in 2005[14]. Widely variable flight payload requirements have driven the cooler design and development in all directions. There has been a push for both lower and higher cooler operating temperatures, with both larger and smaller scale heat lift capability, and the miniaturization of cryocoolers with the incorporation of MEMs technology. Active development work continues for improved mechanical cooler and electronics efficiencies, reduced mechanical cooler parts count for lowering cost, building larger drive electronics towards the kilowatt size, and also simplifying electronics by removing unnecessary control loops for small coolers.

The proven high reliability and the increased efficiency of the mechanical cryocoolers continue to generate new uses for active refrigeration, including vapor shield cooling of cryostats, zero boil-off for stored cryogens, LOX production for planetary missions, and sunshade and telescope cooling applications using active coolers. The push in mechanical cryocooler development has been broad, with a mix and match of cooler types (Stirling/pulse tube, J-T, reverse turbo-Brayton) to fill in the parameter space (temperature, size, capacity and cost) of mechanical cryocoolers.

The intention of this paper is to identify some of the more recent technology development areas that may prove beneficial to future Astrophysics missions such as CMBPol. This paper is not intended to identify and include all cooler developments, and no slight is intended if some significant cooler development efforts have been unintentionally omitted.

5. Cryocooler systems on present space missions

There are currently more than 31 known long-life coolers successfully operating in space [14-16]. Three additional coolers were still operating when the mission ended. The high reliability of cryocoolers has allowed many missions to be extended far beyond their original intended life. Among them, the Rutherford Appleton Lab (RAL) cooler on ATSR-2, which launched in 1995, has been operating for 13 years. The two TRW (NGST) mini pulse tube coolers have been operating for more than 10 years on the C/X mission since its launch in 1998. The majority of the flight instruments to date require cooling at medium to high cryogenic temperatures (50-150K) which has been accomplished using single stage coolers.

The more recent launch of the Japanese AKARI (ASTRO-F) infrared astronomy satellite in February 2006 has been flying two 2-stage Stirling coolers to cool the inner radiation shields for the AKARI cryostat to prolong the superfluid helium supply. Each cooler was providing 350mW of refrigeration at 40K using 50W input power from the bus [17]. While AKARI ran out of its cryogen in August, 2007, the 35-K Stirling coolers have enabled the mission observations to continue at the shorter wavelengths.

6. Cryocooler development for near-term missions

Building on the cryocoolers used in past missions, a large number of new or enhanced cryocoolers- some with multiple coldheads and many with lower operating temperatures ranges- have been developed, or are under development and test, for future missions. The majority of these are focusing on the lower operating temperatures from 6-20K.

6.1. SMILES

Sumitomo Heavy Industries built the 4.5K Joule-Thomson (J-T)/Stirling cooler for the Superconducting Submillimeter-Wave Limb-Emission Sounder (SMILES), to cool SIS mixers and low-noise HEMT amplifiers [18]. The SMILES 4.5-K cooler consists of a ^{4}He Joule-Thomson lower stage to produce 20mW of cooling at 4.5K and a 2-stage Stirling precooler at 20K that also cools HEMT amplifiers at 20K and at 100K. (The 2-stage Stirling cooler is the same as the coolers used on AKARI). The total system input power for the cooler is on the order of 140W. SMILES is to be operated aboard the Japanese Experiment Module (JEM) of the International Space Station (ISS). This instrument was to be delivered to the ISS in 2007, but shuttle delays forced the postponement of the mission launch to 2009.

6.2. MIRI

The NGST hybrid J-T/Pulse Tube cryocooler for the Mid Infrared Instrument (MIRI) on the James Webb Space Telescope (JWST) is designed to cool the 90-kg MIRI instrument to 7K [19]. Current cooling requirements for the cryocooler are for 70mW @ 6K along with 70mW @ 18K for 400W of system input power. The cryocooler consists of a 3- stage pulse tube cooler to reach 17K, with a J-T circuit to provide the 6-K refrigeration to the MIRI instrument. The pulse tube cooler is located on the JWST bus, with the MIRI instrument located within the JWST Instrument Module behind the telescope, about 8 meters from the precooler. MIRI will fly a single mechanical cryocooler system with block-redundant drive electronics. This cooler has been demonstrated to a TRL level of 6 in 2007. JWST is scheduled for launch in 2013.

The MIRI cryocooler was selected from the coolers being developed in the Advanced Cryocooler Technology Development Program (ACTDP) for TPF, JWST and Con-X [20]. Three ACTDP cooler development efforts were being conducted: two hybrid systems using J-T/Stirling and J-T/pulse tube combinations, and a four-stage pulse tube with an optional integral flow loop. The two hybrid systems used a 3-stage Stirling (or pulse tube) cooler to precool the J-T circuit to less than 18K, and a single stage of compression for the J-T circuit provided a sufficient compression ratio to achieve the required 6K refrigeration. All three ACTDP vendors demonstrated the feasibility of operating the ACTDP coolers at 4K through the use of ^{3}He as the working fluid in place of ^{4}He. The ACTDP program ended with the transition into the flight cooler program for JWST/MIRI.

6.3. Planck 20K Sorption Cooler

This hydrogen sorption cryocooler has been designed to deliver approximately 1 watt of refrigeration at 18-20K for the Planck spacecraft to cool the 20K HEMT amplifiers for the Planck Low Frequency instrument (LFI) (~80% of the total heat lift) and to simultaneously provide the pre-cooling for the RAL 4.5K J-T refrigeration stage in the Planck High Frequency Instrument (HFI) m(~20% of the total heat lift). The 4.5K J-T cooler is the upper stage precooler for the HFI 0.1K dilution refrigerators. The Planck sorption cooler requires a total input power of 470W (End of Life), and relies on passive cooling support at three passive radiators (170K, 100K and 60K) to provide pre-cooling to 60K prior to a final counterflow heat exchange and expansion through the J-T valve to produce the 18K refrigeration. The redundant Planck sorption cryocoolers have been integrated in to the Planck spacecraft in preparation for the 2008 launch [21].

6.4. GIFTS

The Geosynchronous Imaging Fourier Transform Spectrometer (GIFTS) instrument, built in partnership between the Space Dynamics Lab (SDL) and the NASA Langley Research Center, is an advanced weather observing instrument that was originally scheduled to be launched on NASA New Millennium Earth Observing (EO3) spacecraft in 2005. It is still looking for an alternate launch opportunity. The GIFTS instrument uses a two-stage Lockheed Martin pulse tube cooler to cool the spectrometer's focal plane to 55K and the optics to 140K. System-level testing of two-stage pulse tube cooler integrated into the GIFTS instrument was performed at SDL [22].

6.5. ABI

The next generation of the NOAA GOES (Geosynchronous) weather satellite will replace the passive cryocooler used on previous NOAA satellites with an active cryocooler. The Advance Baseline Imager (ABI) uses a pulse tube cooler manufactured by NGST with two separate cold heads. One cold head cools the Mid-Wave/Long-Wave IR focal plane to 60K, and the remote second coldhead is used to the optics and the Visible/Near IR focal plane to 200K. The Advance Baseline Imager will use redundant cryocoolers with each of its instrument packages [23].

6.6. AMS-02

The Alpha Magnetic Spectrometer-02 (AMS-02) instrument will be flown as an attached payload to the International Space Station. The instrument uses 4 Sunpower M87 Stirling coolers to maintain the outer vapor-cooled shield of the 2500 liter helium dewar at a temperature of 80K to extend the mission duration. The baseline performance of the coolers is a total of 16 watts of heat lift at 80K for 400 W of input power [24,25].

7. Current cryocooler research and development efforts

7.1. 2K to 6K coolers

7.1.1. Hypres. There have been recent advances in 4K coolers for cooling ground-based superconducting electronics for high speed digital communications in the wireless communications field. The Hypres cooler built by Lockheed Martin presently provides 35mW of refrigeration at 4.5K with additional refrigeration for thermal shielding available at the upper stages: 50mW @ 10K, 150mW @ 25K and 5W @ 70K for 725 W of compressor power [26]. The Hypres refrigeration goal of 100mW @ 4.2K will require further technology advances to improve the cooler efficiency. While the Hypres program is a US Navy and DARPA program and not a space flight program, the development of this cooler will be a benefit to space cooler programs as well.

7.1.2. *University of Twente 4.5 K sorption cooler for DARWIN.* This sorption cooler consists of a cold hydrogen J-T stage operating from 80K to 14.5K and a helium J-T stage operating from 50K and delivering 5mW @ 4.5K. The 2 cooler stages need about 3.5W of total power and are sunk at two passive radiators at temperatures of about 50 and 80K [27]. The 50-K radiator area needed for the sorption cooler is about $2m^2$/mW of cooling at 4.5K [28]. The sorption cooler was tested in the laboratory using a Thales 5W at 50-K Stirling cooler to represent the passive cooler. The total refrigeration package including the 50-K cooler required less than 150W of input power and had a mass of less than 20kg. This laboratory test demonstrated that the cooler package could meet the Darwin power requirements of 200 W.

Journal of Physics: Conference Series **155** (2009) 012008 doi:10.1088/1742-6596/155/1/012008

7.1.3. SPICA. The Japanese are working to develop the 20K, 4.5K and 1.7K coolers needed for the SPICA astronomical mission. The Japanese have made performance enhancements to the SMILES 4.5K cooler to increase the 4.5K capacity to 50mW (compared to 20 mW for SMILES) for a system input power of 145W. Much of the performance improvement was a result of the improvements to the 20K precooler; the cooler is now capable of producing 200mW of refrigeration at 16K (or 325mW @ 20K) plus 1W at 80K for under 90W of input power [29]. Substitution of ^{3}He in the J-T circuit has enabled them to demonstrate 16mW of cooling at 1.7 K [30].

7.2. 10K to 80K Multi-Stage Coolers

The development effort continues for a number of high capacity multi-stage cryocoolers for cooling VLWIR sensors planes at 10K, LWIR sensors at 35K or even MWIR sensors above 50K, while having a large capacity upper refrigeration stage for supporting thermal shielding and/or optical bench cooling. The 10K coolers have also become the precooler for 4K refrigerators. The large refrigeration requirements have led to the manufacture of large input capability of the compressors (up to 800W), necessitating the need for large throughput cooler drive electronics as well. A number of these cooler developments have been listed in the table below.

1.1. 80K-100K Coolers

The use of large scale refrigerators will be required for cryogen storage on long duration space missions. High capacity mechanical coolers will be needed to prolong mission life by minimizing boil- off and providing stable pressurization of the cryogen tank. Large scale refrigeration on the order of 10s of watts may be needed at 25K for hydrogen and at 90K for oxygen. Planetary or lunar in situ oxygen production may also require large scale refrigerators.

Lockheed Martin is working on a several programs for large scale coolers. One cooler program is for a zero boil-off of propellant tanks that has demonstrated 30W @ 95K [31]. This cooler has also been suggested as a lunar LOX liquefier that can provide 40W@95K. Lockheed Martin is in the build phase of another cooler that will provide 25W of refrigeration at 70K [32]. Both coolers are presently in the build phase.

NASA has had a Small Business Innovative Research (SBIR) Subtopic call for Zero Boil-Off Cryocoolers capable of providing 10W of refrigeration at 20K. Creare, in a Phase 1 SBIR effort, has provided some initial design work for a 10W @ 20K reverse turbo-Brayton cooler [33].

Thales Cryogenics (Eindhoven) is developing a 18W @ 80K (or 5W @ 50K) pulse tube cooler for future ESA projects [34]. Thales has redesigned their military tactical coolers into long-life coolers for space applications.

8. Cooler performance summary

Table 2 below summarizes most of the cryocoolers that have been discussed above, and in addition includes some of the larger capacity cryocoolers that have flown or that are in development. The Compressor Power is the input power to the cooler compressor from the drive electronics; the System Power is the input power to the cooler drive electronics. The cooler mass is provided when known. TMU refers to the thermal mechanical unit. The estimated TRL level is also given for the coolers under development. As is often the case, the cooler customer (particularly NASA or the DoD) will fund several cooler manufacturers simultaneously to develop coolers up to a TRL level of 4 or 5, readying the coolers for selection to a particular flight program. Once selected, the flight program will carry the cooler development through a final design modification as necessary to meet the flight program cooling requirements, build the cooler flight electronics and flight qualify the cooler subsystem.

Journal of Physics: Conference Series **155** (2009) 012008 doi:10.1088/1742-6596/155/1/012008

Table 2. Summary of performance characteristics for selected cryocoolers.

Cooler [Ref]	Type	Cold Stage	Upper stages	Input Power (W)		Mass (Kg)	TRL
				Compressor	System		
JEM/SMILES [18]	JT/Stirling	.030W @ 4.5K	200mW @ 20K, 1W @ 100K	140		65 TMU	8
JAXA/ SPICA [29, 30]	JT/Stirling	.050W@4.5K	N/A	145			5
U of Twente DARWIN [27, 28]	He/H_2 sorption	.005W at 4.5K	N/A	4 at 50K		8.3 TMU	4
LM Hypres [26]	4-stage pulse tube	.035W @ 4.5K	50mW@10K, 160mW@25K, 5.2W@66K	725		21 TMU	5
MIRI [19]	JT/ pulse tube	.070W@6K	70mW@18K	320	400	79 TMU + elec	6
LM 10K [35]	3-stage pulse tube	.260W@10K	10.4W@85K	600		16 TMU	4
NGST 10K [36]	3-stage pulse tube	.250W@10K, or .525W@15K	1W@51K	370		18.7 TMU	6
JPL Planck Sorption [21]	JT/ H2 sorption	.963W@18K	Heat rejected to radiator <60K	426			7
JAXA AKARI [17]	2-stage Stirling	.200W@20K	N/A	90			9
LM SBIRS Low [37]	2-stage pulse tube	1.7W@35K	17W@85K	610		23.8 TMU	5
Creare [38]	1-stage reverse turbo-Brayton	3W @ 35K		350		10 TMU	4
Creare HCC [38]	2-stage reverse turbo-Brayton	2W @ 35K	20 W @ 85K	490		18 TMU	4
Ball 35K HCC [39]	2-stage Stirling	1.5W@35K	8.5W@85K	200		14.4 TMU	4
NGST 35K HCC [40, 41]	2-stage linear pulse tube	2.25W@35K	17.4W@85K	500		14.3 TMU	6
Raytheon RSP2 [42]	2-stage Stirling/ pulse tube	2.6W @ 35K	16.2W @ 85K	513		20 TMU	4
NGST ABI [23]	2 parallel pulse tube stages	2.3W@53K	6.7W@183K		170	5.5 TMU, 3.8 elec	7
LM GIFTS [22]	2-stage pulse tube	1.5W@53K	8W@140K		180	6.3 TMU, 4.2 elec	8

LM DARPA 10W [43]	1-stage pulse tube	9.6W@ 77K		220	12 TMU	4
Thales [34]	1 stage Stirling	5W@50K or 18W@80K	350			4
Creare NICMOS [44]	1-stage reverse turbo-Brayton	7W@72K		375	18.5 TMU	9
NGST HEC [45]	1-stage pulse tube	10W@95K	123		4.1 TMU, 3.3 elec	9

Figures 5 and 6 show the performance characteristics of a variety of coolers that have been developed and reported on over the years. Attempts have been made to include only the most recent and best data for each of the coolers. The data points/curves generally represent published data, although a few data points were provided through private communication. Labeling of the data points with the identification of the coolers manufacturers were not included to help keep the plot uncluttered. Figure 5 shows the specific power (the compressor input power required to provide a watt of cooling) for the cold stage only, even if the cooler has multiple stages. The Specific Power curve fit has been biased towards the low values to reflect this. The specific power for multi-stage coolers is generally higher because a significant fraction of the compressor power may be used to provide additional heat lift at the upper stages. It needs to be pointed out that the specific power refers to the power into the cooler compressor itself, and not the power into the cooler drive electronics. Unless the cooler is carried into a flight build, there is generally only laboratory electronics to drive the cooler. To get an estimate of the system level specific power one can assume the cooler flight electronics is 80-85% efficient, with a few additional watts of overhead (or tare)

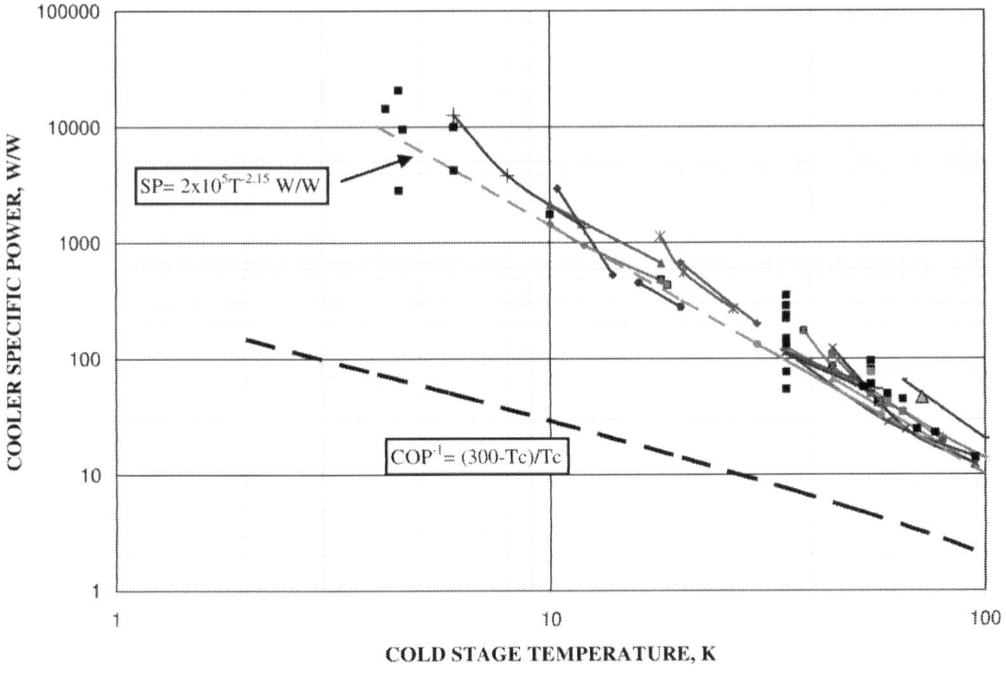

Figure 5. Specific power of single and multi-stage coolers based on the total power into the cooler but only on the heat lift at the cold stage. The ideal Coefficient of Performance (COP) is shown

power to power the control circuits.

Figure 6 shows the cold stage refrigeration capacity as a function of operating temperature for a number of the coolers. There has been a general trend towards larger heat lift capability as the cooler efficiencies have improved.

9. Interfacing active coolers with passive cryocoolers

Cooling chains have long been the subject of study for all future astronomical missions. The requirement for the spacecraft to provide kilowatts of electrical power to handle the cooling of bolometers necessitates the use of passive cooling to ease the power burden. However passive cooling by itself is insufficient if temperatures below about 7K are desired, and so a combination of active and passive cooling must be used. The improved efficiency and reliability of the coolers are providing more options to ease the decision making process. The mechanical coolers can be used to cool telescopes and sunshades in addition to the detectors. And vice versa, the sunshades can be used as passive radiators to enhance the cold-end performance of mechanical cooler.

9.1. Heat Intercept

A cryocooler's refrigeration capacity and operating temperature can be enhanced significantly by using passive cooling to remove heat from the coldfinger of the cooler. The intermediate-temperature radiator is attached to a point partway down the length of the coldfinger regenerator, for example at one of the upper stages of a multi-stage coldfinger, to carry away a large fraction of the heat that comes from the warm end of the coldfinger. The location of the coldfinger intercept point, and the temperature of the radiator, determines the performance improvements realized. Early laboratory heat intercept tests with single stage coolers demonstrated a doubling of the cooler performance, either by doubling the heat life capability or by halving the input requirement for the original heat lift, if the coldfinger were tied to an intermediate-temperature radiator at ~120-160K [46-49]. The Planck hydrogen sorption cooler relies on this technique to remove heat at temperatures of 170K, 100K and

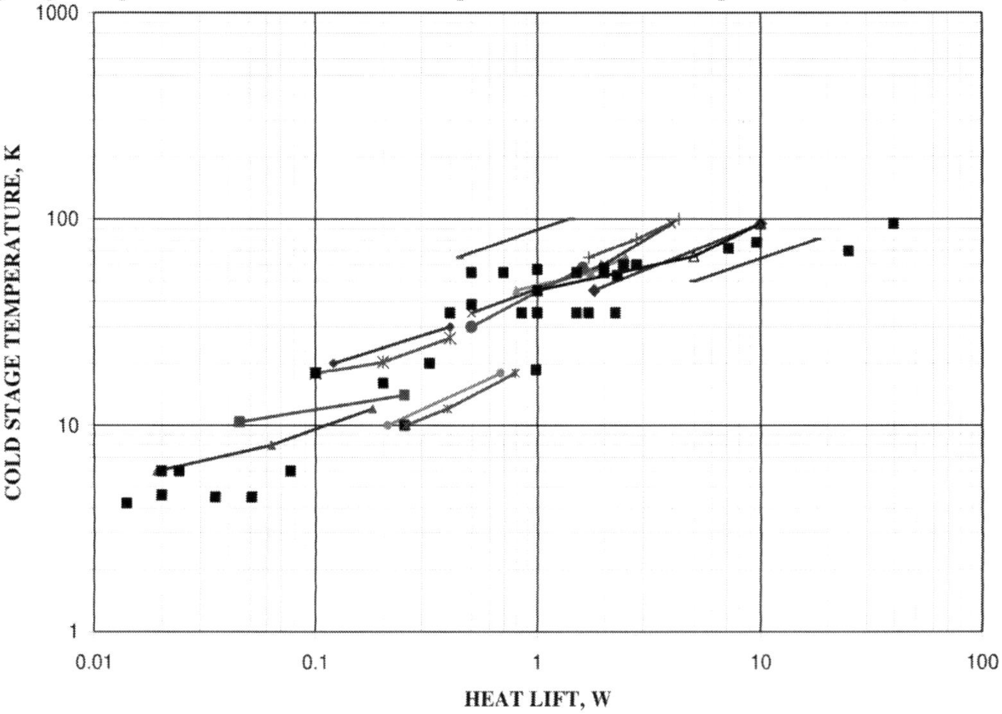

Figure 6. Heat lift capacity of the coldest stage of selected single and multi-stage coolers.

Journal of Physics: Conference Series 155 (2009) 012008 doi:10.1088/1742-6596/155/1/012008

60K to enable it to produce the necessary 18-K refrigeration for the Planck mission.

Two independent studies have been conducted on the effect of a heat interceptor on a two-stage cooler. RAL studies on a 2-stage BAe (now Astrium) Stirling cooler found they were able to increase the 20K refrigeration from 140mW to 220K when intercepting heat at the 1st stage of the coldfinger [50]. They were also able to simultaneously reduce the compressor power by 11%. However they found little performance improvement in reducing the 1st stage temperature below 100K from its nominal temperature of 130K without the intercept. In the other study, CEA/SBT studied the effects of using passive radiators at 50K and 80K for heat intercepts on a 20K 2-stage pulse tube [51]. A G-M cooler was used in laboratory tests to simulate the radiators. The G-M cooler was able to remove 1.7W of heat from the pulse tube regenerator at 80K, or 2.7W at 50K, which resulted in lowering the ultimate no-load temperature of the coldtip (to 13.5K with the 50K heat intercept), but there was little net improvement in refrigeration capacity at the 20K coldtip temperature. This seems to corroborate the RAL results that state there is little improvement in trying to cool the 1st stage below 100K. Further studies are needed to better understand and characterize the cooler performance results.

9.2. Distributed cooling

The need to cool broad areas of sensor arrays, multi-segmented telescopes or sunshades will require the need for distributed cooling either through heat spreading with high conduction materials or by using the helium fluid flow circulation techniques. Development of distributed cooling techniques has been getting much attention from the cryocooler community as they work to satisfy the needs of future space missions. Several different techniques are being pursued which deliver a rectified flow gas stream to the device being cooled. Both single phase and two-phase fluid flows are being considered. Ball Aerospace uses a neon J-T circuit at the base of their 2-stage Stirling cooler to provide 2-phase flow to cool a remotely located device [39]. This circulating flow loop also utilizes a thermal storage unit in the JT circuit to enable extra cooling during peak load periods. Creare and Raytheon teamed together to place a Reverse Turbo Brayton cooler at the cold end of the 2-stage Raytheon cooler to provide either a cold circulating flow loop and/or an additional stage of refrigeration [52]. U of Wisc, in collaboration with Atlas Scientific, is developing a cold rectified-flow loop that siphons part of the cold helium gas from the end of a 2-stage pulse tube to provide cold circulating gas [53]. These uni-directional fluid flow loops can deliver refrigeration over long distances to allow a vibration-free environment where required, keeping the mechanical cooler far removed from the sensor area. The fluid loops can adsorb the heat either via long heat exchanger lengths or through multiple parallel flow paths to minimize the temperature gradients in the object being cooled.

10. Sub-Kelvin Cooling - Background

Cooling to sub-Kelvin temperatures – typically below 300 mK, but increasingly to 50 mK or below – is needed for many detector technologies to reach their measurement goals. The trend for future missions is to use progressively larger detector arrays, with a correspondingly larger heat lift requirement. From the modest array size of the XRS instrument on Astro-E/2 of 32 pixels, which required a cooling power of only 0.3 μW at 60 mK, future missions will implement 1000+ pixel arrays. Even with multiplexing techniques, cooling powers of several μW at 50 mK will be needed.[54]

A number of methods exist for cooling to temperatures below 1 K, and considerable effort has been devoted over the last two decades to adapt them for use in space. These include both solid state (e.g., Adiabatic Demagnetization Refrigerators (ADRs)[55] and Normal-Insulator-Superconductor (NIS) tunnel-junction coolers[56]) and fluid-based systems (e.g., ^{3}He sorption[57] and dilution refrigerators[58]). These techniques vary considerably in their cooling capacities, operating temperature range, heat sink temperature range, and requirements for inclusion with various detector and instrument components. In some cases, they are viable over only part of the necessary temperature range – between the detector and heat sink – and therefore must be used as part of a larger cooling chain. In this paper we present a basic description of each technology that has been

Journal of Physics: Conference Series 155 (2009) 012008 doi:10.1088/1742-6596/155/1/012008

demonstrated for use in space, with a goal of identifying each one's capabilities, advantages and challenges. We conclude with a discussion of future development and cooler architectures that could significantly improve performance (cooling power, mass, size, input power, etc.)

One of the most significant constraints on the design and operation of any cooler is the heat sink to which it rejects heat. At least in their early development, the majority of sub-kelvin refrigerators staged their operation from a superfluid helium bath. Superfluid helium dewars were readily available, and the low, ~1 K, starting temperature was convenient for minimizing parasitic heat inputs. Moreover, the development of helium dewars for missions like IRAS and COBE extended that capability, and that of sub-Kelvin coolers, into space instruments. For a variety of reasons, cryocoolers are being developed to replace stored cryogen dewars for future missions. Offsetting the (significant) advantages of these coolers are their generally higher operating temperature and limited cooling power. For sub-Kelvin coolers, this is providing a bias away from single-shot coolers and toward those that operate continuously, to provide a better match between their peak heat rejection rates and the cryocoolers cycle. Since this has an inherent advantage in lowering the power requirements for the cryocooler, and therefore system mass, it is likely that continuous coolers will become more prevalent in future mission designs.

11. Flight-Capable Sub-Kelvin Cooling Technologies

Table 3 provides a summary of the sub-Kelvin coolers that are detailed in the following section. These are systems that 1) have flight heritage, 2) are currently being readied for flight instrument, or 3) are maturing to the point that they could be considered for future instrument. Cooling powers that have been demonstrated in flight or flight-like coolers are shown in figure 7 over the temperature ranges where these coolers are optimized for use.

Table 3. Demonstrated Performance and Heritage of Flight-Capable Sub-Kelvin Coolers

Technology	Configuration	Operating Temperature	Cooling Power	Heat Sink	Avg. Heat Rejection	TRL	Heritage
Sorption Cooling	^{3}He, single-stage	300 mK	10 μW	1.5 K	3.7 mW	9	IRTS [59], Herschel [57]
	^{3}He, two-stage	300 mK, 240 mK		2.5 K		6	
	^{3}He/^{4}He three-stage	2.5 K, 300 mK, 240 mK		5 K		6	
ADR	Single-stage	60 mK	0.3 μW	1.3 K	0.15mW	9	Astro-E2 [60]
	Two-stage	50 mK	0.4 μW	1.8 K	0.12 mW	5	SRG, Astro-H [62]
	Continuous, Four-stage [61]	50 mK	6 μW	5 K	3.0 mW	4	
Dilution	Open-cycle	100 mK	0.1	4.5	2.0 mW	7	Planck

Refrigeration		µW	K	[58]
NIS Coolers	190 mK		0.3 K	

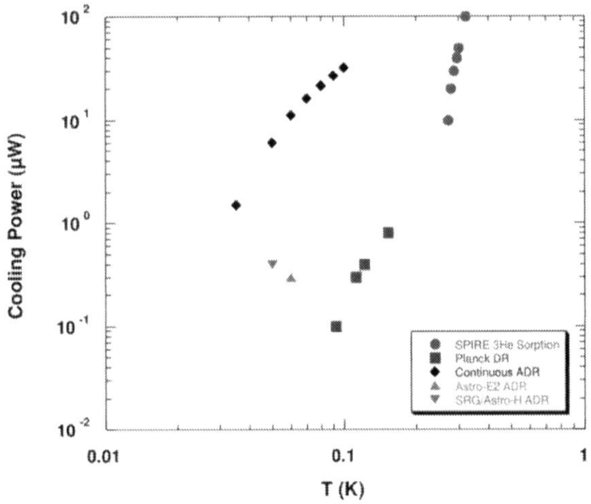

Figure 7. Demonstrated cooling power and operating temperatures for selected coolers.

11.1. ^{3}He Sorption Refrigeration

Sorption refrigerators produce cooling by evaporating liquid from an evaporator using a sorption pump. The sorption pump contains a quantity of material that has a large surface area and large binding energy for, in this case, ^{3}He. When this "getter" is cold, ^{3}He is strongly attracted to its surface, resulting in a very low equilibrium vapor pressure above the material. This allows it to act as a pump, and thereby drive large-scale evaporation of liquid from the evaporator. The remaining liquid is cooled until the evaporation rate balances the heat load. In principle, that point can be varied by changing the surface area and material (binding energy), but in practice it is necessary for the surface area to be so large that the helium, when fully adsorbed, occupies only a fraction of an atomic layer so that the equilibrium pressure remains low throughout the pumping phase. Consequently, the evaporator is typically pumped close to or moderately below the limiting temperature of 300 mK for ^{3}He (1 K for ^{4}He refrigerators). The base temperature can be stabilized over long time scales by regulating the sorption pump temperature, but short-term control is accomplished by heater control of loads coupled to the evaporator. By careful design (or the use of multiple stages) to minimize parasitic heat loads and ensuring a low impedance pumping path, ultimate temperatures of 200 mK or slightly lower have been demonstrated.

During the hold time of the refrigerator, liquid is evaporated as heat is absorbed. When the liquid is depleted, the sorption pump is warmed to desorb and re-condense the ^{3}He gas. For a typical sorption pump made using charcoal, the temperature must be raised to about 50 K to desorb the gas. In ground-based coolers, one can use a condenser anchored to the heat sink, from which the liquid drips back into the evaporator; for space coolers it is necessary to re-condense it in the evaporator. The inside of the evaporator is designed so that surface tension will confine the liquid and ensure it makes thermal contact with the walls. A heat switch is used to thermally connect the evaporator to the heat sink during the condensation phase and thermally isolate it during the cooldown and operational phase. The heat sink can be as warm as the critical point of ^{3}He (3.2 K), but this requires the sorption pump to be heated to high temperature to raise the ^{3}He pressure above the critical value – at a considerable cost in waste heat. Instead, heat sinks are typically 2.5 K or lower to reduce the power needed and to reduce the amount of helium that must be evaporated to cool to the operating point.

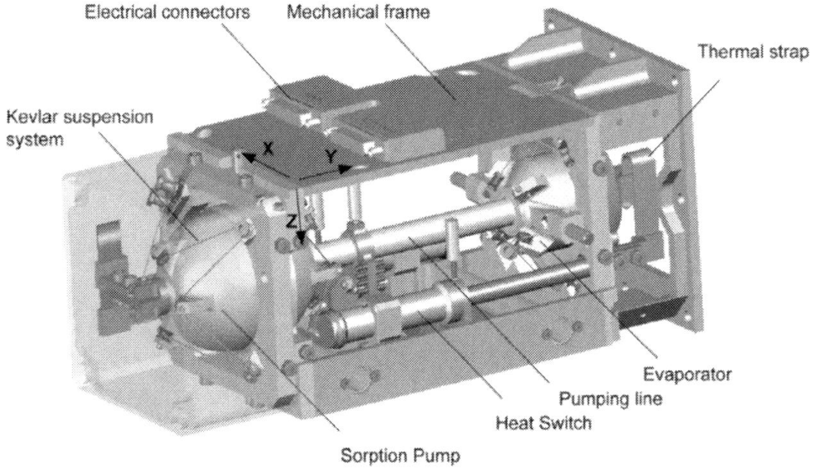

Figure 8. A single-stage ^{3}He cooler will be use to cool the SPIRE instrument on the Herschel mission.

The hold time of a sorption cooler is determined by the detector heat load and the volume of liquid remaining after pumping to the operating point. The design values for Herschel [57] (Figure 8) are 10 μW at 290 mK, and a hold time of 70 hours. The cooler generates waste heat during operation due to the heat of adsorption in the pump, and during recycling due to the heat of condensation of the ^{3}He charge, the heat required to warm the sorption bed, and heat required to operate the heat switch. For the Herschel cooler, operating at 277 mK with a 10 μW load, the total heat rejected is 927 J into the superfluid helium at ~1.5 K. From this, one can see that one of the main disadvantages of sorption refrigerators is their low overall efficiency (~1.5% of Carnot).

11.2. Adiabatic Demagnetization Refrigeration

ADRs make use of the magnetocaloric effect: an increasing magnetic field in a paramagnetic material causes the magnetic moments to increasingly align with the field, decreasing their entropy. This appears as heat in the lattice, causing the material to warm. Conversely, a descreasing field produces cooling in the material. What is notable about this process is that the heating and cooling are essentially reversible, leading to extremely high efficiencies. A typical single-shot ADR consists of a capsule of paramagnetic material (usually called a "salt pill"), a superconducting magnet, and a heat switch. The heat switch is used to thermally connect the salt pill to a heat sink while it is being magnetized, and to thermally isolate it after reaching full field. Subsequent demagnetization then causes the salt pill to cool to its operating temperature. Depending on starting and ending temperatures, and the peak field, this point will be reached with some magnetic field remaining. This provides a means of regulating temperature – with no waste heat input – as the salt pill absorbs heat both from the load (detectors or other experiment components) and parasitic sources. As heat is absorbed, the field is continually reduced until it reaches zero. The ADR must then be recycled. For space missions, the hold time requirement is typically 24 hours, followed by a recycling period of no more than 1 hour.

While the magnetocaloric effect is usable over a very wide range of temperatures (high and low), there are practical limitations for a single stage device. First, for a single stage to reach very low temperature the magnetic field applied must be roughly proportional to the heat sink temperature (1 T for a 1 K sink, and 4 T for a 4 K sink for operation at 100 mK). At present, high field to current ratio magnet technology is limited to NbTi, and practically limited to heat sinks at 5 K or lower and fileds of 4 T or lower. Single-stage ADRs operating at, for example, 50 mK are therefore limited to heat sinks up to perhaps 2 K. With colder heat sinks, temperatures as low as 20 mK are readily achievable. However, ADRs are easily adaptable to multi-stage configurations which also provide size and mass benefits. As an example, the single-stage ADR on Astro-E2 [60, 61] (Figure 9) operating at 60 mK used 920 g of ferric ammonium alum refrigerant, a 2 T NbTi magnet and a 1.3 K superfluid helium bath as a heat sink. Its net (detector) cooling power was 0.3 microwatts, and the total mass was 15 kg. For the upcoming Astro-H [62] mission with a 1.8 K heat sink, a two-stage design was chosen (240 g chrome potassium alum with a 2 T NbTi magnet, and 72 g gadolinium lithium fluoride with a 3 T NbTi magnet) that will provide 0.4 μW of detector cooling at 50 mK with a mass of only 8 kg.

Figure 9. A single-stage ADR (salt pill/heat switch assembly at left; magnet at right) was flown on the Suzaku (Astro-E2) mission to cool the X-Ray Spectrometer detector array to 60 mK.

The heat rejected by the ADR comes from two main sources – the heat of magnetization of the spins and hysteresis heating in the superconducting magnet. However, as in the sorption cooler, significant heating can come from operation of the heat switches, especially active gas-gap switches. Generally, though, magnet hysteresis and heat switch dissipation are at most comparable to the heat of magnetization, giving overall efficiencies that are only about half of the efficiency of the salt pills in absorbing heat, which can be as high as 70% (of Carnot) at 50 mK.

ADRs have the disadvantage of needing current leads for the magnets – on the order of a few amps – but recent development in high-Tc materials allows these currents to be introduced to low temperature with a quite manageable impact on the thermal design. Materials such as MgB_2 and YBCO deposited on low conductance tape can be heat sunk to cryocooler stages or vapor cooled with venting cryogen to reduce heat loads on the cold stage to a small fraction of a mW.

11.3. Dilution Refrigeration

Dilution refrigerators (DRs) were developed in the 1960s and quickly became the standard laboratory sub-Kelvin cooling. In comparison to other techniques (principally ADR and ^{3}He sorption), DRs operated continuously, had very large cooling power, and could reach temperatures of a few mK. These systems relied critically on gravity for phase separation of the two helium isotopes, and it is this fact that has precluded their direct application in space, as no other mechanism has yet succeeded in providing the required ^{3}He/^{4}He phase separation and liquid confinement.

The Planck mission [58], however, developed an open-cycle DR (Figure 10) that has achieved cooling to 100 mK. The system feeds gaseous ^{3}He and ^{4}He from external storage tanks, condenses them at 4.5 K with a sequence of radiative and Joule-Thomson (JT) coolers, and pre-cools the individual streams to 1.6 K with a JT cooler which uses the exiting ^{3}He/^{4}He mixture as its expansion fluid. The two streams are further pre-cooled in counterflow heat exchangers with the exiting mixture.

When the two streams are finally mixed, they cool as the pure ^{3}He dilutes the ^{4}He. Because the ^{4}He flow must be cooled by the dilution process, the net cooling is much less than for a conventional DR cycle where (ideally) only ^{3}He is circulated. Nevertheless, by careful minimization of parasitic heat loads, the 100 nW cooling power is sufficient to cool the bolometers to 100 mK. The external tanks contain enough ^{3}He and ^{4}He to achieve a lifetime of 24 months.

The pre-cooling requirement from auxiliary cryocoolers for the dilution unit is 2 mW at 4.5 K. Like sorption coolers, this results in a low overall efficiency, less than 1% of Carnot. To improve on this performance, plans are underway to convert the operation to a closed-cycle. This has the potential to increase the cooling power to into the range of 1 μW at 50 mK, and remove the lifetime limitation.

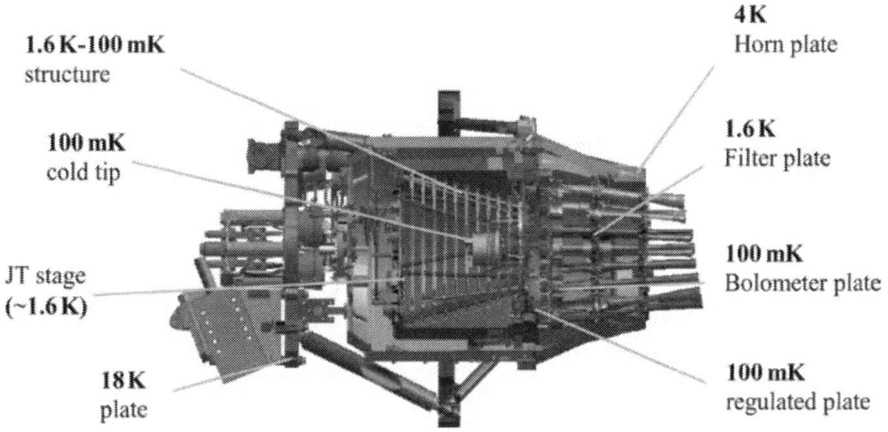

Figure 10. Schematic of the Planck focal plane with the dilution refrigerator.

11.4. NIS Coolers

The most recently developed technique, shown in figure 11, for cooling at low temperature is the concept of using NIS tunnel junctions [56], which have the potential to cool from 300 to 100 mK. Biasing an NIS junction causes the hottest electrons in the normal metal to tunnel across the insulator barrier into the superconductor, which cools the remaining normal-metal electrons. Extending the normal metal onto a membrane as a cold finger cools both the electrons and lattice of the electrically separate membrane. Although the magnitude of the cooling power is small, the virtue of this technique is that the cooler can be integrated directly with the detector pixels, and thus cool only the pixels suspended on insulating supports. This is done instead of cooling the frame surrounding the

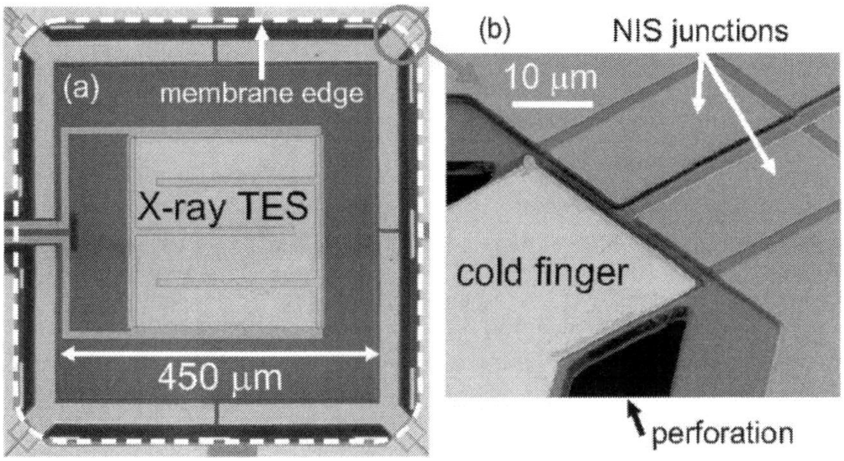

Figure 11. Micrograph of an NIS cooler integrated with a TES x-ray microcalorimeter.

detector array and thus having to absorb parasitic heat from lead wires and other structures. Figure 11 shows a test device in which an NIS cooler being used to cool an x-ray transition-edge sensor (TES). The heat sink for the NIS cooler is an ADR stage that can provide base temperatures from 300 mK to below 50 mK. In this case, cooling from 300 mK to 190 mK was demonstrated. The technique is in principle capable of cooling into the 50 mK range from 300 mK.

12. Future Developments and Cooler Architectures

12.1. Single-Shot versus Continuous Coolers

Single-shot coolers are well-suited to operation with stored cryogens, since only the total heat transferred to the cryogen is a concern. However, for cryocooler-based operation, the peak rate at which heat is rejected during recycling becomes the most significant concern, although load-leveling techniques can be applied which refocus attention toward the average rate. Consequently, adapting single-shot coolers to operate continuously can be very beneficial (in reducing the mass and simplifying the requirements for both the sub-Kelvin refrigerator and the cryocooler) by removing the need to store heat for long periods of time and reject it quickly in a short recycle operation. DR and tunnel-junction coolers are inherently continuous techniques, so this conversion applies mainly applies to sorption refrigerators and ADRs. However, since continuously operating versions of the these coolers can be used, for example, as pre-coolers for DR or NIS coolers, there is potential benefit for a wide range of cooler combinations.

Sorption refrigerators and ADRs can employ multiple stages either in parallel or in series to accomplish continuous cooling. The simplest, but by no means most efficient or least complex, way is to use two separate coolers in parallel, using one to cool the load while the other is recycling. It requires two heat switches to couple the coolers to the load. Two key requirements for these switches are that they conduct well at the cooler's operating temperature, and have sufficiently low heat leak when one end is raised to the heat sink temperature. For ^{3}He sorption coolers at ~250 mK, gas-gap

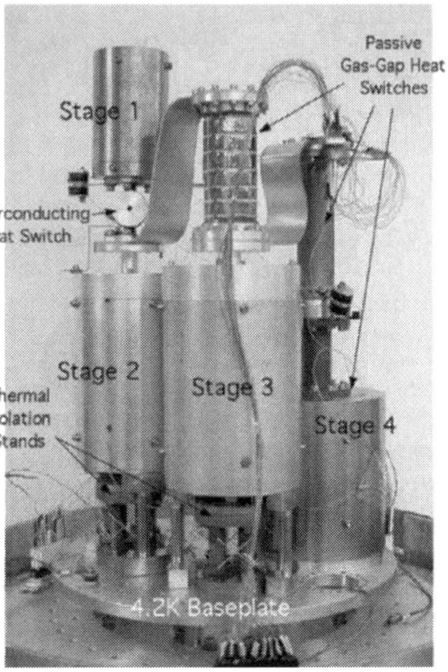

Figure 12. A continuous ADR capable of operating at 50 mK and below with a 5 K heat sink.

switches are suitable. For an ADR at 50 mK, only switches using metal as the conductive element will be functional, limiting the choices to mechanical, magnetoresistive, and superconducting switches. Each has significant limitations, making the series arrangement of stages preferable for a continuous ADR [63, 64]. Such a system has been developed. Shown in figure 12, it uses 4 stages in which heat is cascaded from a stage that is essentially a thermal buffer, sequentially through the upper stages to a heat sink. It has demonstrated continuous cooling as low as 35 mK with a 4.2 K heat sink. With a mass of 8 kg, produces net cooling powers of 6 μW at 50 mK and more than 30 μW at 100 mK.

12.2. Sub-Kelvin Cooler Combinations

As seen in the companion papers on flight cryogen and cryocooler systems, one can choose from a rather wide range of pre-coolers as a basis for sub-Kelvin coolers. In the stored cryogen category are superfluid helium at 1.3-1.7 K, and solid hydrogen at 6 K. In the cryocooler category, pulse-tube and Stirling cycle coolers can reach temperatures near 4 K with cooling powers in the 10-100 mW range, and JT systems can cool below 2 K with up to 10 mW of cooling. This opens a large parameter space for optimizing system design, and/or for developing cooler systems that meet unique constraints such as low cold mass or low magnetic field requirements.

ADRs have the widest operating temperature range of the coolers discussed. Temperatures of 50 mK and below can be reached from heat sink of 4-5 K. At the warm end, one can in principle add stages to arbitrarily extend an ADR's operating range; the present limitation is NbTi magnet technology which requires temperatures below 6 K. Other magnet technologies are emerging, such as Nb_3Sn [65], which show promise for enabling low (<5 A) current, high (>3 T) field magnets to operate at temperatures up to 10 K. As this becomes a reality, it will enable the use of multi-stage ADRs with smaller, more efficient solid hydrogen dewars, or with simpler cryocoolers having fewer stages. An ADR is the only sub-Kelvin technology that can be extended for use above 5 K, and its high efficiency suggests that that there would be system-level advantages to doing so.

In the low temperature limit, there are combinations of coolers that could offer mass or size advantages, particularly in lowering the cold mass. For example, a ^3He refrigerator can produce substantial cooling at 250-300 mK. Attaching a single ADR stage would be a very simple way to extend cooling to 50 mK or below, and by adding two stages [66], continuous cooling at 50 mK could be realized. The advantage here is that these stages, like the ^3He cooler, would be relatively small and lightweight, as they would require magnetic fields of only a few tenths of a T. Such fields would be easy to shield from detectors and require low (<1 A) currents. The ^3He cooler could be a single-shot version and coupled to a ^4He single-shot refrigerator to operate from a 5 K sink.

A variation on this would be to couple a dilution unit to a ^3He sorption cooler. While a cooler using this concept [67] is already being developed for use in 1 g, the concept could improve the performance of the open-cycle DR by pre-cooling the mixture below 1.5 K, and reducing the fraction of dilution cooling used to cool the incoming mixture.

13. Summary

A number of far infrared/submillimeter/CMB missions are enabled by advancing the state of the art in efficiently designed space flight cooling systems. There are a variety of cryogenic cooling techniques available that span a wide range of temperatures and heat lift capabilities from which to select. Multi-stage stored cryogen-free cooling systems have been developed by Planck and JWST. The coolers will be able to cool sunshades to maintain low and steady temperatures, to cool telescopes, and to provide the pre-cooling necessary for the sub-Kelvin coolers. Closely related to the cooler development for a specific mission will be the need to validate the integration technology readiness for interfacing the coolers with large sensor arrays, telescopes and/or sunshades. Verification of the thermal performance of these high performance systems with an open geometry operating over a large temperature range is difficult and potentially expensive. A combination of conservative design and subscale tests to validate the thermal models is warranted. Trades for use of active coolers versus stored cryogen systems have been made. Over the last several years significant advances have been made in mechanical cryocoolers operating below 10 K and these look to be the choice over stored cryogens for the future. A variety of sub-Kelvin cooling technologies have been brought to flight-ready or flight-proven status in order to meet the demand for very low temperature detector cooling. There are still challenges to their use on future missions, including lower temperature operation and higher cooling power requirements for larger detector arrays, the need to interface these coolers with cryocoolers operating at higher temperature, and longer mission lifetimes than has been demonstrated on any previous flight instrument. There are, however, coolers under development whose performance envelopes the estimated requirements for these missions, and which can operate with existing cryocoolers. What they, and other promising technologies, need most is further investment to achieve sufficiently high TRLs in advance of proposing their use on future missions.

14. Acknowledgments

Part of the work described in this paper was carried out at NASA/Goddard Space Flight Center. Part of the work was also carried out at the Jet Propulsion Laboratory, California Institute of Technology through an agreement with the National Aeronautics and Space Administration.

15. References

[1] see for instance, Bard S, Stein J, and Petrick W, "Advanced Radiative Cooler with Angled Shields." *Progress in-Astronautics and Aeronautics* **83** AIAA. New York, 1982, pp. 249-258.

[2] Tauber J, "The Planck mission", *Advances in Space Research*, Elsevier, (2004).

[3] Cleveland P and Parrish K, "Thermal system verification and model validation for NASA's cryogenic passively cooled James Webb Space Telescope (JWST)", Proc. Soc. Automotive Engineers, 05-CES-236 (2005).

[4] see for instance, Bennett C, et al., "First-year Wilkinson Microwave Anisotropy Probe (WMAP) observations: preliminary maps and basic results", *Astrophysical Journal Supplement Series*, **148**:1–27, 2003 September

[5] Pilbrat G., "Herschel mission overview and key programmes", http://herschel.esac.esa.int/Publ/2008/SPIE2008_Herschel_paper.pdf

[6] Finley P. and Schweikart R., "Mid-mission update of Spitzer Space Telescope cryogenic performance", *Adv. Cry. Eng.* **51**, pp. 1295-1302 (AIP Conf. Proc. 823), New York, (2006) Weisend II J, Ed.

[7] Kelsall T, et al. "The *COBE* Diffuse Infrared Background Experiment Search for the Cosmic Infrared Background. II. Model of the Interplanetary Dust Cloud", *Astrophysics Journal*, **508**:44-73 (1998).

[8] DiPirro M, et al., "High fidelity cryothermal test of a subscale large space telescope", Proc. of SPIE **6692**, 669202-1.

[9] Collaudin B, "ISO, cryogenic aspects", *Proc. 6th European Symposium on Space Environmental Control Systems*, Noordwijk, Netherlands 20-22 May 1997. (ESA SP-400, August, 1997.

[10] Murray D, Taber M, and Burns K, "Flight performance of the Gravity Probe B Cryogenic System", *Adv. Cry. Eng.* **51**, pp. 1303-1314 (AIP Conf. Proc. 823), New York, 2006. Ed. Weisend II J.

[11] Swift W, McCormick J, Breedlove J, Dolan F, Sixsmith H, "Initial operation of the NICMOS cryocooler on the Hubble Space Telescope", Cryocoolers 12, Ross Jr R, ed. 2003, pp. 563-570.

[12] Shirron P and DiPirro M, "Suppression of Superfluid Film Flow in the XRS Helium Dewar", *Adv. Cryo. Eng.*, **43** (1998) 949; Kelley R, et al.," The Suzaku High Resolution X-Ray Spectrometer", PASJ: Publ. Astron. Soc. Japan **59**, S77–S112, 2007 January 25

[13] Donabedian M, Ed., *Spacecraft Thermal Control Handbook Vol II: Cryogenics*, The Aerospace Press, (2003) Chapter 19, pp. 469-480.

[14] Ross R G, Jr 2007 Aerospace coolers: a 50-year quest for long-life cryogenic cooling in Space Cryogenic Engineering: Fifty Years of Progress ed K Timmerhaus and R Reed (New York: Springer Publishers) chapter 11 pp 225-84
[15] Ross R G Jr. and Boyle R F 2007 An overview of NASA space cryocooler programs – 2006," Cryocoolers 14 ed S. Miller and R G Ross, Jr (Boulder, CO: ICC Press) pp 1-10
[16] Gibson A F, Bradshaw T W and Linder M 2008 Heritage overview: 20 years of commercial production of cryocoolers for space Adv. In Cryogenic Engineering: Transactions of the Cryogenic Engineering Conference - CEC, Vol. 53 (Chattanooga, TN, 16-20 July 2007) (AIP Conference Proceedings vol 985) ed J G Weisband, et al (New York: Amer. Inst. of Physics) pp 493 --505
[17] Hirabashi M, et al 2008 Thermal design and it on-orbit performance of the AKARI cryostat Cryogenics 48 189 -97
[18] Narasaki K, et al 2004 Development of cryogenic system for SMILES Adv. In Cryogenic Engineering: Transactions of the Cryogenic Engineering Conference – CEC, Vol. 49 (Anchorage, AK, 22-26 September 2003) (AIP Conference Proceedings vol 710) ed J Waynert, et al (New York: American Institute of Physics) pp 1785--94
[19] Durand D, Colbert R, Jaco C, Michaelian M, Nguyen T, Petach M and Tward E 2007 NGST Advanced Cryocooler Technology Development Program (ACTDP) cooler system Cryocoolers 14 ed S Miller and R G Ross, Jr (Boulder, CO: ICC Press) pp 21 -5
[20] Ross R G, Jr and Johnson D L 2006 NASA's advanced cryocooler technology development program (ACTDP) Adv. in Cryogenic Engineering: Transactions of the Cryogenic Engineering Conference – CEC,: Vol. 51 (Keystone, CO, 28 Aug-2 Sept, 2005) (AIP Conference Proceedings vol 823) ed J G Weisend, et al (New York: American Institute of Physics) pp 607 --14
[21] Pearson D, Zhang B, Prina M, Paine C, Morgante G, Bhandari P, Bowman R and Nash A 2007 Flight acceptance testing of the two JPL Planck sorption coolers Cryocoolers 14 ed S Miller and R G Ross, Jr (Boulder, CO: ICC Press) pp 497-504
[22] Jensen S M, Hansen G, Nast T, Roth E and Clappier R 2007 Thermal/mechanical system level test results of the GIFTS 2-stage pulse tube cryocooler Cryocoolers 14 ed S Miller and R G Ross, Jr (Boulder, CO: ICC Press) pp 65-74
[23] Colbert R, Pruitt G, Nguyen T, Raab J, Clark S and Ramsey P 2009 ABI cooler system protoflight performance Cryocoolers 15 (Long Beach, CA, 9-12 June 2008) ed S Miller and R G Ross,Jr (Boulder, CO: ICC Press) (In Press).
[24] Breon S R, Shirey K A, Banks I S, Warner B A, Boyle R F andMustafi S 2003 Operation of a Sunpower M87 cryocooler in a magnetic field Cryocoolers 12 ed R G Ross, Jr (New York: Plenum Publishers) pp 761-9
[25] Banks S, Breon S and Shirey K 2004 AMS-02 cryocooler baseline configuration and EM qualification program Cryogenics 44 551 -7
[26] Nast T, Olson J, Champagne P, Mix J, Evtimov B, Roth E and Collaco A 2008 Development of a 4.5K pulse tube cryocooler for superconducting electronics Adv. In Cryogenic Engineering: Transactions of the Cryogenic Engineering Conference – CEC, Vol. 53 (Chattanooga, TN, 16-20 July 2007) (AIP Conference Proceedings vol 985) ed J G Weisband, et al (New York: American Institute of Physics) pp 881 – 6
[27] Doornink D J, Burger J F and ter Brake H J M 2008 Sorption cooling: a valid extension to passive cooling," Cryogenics 48 272-9
[28] Burger J, Holland H, Meijer R, ter Brake H J M, Doornink J and Sirbi A 2009 Further developments on a vibration-free helium-hydrogen sorption cooler Cryocoolers 15 (Long Beach, CA, 9-12 June 2008) ed S. Miller and R G Ross, Jr (Boulder, CO: ICC Press) (in press)
[29] Sugita H et al 2006 Cryogenic infrared mission "JAXA/SPICA" with advanced cryocoolers Cryogenics 46 149-157
[30] Sugita H et al 2008 Development of mechanical cryocoolers for the Japanese IR space telescope SPICA Cryogenics 48 258-66

[31] Frank D, Roth E, Olson J, Evtimov B, Nast T, Sompayrac B and Clark L D 2007 Development of a cryocooler to provide zero boil-off of a cryogenic propellant tank Cryocoolers 14 ed S Miller and R G Ross, Jr (Boulder, CO: ICC Press) pp 583-8
[32] Nast T C, private communication, 2008.
[33] Zagorola M V, private communication, 2008.
[34] van de Groep W, Mullie J, Willems D and Benschop T 2009 Development of a 15W coaxial pulse tube cooler Cryocoolers 15 (Long Beach, CA, 9-12 June 2008) ed S. Miller and R G Ross, Jr (Boulder, CO: ICC Press) (in press)
[35] Olson J R and Davis T 2006 Development of a 3-Stage pulse tube cryocooler for cooling at 10K and 75K Adv. In Cryogenic Engineering: Transactions of the Cryogenic Engineering Conference – CEC, Vol 51 (Keystone, CO, 28 Aug-2 Sept, 2005) (AIP Conference Proceedings vol 823) ed J G Weisend, et al (New York: Amer. Inst. of Physics) pp 1885-92
[36] Jaco C, Nguyen T and Raab J 2009 10K Pulse tube cooler performance data Cryocoolers 15 (Long Beach, CA, 9-12 June 2008) ed S. Miller and R G Ross, Jr (Boulder, CO: ICC Press) (in press)
[37] Foster W G, Olson J, Champagne P, Evtimov B, Will E, Collaco A, Nast T, Clappier R, Mitchell A, Jungkman D, Radebaugh R and Curran D G T 2003 Development of a high capacity two-stage pulse tube cryocooler Cryocoolers 12 ed R G Ross, Jr (New York: Kluwer Academic/Plenum Publishers) pp 225-32
[38] Zagarola M V, Dietz A J, Swift W L and Davis T M 2004 35K Turbo-Brayton cryocooler technology Adv. In Cryogenic Engineering: Transactions of the Cryogenic Engineering Conference – CEC, Vol. 49 (Anchorage, AK, 22-26 September 2003) (AIP Conference Proceedings vol 710) ed J Waynert, et al (New York: American Institute of Physics) pp 1635-42
[39] Gully W, Glaister D, Hendershott P, Marquardt E, Lester J and Levenduski R 2009 Initial test results for a 35K Cryocoolers 15, Cryocoolers 15 (Long Beach, CA, 9-12 June 2008) ed S Miller and R G Ross, Jr (Boulder, CO: ICC Press) (in press)
[40] Jaco C, Nguyen T, Colbert R, Pietrzak T, Chan C K. and Tward E 2004 High capacity staged pulse tube cooler Adv. In Cryogenic Engineering: Transactions of the Cryogenic Engineering Conference – CEC Vol. 49 (Anchorage, AK, 22-26 September 2003) (AIP Conference Proceedings vol 710) ed J Waynert, et al (New York: American Institute of Physics) pp 1263-8
[41] Jaco C, Nguyen T and Tward E 2008 High capacity two-stage coaxial pulse tube cooler Adv. In Cryogenic Engineering: Transactions of the Cryogenic Engineering Conference – CEC, Vol 53 (Chattanooga, TN, 16-20 July 2007) (AIP Conference Proceedings vol 985) ed J G Weisband, et al (New York: American Institute of Physics) pp 530 --7
[42] Kirkconnell C S, Hon R C and Roberts T 2009 Raytheon Stirling/pulse tube cryocooler maturation programs Cryocoolers 15 (Long Beach, CA, 9-12 June 2008) ed S Miller and R G Ross, Jr (Boulder, CO: ICC Press) (in press).
[43] Kotsubo V, Olson J R, Champagne P, Williams B, Clappier B and Nast T C 1999 Development of pulse tube cryocoolers for HTS satellite communications Cryocoolers 10 ed R G Ross, Jr (New York: Kluwer Academic/Plenum Publishers) pp 171-9
[44] Swift W L, Dolan F X and Zagorola M V 2008 The Nicmos cooling system – 5 years of successful on-orbit operation Adv. In Cryogenic Engineering: Transactions of the Cryogenic Engineering Conference – CEC, Vol. 53 (Chattanooga, TN, 16-20 July 2007) (AIP Conference Proceedings vol 985) ed J G Weisband, et al (New York: American Institute of Physics) pp 799 --806
[45] Tward E, Chan C K, Colbert R, Jaco C, Nguyen T, Orsini R and Raab J 2002 High efficiency cryocooler Adv. In Cryogenic Engineering: Transactions of the Cryogenic Engineering Conference – CEC, Vol. 47 (Madison, WI, 16-20 July 2001) (AIP Conference Proceedings 613) ed M Adams, et al (New York: American Institute of Physics) pp 1077 --84
[46] Johnson D L and Ross R G, Jr 1995 Cryocooler coldfinger heat interceptor Cryocoolers 8 ed R G Ross, Jr (New York: Plenum Press) pp 709-17

[47] Johnson D L and Ross R G, Jr 1994 Spacecraft cryocooler thermal integration Proc. of the Spacecraft Thermal Control Symposium (Albuquerque, NM, November 1994) (Abuquerque, NM: AirForce Research Laboratory)
[48] Gillman D C 1997 Cryocooler heat interceptor test for the SMTS program Cryocoolers 9 ed R G Ross, Jr (New York: Plenum Press) pp 783-93
[49] Gully Ph, Charles I, Durand J L and Briet R 2008 A new mini pulse tube with a heat interceptor Adv. In Cryogenic Engineering: Transactions of the Cryogenic Engineering Conference – CEC, Vol. 53 (Chattanooga, TN, 16-20 July 2007) (AIP Conference Proceedings vol 985) ed J G Weisband, et al (New York: American Institute of Physics) pp 1651—58
[50] Bradshaw T W, Orlaska A H, Jewell C, Jones B G and Scull S 1997 Improvements to the cooling power of a space qualified two-stage stirling cycle cooler Cryocoolers 9 ed R G Ross, Jr (New York: Plenum Press) pp 79-88
[51] Charles I, Golanski L, Gauthier A, Coynel A and Duval J M 2008 20K Coaxial pulse tube using passive pre-cooling Adv. In Cryogenic Engineering: Transactions of the Cryogenic Engineering Conference – CEC, Vol. 53 (Chattanooga, TN, 16-20 July 2007) (AIP Conference Proceedings vol 985) ed J G Weisband, et al (New York: American Institute of Physics) pp 887--94
[52] Kirkconnell C S, Zagarola M V and Russo J T 2006 Hybrid Stirling/reverse Brayton and multi-stage Brayton cryocoolers for space applications Adv. In Cryogenic Engineering: Transactions of the Cryogenic Engineering Conference – CEC, Vol 51 (Keystone, CO, 28 Aug-2 Sept, 2005) (AIP Conference Proceedings vol 823) ed J G Weisend, et al (New York: American Institute of Physics) pp 1489 -- 96
[53] Feller J, Salerno L, Maddocks J, Helvensteijn B, Kashani A, Nellis G and Gianchandani Y 2009 Distributed cooling techniques for space applications Cryocoolers 15 (Long Beach, CA, 9 -12 June 2008) ed S Miller and R G Ross, Jr (Boulder CO: ICC Press) (in press).
[54] Whitehouse PL, Shirron PJ and Kelley RL 2004 The X-ray Microcalorimeter Spectrometer (XMS): a Reference Cryogenic Instrument Design for Constellation-X Cryogenics 44 543.
[55] Serlemitsos AT, SanSebastian M and Kunes E 1992 Design of a spaceworthy adiabatic demagnetization refrigerator Cryogenics 32 117.
[56] N. A. Miller, G. C. O'Neil, J. A. Beall, G. C. Hilton, K. D. Irwin, D. R. Schmidt, L. R. Vale and J. N. Ullom 2008 High resolution x-ray transition-edge sensor cooled by tunnel junction refrigerators Appl. Phys. Lett. 92 163501.
[57] Duband L, Clerc L, Ercolani E, Guillemet L, and Vallcorba R 2008 Herschel flight models sorption coolers Cryogenics 48 95.
[58] Triqueneaux S, Sentis L, Camus P, Benoit A, Guyot G 2006 Design and performance of the dilution cooler system for the Planck mission Cryogenics 46 288.
[59] Freund MM, Duband L, Lange AE, Murakami H, Hirao T, Matsumoto T and Sato S 1996 Design and Flight Performance of a Space-borne 3He Refrigerator for the Infrared Telescope in Space Cryogenics 38 435.
[60] Serlemitsos AT, SanSebastian M and Kunes E 1998 Final Design of the Astro-E/XRS ADR Adv. Cryo. Eng. 43 957.
[61] Kelley RL, et al. 2007 The Suzaku High Resolution X-Ray Spectrometer Publ. Astron. Soc. Japan 59 S77-S112.
[62] Mitsuda K, et al. 2008 The X-Ray Microcalorimeter on the NeXT Mission, submitted for publication in Proc. SPIE 7011.
[63] Shirron PJ, Canavan ER, DiPirro MJ, Tuttle JG and Yeager CJ 2000 A Multi-Stage Continuous Duty Adiabatic Demagnetization Refrigerator Adv. Cryo. Eng. 45 1629
[64] Shirron PJ, Wegel D and DiPirro MJ 2006 A Continuous Adiabatic Demagnetization Refrigerator for Cooling to 10 mK and Below AIP Conference Proceedings 850 1573.
[65] Tuttle JG, Pourrahimi S, Canavan ER, DiPirro MJ and Shirron PJ 2008 Development of a Lightweight Low-Current 10 K 4 T Magnet for Space-Flight ADRs Cryogenics 48 248.
[66] Duval JM, Luchier N, Duband L and Sirbi A 2008 50 mK Continuous Cooling with ADRs

Journal of Physics: Conference Series **155** (2009) 012008 doi:10.1088/1742-6596/155/1/012008

Coupled to Helium3 Sorption Cooler, submitted for publication in the proceedings of the 15th International Cryocooler Conference, Long Beach, CA.
[67] Prouvé T, Luchier N and Duband L 2008 Pocket Dilution Cooler, submitted for publication in the proceedings of the 15th International Cryocooler Conference, Long Beach, CA.

AUTHOR INDEX

Institute of Physics
Dirac House, Temple Back
Bristol BS1 6BE UK